SHIFTING AGRICULTURE AND SUSTAINABLE DEVELOPMENT

An Interdisciplinary Study from North-Eastern India

MAN AND THE BIOSPHERE SERIES

Series Editor J.N.R. Jeffers

VOLUME 10

SHIFTING AGRICULTURE AND SUSTAINABLE DEVELOPMENT

An Interdisciplinary Study from North-Eastern India

P. S. Ramakrishnan

School of Environmental Sciences
Jawaharlal Nehru University
New Dehli, 110067, India

PUBLISHED BY

UNESCO

PARIS

AND

The Parthenon Publishing Group

International Publishers in Science, Technology & Education

Published in 1992 by the United Nations Educational, Scientific and Cultural Organization,
7 Place de Fontenoy, 75700 Paris, France—UNESCO ISBN 92-3-102749-2
and
The Parthenon Publishing Group Limited
Casterton Hall, Carnforth,
Lancs LA6 2LA, UK—ISBN 1-85070-383-3
and
The Parthenon Publishing Group Inc.
120 Mill Road,
Park Ridge
New Jersey, 07656, USA—ISBN 1-85070-383-3

Typeset by Lasertext Ltd, Stretford, Manchester
Printed and bound in Great Britain by
Butler and Tanner Ltd., Frome and London

British Library Cataloguing in Publication Data

Ramakrishnan, P.S.
Shifting Agriculture and Sustainable Development. — (Man & the Biosphere Series; v. 10).
I. Title. II. Series
304.20954

ISBN 1-85070-383-3

Library of Congress Cataloging-in-Publication Data

Ramakrishnan, P.S.
Shifting agriculture and sustainable development / P.S. Ramakrishnan.
p. cm. — (Man and the biosphere series : v. 10)
Includes bibliographical references and index.
ISBN 1-85070-383-3
1. Shifting cultivation—Environmental aspects—India, Northeastern. 2. Agricultural ecology—India, Northeastern. 3. Forest ecology—India, Northeastern. 4. Human ecology—India, Northeastern. I. Title. II. Series.
S602.87.R36 1992
333.76′13′09541—dc20 92-5059
CIP

PREFACE

UNESCO's Man and the Biosphere Programme

Improving scientific understanding of natural and social processes relating to man's interactions with his environment, providing information useful to decision-making on resource use, promoting the conservation of genetic diversity as an integral part of land management, enjoining the efforts of scientists, policy-makers and local people in problem-solving ventures, mobilizing resources for field activities, strengthening of regional co-operative frameworks, these are some of the generic characteristics of UNESCO's Man and Biosphere (MAB) Programme.

The MAB Programme was launched in the early 1970s. It is a nationally based, international programme of research, training, demonstration and information diffusion. The overall aim is to contribute to efforts for providing the scientific basis and trained personnel needed to deal with problems of rational utilization and conservation of resources and resource systems, and problems of human settlements. MAB emphasizes research for solving problems: it thus involves research by interdisciplinary teams on the interactions between ecological and social systems; field training; and applying a systems approach to understanding the relationships between the natural and human components of development and environmental management.

MAB is a decentralized programme with field projects and training activities in all regions of the world. These are carried out by scientists and technicians from universities, academies of sciences, national research laboratories and other research and development institutions, under the auspices of more than a hundred MAB National Committees. Activities are undertaken in co-operation with a range of international governmental and non-governmental organizations.

Man and the Biosphere Book Series

The Man and the Biosphere Series was launched with the aim of communicating some of the results generated by the MAB Programme and is aimed primarily at upper level university students, scientists and resource managers, who are not necessarily specialists in ecology. The books are not normally suitable for undergraduate text books but rather provide additional resource material in the form of case studies based on primary data collection and written by the researchers involved; global and regional syntheses of comparative research conducted in several sites or countries; and state-of-the-art assessments of knowledge or methodological approaches based on scientific meetings, commissioned reports or panels of experts. The Series Editor is John Jeffers, formerly Director of the Institute of Terrestrial Ecology, in the United Kingdom, who has been associated with MAB since its inception.

Shifting Agriculture and Sustainable Development

The overall objective of work within the MAB Programme in the humid tropics is to contribute to the development of sustainable land-use systems that are in tune with the social, cultural and biological characteristics of the peoples and ecological systems of these regions in a time of rapid and far-reaching change. Research activities in these fields are integrally linked with efforts to synthesize and diffuse information on tropical ecology and resource management, and with a programme of training and institutional development.

Field research activities are undertaken by scientists from the tropical countries themselves, sometimes in collaboration with researchers from institutions in temperate regions and with the support of UNESCO, UNEP, FAO and other cooperating international organizations. More than twenty major field projects have been completed or are underway, dealing with such topics as people–forest interactions in East Kalimantan (Indonesia), rehabilitation of degraded forest areas in Guangzhou (China), the effects of increasing human activities on tropical forest ecosystems at Tai (Côte d'Ivoire) and learning from traditional agricultural systems in lowland wetland areas in Mexico. Two projects have generated case studies which have previously been published in the Man and the Biosphere Book Series – on the structure and the function of a nutrient stressed forest system in San Carlos de Rio Negro in Venezuela (Volume 2) and on the social and environmental effects of pulpwood logging in the Gogol Valley in Papua New Guinea (Volume 3).

The present volume presents a wide ranging synthesis of a long-term ecological study of shifting cultivation in north-eastern India, supported by India–MAB, the Department of Environment and Forests, the Department of Science and Technology, the University Grants Commission and other national institutions. The work has been carried out by a team of scientists headed by P.S. Ramakrishnan, Professor at the School of Environmental Sciences at the Jawaharlal Nehru University, New Delhi.

The project in north-eastern India has been wide ranging in scope and content, covering such topics as nutrient cycling, hydrology, plant succession, soil microbiology and socio-economics. The project has sought to combine detailed ecological studies on the dynamics of rural ecosystems with practical suggestions for improving the systems of land use and land management in the region. It has also placed studies on rain forest conservation and management within the broader human ecological context of village function and its redevelopment. Though the present case study is focused on upland areas of north-eastern India, it is hoped that the approaches and concepts set out in this volume will be of wider interest to those interested in the ecological systems and cultures of the humid tropics, as an example of the sort of contribution that scientists can make in assisting tropical peoples to adapt to changing social and economic circumstances.

It is a pleasure for UNESCO to thank Professor Ramakrishnan and his gifted team of Indian scientists for their foresight in undertaking the ambitious research effort in north-eastern India and the leader of the team for his diligence and perseverance in seeing the present volume into print.

MAN AND THE BIOSPHERE SERIES

---

1. The Control of Eutrophication of Lakes and Reservoirs. S.-O. Ryding; W. Rast (eds.), 1989.
2. An Amazonian Rain Forest. The Structure and Function of a Nutrient Stressed Ecosystem and the Impact of Slash-and-Burn Agriculture. C.F. Jordan, 1989.
3. Exploiting the Tropical Rain Forest: An Account of Pulpwood Logging in Papua New Guinea. D. Lamb, 1990.
4. The Ecology and Management of Aquatic-Terrestrial Ecotones. R.J. Naiman; H. Décamps (eds.), 1990.
5. Sustainable Development and Environmental Management of Small Islands. W. Beller; P. d'Ayala; P. Hein (eds.), 1990.
6. Rain Forest Regeneration and Management. A. Gómez-Pompa; T.C. Whitmore; M. Hadley (eds.), 1991.
7. Reproductive Ecology of Tropical Forest Plants. K. Bawa; M. Hadley (eds.), 1990.
8. Biohistory: The Interplay between Human Society and the Biosphere – Past and Present. S. Boyden, 1992.
9. Sustainable Investment and Resource Use: Equity, Environmental Integrity and Economic Efficiency. M.D. Young, 1992.

CONTENTS

FOREWORD

Agriculture is an important economic activity for a large population of the developing world. The view held a few decades ago that over-emphasized industrialization as the main hope of the developing countries has undergone remarkable changes in more recent times. Today, economists emphasize agriculture and rural development as the *sine qua non* of national development. Indeed, such a reorientation wherein rural development is well balanced with industrial growth is now being considered essential to correct serious imbalances in the economy, for the equitable distribution of wealth and for social justice.

It is against this background that increases in agricultural production in the developing world need to be viewed and evaluated. While the 'green revolution' in India, for instance, had positive repercussions in terms of general self-sufficiency in food production, it brought with it the much larger problems of environmental degradations and unequal distribution of wealth. Being largely confined to a relatively small section of the society, its wider inapplicability has led to a renewed scientific interest in traditional systems of agriculture which presumably offer ecological efficiency and sustainability, together with social justice. Sustainable agriculture not only demands efficient use of water and nutrients based on recyling of locally available resources, but also regulated cropping done in a manner that would contribute towards sustaining soil fertility. Application of wrong technology and/or over-exploitation of natural resources may have short-term gains but could often lead to ecological degradation, no longer sustainable by man. Viewed in this context, the north-east Indian case study is significant.

It is well recognized now that conservation and sustainable development are two sides of the same coin, closely interlinked in that one cannot be achieved at the expense of the other. From a human angle, such an integrated approach demands satisfying basic human needs in an equitable manner, maintenance and indeed promotion of social, cultural and biological diversity, and ecological integrity of the system. The north-eastern region of India, with a complex and large natural resource base and a hilly terrain, provides the ecological diversity. The people of the region, comprising many tribes, provide the socio-cultural diversity. To build upon such a diversity profile on a sustainable basis is a challenging opportunity; the traditional knowledge and technology obviously have to be the starting points if peoples' involvement is to be ensured and development sustainable. Whether it be the highly complex jhum (shifting agriculture) systems of the different tribal communities, or the equally fascinating sedentary agricultural systems in the valleys, or the multilayered home gardens, they all link the traditional animal husbandry or domestic sub-systems in the context of the forest ecosystem in which they are set. Any meaningful developmental effort needs to have this traditional knowledge as its basis. While imported technology from outside the region may sometimes be unavoidable in a larger context, its sustainability should be ensured. The burden of proof that such introductions are not detrimental to conservation has to rest on those developmental agencies seeking to introduce the technology. Thus, a holistic approach to conservation and development based on ecological, economic and social considerations has to be worked out. This volume thus raises: (a) issues related to human ecology in the context of village ecosystem function and its redevelopment, (b) bio-ecological issues concerning rain forest conservation, management and redevelopment, and (c) the obvious linkage between the two. Such relationships have therefore been drawn between the natural and social sciences.

The book has three sections : 1, Agroecosystem and village ecosystem function; 2, Secondary successional patterns and processes; and 3, Management implications. While the nine chapters of the first two sections discuss applications of a specific ecological study and its obvious social linkages, the two chapters of the third section consider general issues of conservation and sustainable development of interest to planners, administrators and non-governmental organizations interested in tribal development. This is significant in the Indian context because the geographical distribution of biological diversity and exploitable natural resources often overlap and are to be found in tribal areas. Though the present study is location-specific to north-eastern hill areas of India, the concepts and principles involved would have wider applicability for redeveloping traditional societies elsewhere in the humid tropics.

I wish to acknowledge the substantial contribution on rain forest ecosystem and restoration ecology by the team of about 25 young students who undertook Ph.D. research and an equal number of those who did post-doctoral work with me over the last 18 years. Without their hard work, enthusiasm, devotion, skills and endurance, this study could not have been brought to this level of completion. All of them had to survive extremely difficult field conditions, sometimes even risking their lives to get the work completed. The end-product is a new breed of ecologists who can handle environmental issues with a keen sensitivity to the social sciences.

For their intense academic involvement throughout this study, and help in many ways I especially thank : U. Baruah, T. Bhadauria, R. Boojh, U.M. Chandrashekara, A.K. Das, M. Deb, P. Dev, B.P. Dey, A.K. Gangwar, U. Gupta, V. Jeeva, R.S. Khiewtam, A. Kumar, S.P.S. Kushwaha, R.K. Maikhuri, B.K. Mishra, K.C. Mishra, S. Patnaik, S.C. Ram, K.S. Rao, S.N. Sharma, K.G. Saxena, R.P. Shukla, G.S. Singh, J. Singh, P.S. Swamy, and O.P. Toky.

The financial and/or institutional support received from the Department of Science and Technology, Department of Environment and Forests, Council of Scientific and Industrial Research, University Grants Commission, North-Eastern Hill University, Jawaharlal Nehru University and the G.B. Pant Institute of Himalayan Environment and Development, in India are acknowledged. The Pitamber Pant Environmental Award that I received from the Ministry of Environment and Forests, Goverment of India facilitated this work. The encouragement and support from my friends in UNESCO, Bernd Von Droste, Malcolm Hadley and John Jeffers led to speedy completion of this volume. Efficient typing of the manuscript by Mr Hari Ram facilitated its completion. I acknowledge the support and indulgence of my family (Poonam, Puneet and Sudha, with substantial field support from the first two) and of those personally dear to me, for the time spent away from them.

GENERAL INTRODUCTION

The north-eastern part of India, covering about 0.26 million km^2 of hills, valleys and plateau (Fig.0.1), is ethnically and culturally very distinct from the rest of the country. The region comprises seven states : Arunachal Pradesh, Assam, Manipur, Meghalaya, Mizoram, Nagaland and Tripura, and is bordered by Bhutan, Tibet, Burma and Bangladesh. Unlike the rest of the country, the people in the area are Tibeto-Mongoloids having migrated into the region from the neighbouring countries and speaking a variety of languages of Mon Khmer and Tibeto-Burman origin. The hills constitute about 70% of the total land area, where shifting agriculture (locally called jhum) is the chief land use. The Brahmaputra valley in Assam is the largest valley land with sedentary wet rice cultivation, and covers an area of about 56 × 10^3 km^2, with a small hill area comprising the Mikir and North-Cachar hills.

LAND

North-eastern India covers 7.76% of the geographical area of the country (Table 0.1). This region has been suggested to have 46.9% of the geographical area under forests, representing about 15.9% of India's forest area (North-Eastern Council, 1984). However, the results obtained by the National Remote Sensing Agency (1983a) suggest a much smaller forest cover for the country as a whole, with a little over 10% good forest cover. Viewed in this context and realizing that much of what is classified as forest in the north-eastern region may be degraded arrested successional stages of forest, these statistical estimates for forest cover of the north-east (National Remote Sensing Agency, 1983b) may need a downward revision.

Of the reported 21,040 km^2 of evergreen rain forest, 8,340 km^2 semi-evergreen rain forest and 102,000 km^2 of moist deciduous forest for India

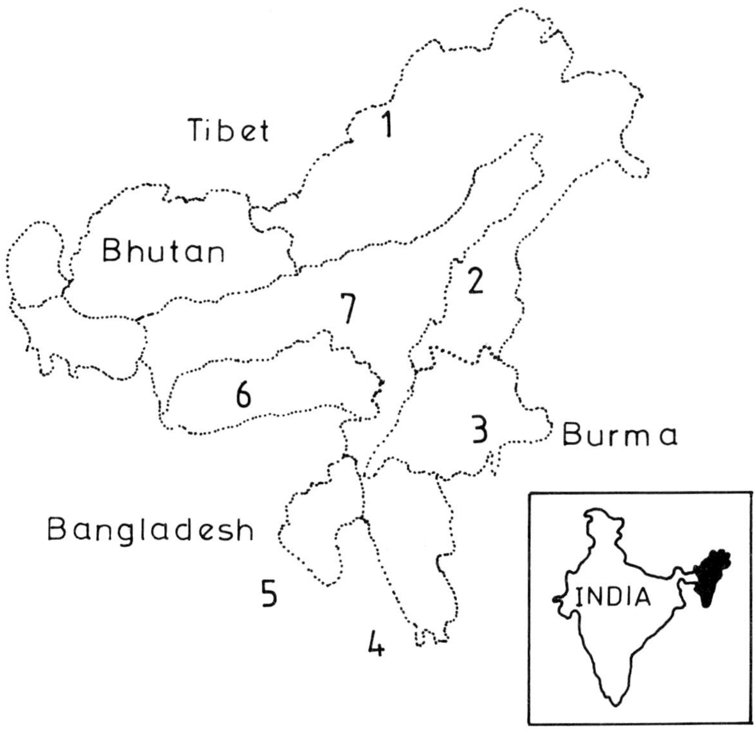

Figure 0.1 The study area of Arunachal Pradesh (1), Nagaland (2), Manipur (3), Mizoram (4), Tripura (5), Meghalaya (6) and Assam (7) in north-eastern India, with an inset map of the country. This is only indicative

Table 0.1 Important land use and population parameters – a comparison of north-eastern region and the nation (from North-Eastern Council, 1984)

Item	*North-east India*	*All India*
Geographical area ($\times 10^3$ ha)	25,505	328,778
Population density (persons km^{-2})	104	201
Decennial growth (%)	35.73	24.43
Area under settled agriculture ($\times 10^3$ ha)	4,090	155,626
Area under forest ($\times 10^3$ ha)	11,956	75,027
Area affected by shifting agriculture ($\times 10^3$ ha yr^{-1})	2,696	NA*

*Not available

(Myers, 1980), a major proportion is confined to north-eastern India, a narrow belt occurs along the western Ghats and some in the Andaman and Nicobar Islands.

VEGETATION

The climax sub-tropical vegetation typically is comprised *Artocarpus chaplasha, Michelia champaca, Mesua ferrea, Alstonia scholaris, Morus laevigata, Castanopsis indica, Ealeocarpus* spp. and *Garcinia* spp. (Puri, 1960; Shukla and Ramakrishnan, 1982b). Secondary successional forests are largely of bamboos dominated by *Dendrocalamus hamiltonii, Bambusa tulda, Bambusa khasiana* and *Neohouzeua dulloa* (Toky and Ramakrishnan, 1983a; Rao and Ramakrishnan, 1989) or comprised broad-leaved species such as *Alnus nepalensis, Vitex peduncularis, Vitex glabrata, Terminalia belerica, Lagerstroemia parviflora, Schima wallichii, Dillenia pentagyna* and *Cedrela toona.*

The sub-temperate climax forest is a mixed broad-leaved forest with species such as *Quercus dealbata, Quercus griffithii, Englehardtia spicata, Acer laevigatum, Betula alnoides, Castanopsis hystrix, Cinnamomum* spp., *Litsea meissneri, Litsea sebifera, Machillus kingii* and *Manglietia insignis* as important members (Boojh and Ramakrishnan, 1985; Kheiwtam, 1986). Early successional forests often are either pure pine (*Pinus kesiya*) forests or mixed with *Schima wallichii, Schima khasiana,* and *Alnus nepalensis* (Mishra and Ramakrishnan, 1983a; Ramakrishnan and Das, 1983).

Arrested succession of weeds which form extensive formations throughout have exotics such as *Eupatorium* spp., *Mikania micrantha,* and natives such as *Imperata cylindrica, Saccharum* spp., *Thysanolaena maxima,* and *Pteridium aquilinum* at different altitudes (Toky and Ramakrishnan, 1983a; Mishra and Ramakrishnan, 1983a).

Forest conversions

Large-scale timber extraction has been carried out in this region during the last few decades (Ramakrishnan, 1985d; Ramakrishnan and Toky, 1983). In fact, a substantial part of the timber needs of the country is met from the north-east. The secondary damage done to the forest due to the falling trees during timber harvest, destroying everything else in their path is also substantial. Bamboo forests, which are seral communities in the region, are being harvested in a major way by the pulping industry.

Much of the fuelwood consumed in the region comes from secondary forests, with about 70% being from the peripheral disturbed zone (Myers, 1980); only a small part comes from the primary forest. However, increased population pressure and rapid depletion of primary forest cover in the region is rapidly changing the situation (Ramakrishnan, 1985d). If about 250 kg of fuelwood is taken as the annual average consumption by an individual, as seen from the study of Sethliew village in Meghalaya (Mishra and Ramakrishnan, 1982), then for a population of 19.58 million (according

Plate 0.1 Fuelwood in the desertified areas of Meghalaya has often to be carried over long distances

to 1971 census data – Anonymous, 1971) the fuelwood need in the region itself would work out to be 4,895 × 10^6 kg. This is apart from the fuelwood extracted for export outside the region. With rapid increase in population and drastic decline in forest cover, the villagers are now forced to trek 10–15 km outside the village boundary (Plate 0.1).

Forest farming variously termed shifting agriculture, slash-and-burn agriculture, rotational bush fallow agriculture, swidden, and locally called 'jhum' has been an important factor for forest conversion. According to 1971 census data (Anonymous, 1971), 455 × 10^3 families are involved in jhum covering an area of 2,696 × 10^3 ha yr^{-1} in north-eastern India (cf Table 0.1). At a population growth rate of 24–30% per decade, this is expected to go up to 766 × 10^3 families by the turn of the century (North-Eastern Council, 1982). About 2.7 million ha of land are considered to be under jhum in the region. About 7.3 million ha of land were considered to be affected by jhum in 1975. The exact area under jhum, however, is still debatable. This is because the calculations are based upon a number of approximations regarding jhum cycle length and the area owned per family. This does not account for the land area already degraded and rendered unfit for cultivation. Sequential and reliable data on land cover following standardized methodology are required (Singh, 1987); otherwise the conclusions could be misleading. Therefore, these estimates of degraded lands can at best be considered as indicative of the magnitude of the problem, and the values are likely to be underestimates.

CLIMATE AND SOIL

The climate ranges from sub-tropical at lower elevations (Toky and Ramakrishnan, 1981a) to sub-temperate in the hills (Kushwaha *et al.*, 1981a). The temperature/ rainfall pattern shown in Fig. 0.2 for Burnihat at lower elevations in Meghalaya (100 m altitude) and for Shillong at higher elevations in Meghalaya (1,500 m altitude) indicates the typical patterns for the sub-tropical and sub-temperate locations, respectively. The climate is typically monsoonic with an average rainfall of about 2,000 mm, which locally may go up to over 10,000 mm as at Cherrapunji (Ramakrishnan, 1985e). Much of the annual rainfall occurs during the monsoon period from May to September. Winter extends from November to February, with occasional showers. March and April represent the brief dry summer and are accompanied by strong winds. The average maximum and minimum temperatures are 25°C and 12°C, respectively for November to February at Burnihat, and the corresponding temperatures at Shillong are 17°C and 8°C. The rest of the year has on average maximum and minimum temperatures of 32°C and 23°C, respectively, at Burnihat; it was 23°C and 12°, respectively at Shillong. At higher elevations frosts are very common during December–January.

The soils of north-eastern hill areas (Anonymous, 1977) are oxisol (red, mixed red and yellow) or laterosol, (comprising laterite and lateritic types) or ultisol (mountain brown). The Assam valley has inceptisol (alluvial–deltaic). The soils are acidic with average pH ranging from 4 to 5.5. The hill soils are sandy loam derived from metamorphic rocks and are low in phosphorus and potassuim. Nitrogen availability is uncertain partly because of steep slopes of 30–40° angles that are susceptible to intense run-off and leaching losses and partly because of volatilization during slash-and-burn operations related to jhum.

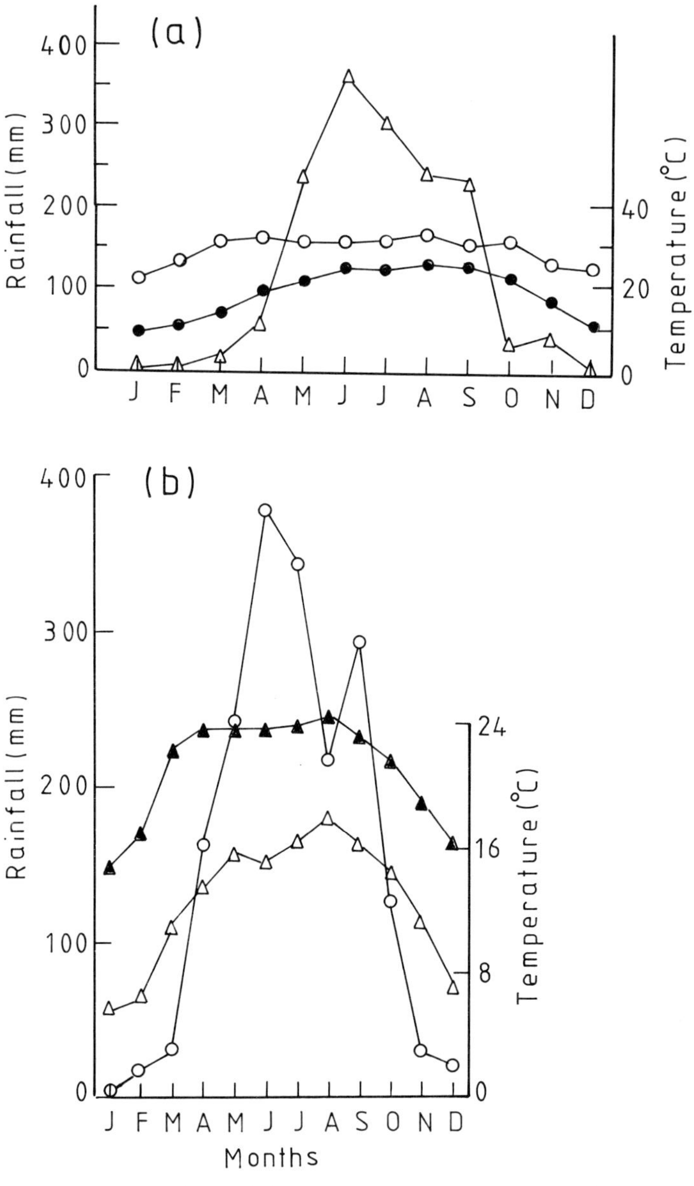

Figure 0.2 Temperature and rainfall data for the study area at Burnihat (100 m altitude) (a), and at Shillong (1,500 m altitude) (b) in Meghalaya, which is typical of the pattern for the north-eastern region, except for local variations (from Toky and Ramakrishnan, 1981a; Kushwaha *et al.*, 1981a). ○, mean max. temperature; ●, mean min. temperature; and △, mean rainfall; (b) ▲, mean max. temperature; △, mean min. temperature; and ○, mean rainfall

Table 0.2 Population structure of the north-eastern region of India (from Anonymous, 1971)

State	*Geographical area* ($km^2 \times 10^3$)	*Population* ($\times 10^3$)	*Tribals* (%)
Arunachal Pradesh	83.5	467.5	79
Assam	78.3	12,911.9	–
Plains	–	12,456.5	11
Hills	–	455.4	58
Manipur	22.4	1,072.8	31
Meghalaya	22.5	1,011.7	80
Mizoram	21.1	321.9	94
Nagaland	16.6	516.5	89
Tripura	10.5	1,556.3	29

PEOPLE

With about 30 million people over a land area of 255×10^3 km^2, jhum is practised by the tribals of the six hill states of Arunachal Pradesh, Manipur, Meghalaya, Mizoram, Nagaland and Tripura and by the localized hill tribes of Assam. The population structure and land area according to 1971 census data (Anonymous, 1971) given in Table 0.2 suggest that the tribal population largely live in the hill areas.

The population growth in the north-eastern hill region is much above the national average, though the population density is lower (cf Table 0.1). The population grew at a rate much faster in the post-independence period than during earlier times. (Fig. 0.3). The trend in growth in the north-east is similar to the rest of the country, but with a faster rate. Such a trend has had its adverse impact on the natural resources.

There are over 100 different tribes in the north-east, differing linguistically and culturally. Often the tribals change culturally and linguistically over very short distances. Some of the tribes are represented only by a few hundred individuals. The 53 more important tribes in the region with their population size according to 1971 census data (Anonymous, 1971) are shown in Table 0.3.

All the tribes practise jhum. The land is owned by the community. Each village operates in a given demarcated area and the land is distributed amongst the families by the village headman or the village council. The area allocated to a family is determined by the number of members in the family. Once allocated, the land is owned by the family as long as jhum is executed. However, there could be many minor variations depending upon the tribal background (UNESCO, 1983). The residence is mostly fixed and the families operate around the village over a radius of 10–15 km.

The jhum procedures are closely linked with socio-cultural practices and religious beliefs (Ramakrishnan, 1983, 1985c) Thus the Garos believe that

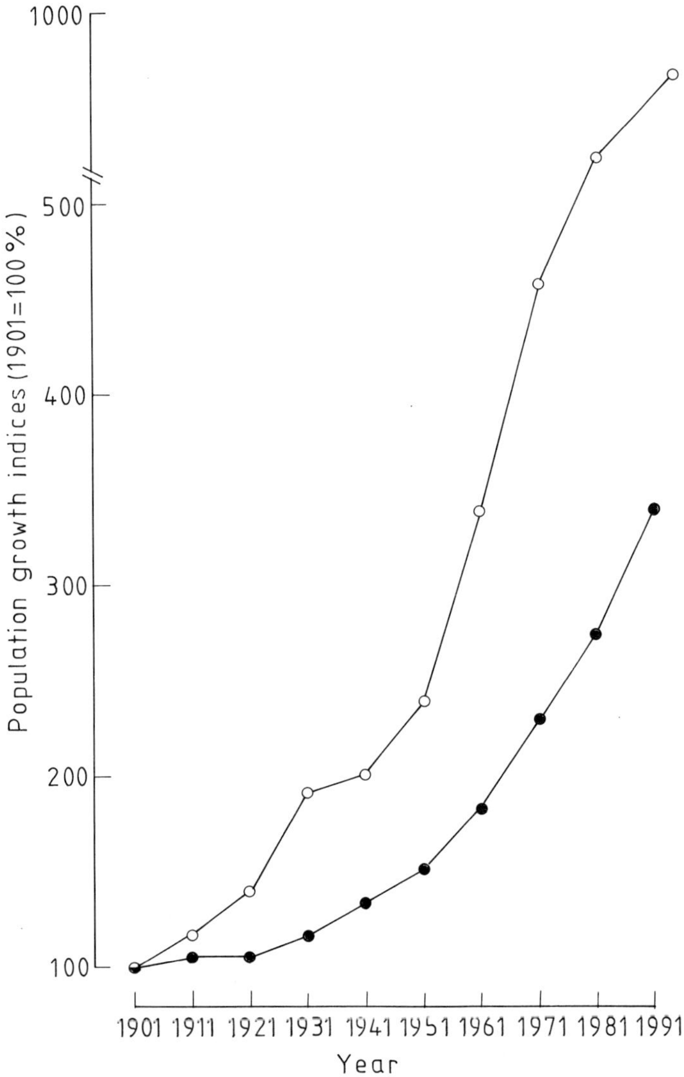

Figure 0.3 Population growth in north-eastern India (○) compared with all India average (●) expressed as growth indices (1901 = 100%)

their deity 'Misipa' helps in obtaining good returns from the jhum plots and therefore jhum operations such as land allocation, felling and burning, and crop harvest are all linked with religious rites and festivities. A detailed account of the cultural profile of many tribes of north-eastern India are given elsewhere (UNESCO, 1983). Indeed, many social anthropologists believe that jhum is a way of life for the tribals and therefore drastic changes

Table 0.3 Some important tribes of north-eastern region of India (from Anonymous, 1971)

Tribe	*Population*	*Tribe*	*Population*
Arunachal Pradesh		**Manipur**	
Adi	99,372	Aimol	108
Aka	2,347	Anal	6,592
Apatani	12,888	Chiru	3,590
Bongro	1,085	Chothe	1,117
Hill Miri	8,174	Hamar	38,207
Khamti	4,078	Kabui	17,360
Miji	3,549	Khoira	406
Mishing	3,359	Maram	19,968
Monpa	28,209	Tangkhul	58,167
Nishi	80,325	**Mizoram**	
Nocte	23,165	Mizo	270,312
Sherdukpen	1,639	Pawi-Lakher	21,427
Singpho	1,567	Ralte	170
Sulung	4,250	**Tripura**	
Tagin	20,377	Chackma	68,711
Wancho	28,650	Jamatia	22,446
Meghalaya		Magh	12,378
Garo	411,532	Riang	74,931
Hajong	23,978	Tripuri	268,948
Jaintia	82,493	Mariang	9,710
Khasi	384,006	**Nagaland**	
Assam		Angami	43,569
Boro-Kachari	543,615	Ao	62,275
Chutiya	9,103	Chang	15,816
Dimasa	37,900	Konyak	72,338
Karbi	17,360	Lhota	36,949
Lalung	10,650	Phom	18,017
Mech	12,919	Rengma	8,578
Mishing	180,684	Sema	65,227

in agricultural technology are considered to cause social disruptions (Ramakrishnan, 1983; Ramakrishnan and Maikhuri, 1988). However, because of drastically shortened jhum cycles (the length of the intervening fallow phase between two successive croppings on the same site) to 5 years or less, distortions in jhum have occurred. Therefore, there is an obvious need for change.

It is in this context of continuity and the obviously unavoidable need for change that are inherent to all social systems that the various socio-economic activities of the tribal village communities have to be understood, evaluated and linked with forest ecosystem function. In an ecological sense, the emphasis has to be towards linking up natural sciences with social sciences, so that the studies reflect on ecosystem function impacted by human activities and problems related to management of human-impacted ecosystems (UNESCO, 1986). We, therefore, consider here village ecosystem function with emphasis on land use, various physical and biological parameters under which tribal societies operate, and with emphasis on the linked forest ecosystem function. We also consider ecosystem redevelopment strategies on a short- and long-term basis so that social disruptions are minimized, if not avoided.

SECTION 1

Agroecosystem and Village Ecosystem Function

CHAPTER 1

CROPPING AND YIELD PATTERNS AND ENERGY BUDGET UNDER JHUM

INTRODUCTION

Agriculture is the predominant activity of a large majority of the world population. Industrialization of agriculture through large fossil fuel energy subsidies, sophisticated chemical control measures for pests and diseases, and high yielding crop varieties has resulted in huge increases in agricultural yields during the last half century. Such agricultural systems are efficient in terms of human time and labour but suffer from many deficiencies. They are highly inefficient from an overall energetic point of view, because five to ten units of fuel energy are required to produce a single unit of food energy (Steinhart and Steinhart, 1974). Apart from the ecological instability of the monoculture of a single high-yielding variety of crop, modern agriculture also causes varied environmental problems related to intensive use of chemical fertilizers and pesticides. The obvious inapplicability of such systems as models for development in an energy limited world has led to a renewed interest in an understanding of traditional agricultural systems, which presumably offer higher ecological efficiency.

The forest farmer in the humid tropics has managed his traditional 'shifting agriculture', variously termed as 'slash-and-burn agriculture', 'rotational bush fallow agriculture' or popularly known in India as 'Jhum', for centuries, with optimum yield on a long-term basis, rather than trying to maximize production on a short-term basis (Watters, 1960; Spencer, 1966; Ruthenberg, 1971; Soemarwoto, 1975; Gliessman *et al.*, 1981; Ramakrishnan, 1984a).

Shifting agriculture systems have been held up as models of productive efficiencies where 5–50 units of food energy are obtained for each unit of energy expended (Rappaport, 1971; Steinhart and Steinhart, 1974). The possibility for increased crop production has been suggested (Greenland, 1975; Revelle, 1976; Mutsaers *et al.*, 1981) without departing too much from these traditional systems that have been considered as the most evolved system for the forested areas of the tropics and the sub-tropics (Conklin, 1957; Carneiro, 1960; Nye and Greenland, 1960; Watters, 1971; Ramakrishnan, 1985a, 1985c, 1988c).

Earlier attempts at an evaluation of the cropping patterns, monetary and energy efficiencies of shifting agriculture elswhere (Ruthenberg, 1971; Watters, 1971) have been on isolated systems, often without fully defining the parameters under which a given system operates and without making adequate comparisons between all available agroecosystem types. Our studies on the jhum systems of north-eastern India have made such comparisons possible, with greater precision. We have been able to look at jhum systems under varied jhum cycles (the length of the intervening fallow period between two successive croppings at the same site) (Toky and Ramakrishnan 1981a; Mishra and Ramakrishnan, 1981; Maikhuri and Ramakrishnan, 1990a,b), under diverse ecological conditions (Ramakrishnan, 1988d, 1990) and under different socio-economic and cultural backgrounds of societies (Ramakrishnan, 1983; Patnaik and Ramakrishnan, 1989; Kumar and Ramakrishnan, 1989a; Maikhuri and Ramakrishnan 1990a,b).

JHUM AGROECOSYSTEM IN NORTH-EASTERN INDIA

Realizing that there are over a hundred different tribal communities which are highly insulated because of language, topography and socio-cultural factors, one could reasonably expect over a hundred different variations in this land use practice. If ecological conditions are superimposed on this then the heterogeneity gets very complex indeed. Therefore, it is important to realize that we are talking of a variety of jhum systems rather than a single, homogeneous system of 'shifting agriculture', as often assumed in the scientific literature.

Jhum is the chief land use practice in the north-eastern hill region of India. The jhum cycle, until recently, was more than 20 years (Ramakrishnan and Toky, 1978; Ramakrishnan *et al.*, 1981a,b) but has now come down to an average of 4–5 years. Long cycles were possible earlier because population pressure was not heavy and land availability did not limit the cycle. In north-eastern India, the average size of a jhum plot varies from 1.0 to 2.5 ha. The average family consists of two adults and three to four children.

Jhum at lower elevations

The low elevation jhum prevalent at Burnihat in Meghalaya (Plate 1.1) is typical of the version found throughout the north-eastern hill areas in that: (1) the cultivation is carried out on slopes of 30–40° angle, (2) the climate

Plate 1.1 A general view of the jhum plot at Burnihat (90 m) in Meghalaya on a steep slope of 30–40° angle. In the foreground is wet rice cultivation in the valley land which is common in the region

is monsoonic with a high rainfall of over 2,200 mm followed by a dry winter and a brief warm summer, supporting a mixed sub-tropical humid forest, (3) the normal jhum cycle is of 4–5 years, but rarely longer, and more importantly the forest is clear-felled before cropping (Toky and Ramakrishnan, 1981a).

The study site at Burnihat in (100 m altitude) the Khasi hills of Meghalaya and about 90 km north of Shillong is located 26°N and 91.5°E. The soil is red, sandy loam and of lateritic origin (oxysol) with a pH ranging from 5 to 7. The climate is monsoonic with an annual average rainfall of 2,200 mm, 90% of which occurs between May to September (Toky and Ramakrishnan, 1981a). The summer temperatures have an average maximum of 33°C and a minimum of 29°C the average maximum winter temperature is 25°C and the minimum is 15°C.

During the winter months (December–January), the undergrowth is slashed, and small trees and bamboos are felled. Short tree stumps and large tree boles are left intact. The underground organs of different species are not disturbed. This laborious process is often completed by the men from two or three families. Such a joint effort is one of the essential ingredients of a well-knit social organization. This effort, along with the process of allotment of sites for jhum by the village headman who is in

Plate 1.2 A jhum plot at Lailad (296 m) in Meghalaya after the slash-and-burn operation. Uncut trees and tree stumps are left *in situ*. A small hut (top left) is for the farmer to live in and protect his field. Note the small heaps of ash left after a second burning of the unburnt biomass

overall control of the village community, helps to promote kinship among the members of the village, and among many communities such as the Garos and the Mikirs (Maikhuri and Ramakrishnan, 1990a).

Towards the end of March or the beginning of April, before onset of the monsoon, the debris is burnt *in situ*. Before burning, a fire line is cleared around the field. Burning is often repeated to destroy any unburnt material that has first been collected in heaps. A bamboo hut is built for temporary living. The family's presence protects the field from wild animals (Plate 1.2).

Sowing is done after the first few monsoon showers. The seed mixtures used for different jhum cycles may vary considerably. Whereas cereals are emphasized under long jhum cycles, perennials and tuber crops are emphasized under short jhum cycles, as by the Garos in Meghalaya (Toky and Ramakrishnan, 1981a), or by the Nishis in Arunachal Pradesh (Maikhuri and Ramakrishnan, 1990b). The details of the crops sown at Burnihat are given in Table 1.1. Some 8–35 crop species are grown together at Burnihat, but the number may be up to 35 species elsewhere in Arunachal Pradesh (Maikhuri and Ramakrishnan, 1990b) or in the Garo and the Naga hills (Kushwaha and Ramakrishnan, 1987). Seeds of pulses, cucurbits, vegetables and cereals are mixed with dry soil from the site to ensure their

Table 1.1 Phytosociology of crops under different jhum cycles of the Garos at Burnihat in Meghalaya (from Toky and Ramakrishnan, 1981a)

	Frequency (%)			*Density* (plants mm^{-2})			*Basal area* (cm^2)			*IVI**		
Crops species	*30-yr*	*10-yr*	*5-yr*	*30-yr*	*10-yr*	*5-yr*	*30-yr*	*10-yr*	*5-yr*	*30-yr*	*10-yr*	*5-yr*
Grain and seed												
Oryza sativa	100	100	100	8.6	12.2	1.5	52.9	23.9	3.2	163.6	175.5	114.2
Sesamum indicum	80	35	5	1.1	0.5	0.1	1.5	1.4	0.1	28.3	19.9	4.6
Zea mays	80	25	15	1.8	1.5	0.2	4.8	1.8	0.2	38.2	23.4	13.2
Setaria italica	90	30	15	2.1	0.5	0.2	0.9	0.2	0.1	36.9	14.9	14.8
Phaseolus mungo	5	–	–	0.1	–	–	0.01	–	–	1.9	–	–
Ricinus communis	20	–	–	0.2	–	–	1.9	–	–	9.0	–	–
Leaf and fruit vegetables												
Hibiscus sabdariffa	10	20	20	0.1	0.2	0.2	0.5	0.8	0.8	3.8	10.7	19.7
Hibiscus esculentus	–	10	–	–	0.1	–	–	0.1	–	–	4.8	–
Capsicum frutesence	–	5	–	–	0.1	–	–	0.02	–	–	2.2	–
Lagenaria leucantha	10	5	–	0.2	0.3	–	0.2	0.2	–	3.7	3.6	–
Cucurbita maxima	5	–	–	0.1	–	–	0.1	–	–	1.6	–	–
Cucumis sativa	10	–	–	0.2	–	–	0.2	–	–	3.7	–	–
Momordica charantia	–	10	–	–	0.2	–	–	0.01	–	–	4.7	–
Musa sapientum	–	10	30	–	0.1	0.3	6.8	–	20.3	–	21.6	95.4
Tuber and rhizome												
Manihot esculenta	5	10	10	0.1	0.1	0.1	0.2	0.7	0.7	1.8	6.3	11.0
Colocasia antiquorum	5	5	15	0.2	0.2	0.2	2.0	3.6	3.6	5.2	12.3	27.0
Zingiber officinalis	5	–	–	0.2	–	–	0.1	–	–	1.7	–	–

* IVI = importance value indices

Plate 1.3 A Garo farmer at Lailad (296 m) in Meghalaya dibbling the seeds of rice along the bottom of the slope using a bamboo stick

uniform distribution, and broadcast soon after the burn. Maize seeds are dibbled at regular intervals amongst other crops. Similarly, rice is sown into the crop mixture by dibbling with a long stick, after the first rainfall in mid-April (Plate 1.3). Semi-perennial and perennial crops such as ginger, colocasia, tapioca, banana and castor are sown intermittently throughout the growing season (Plate 1.4). The leaves of *Ricinus communis* are used for rearing young silk worm caterpillars. Crops are harvested as they mature (Table 1.2) making room for the other crops left behind. Other crop species could be *Coix lacryma – jobi*, *Eleusine coracana*, *Ipomoea batatus*, *Dioscorea alata*, etc. (Maikhuri and Ramakrishnan, 1990b).

Throughout the cropping period, weeds pose a problem. The most common weeds are root sprouts, rhizome sprouts, stump sprouts, tree seedlings, grasses and herbs. Under long jhum cycles, the problem is less severe than under short jhum cycles (Saxena and Ramakrishnan, 1984a), where many weeds, particularly *Imperata cylindrica*, keep sprouting from underground rhizomes and are difficult to eradicate. Others, like *Eupatorium odoratum* are kept under control through frequent slashing. Hand-hoeing is usually done twice (3–4 times under shorter cycles) during the cropping, mainly by women.

Table 1.2 Sequential harvesting of crops on jhum plots under a 30-year cycle of the Garos at Burnihat in Meghalaya (from Ramakrishnan *et al.*, 1981a)

Species	*Harvesting time*
Setaria italica	Mid-July
Zea mays	Mid-July
Oryza sativa	Early September
Lagenaria spp.	"
Cucumis sativa	"
Zingiber officianalis	Early October
Sesamum indicum	"
Phaseolus mungo	"
Cucurbita spp.	Early November
Manihot esculenta	"
Colocasia antiquorum	"
Hibiscus sabdariffa	Early December
Ricinus communis	(Perennial crop)

All the seeds were sown in April

Jhum at higher elevations

The jhum done by the Khasis at higher elevations (1,500 m altitude and above) in Meghalaya is a modified version of the typical one outlined above. It is commonly practised around Shillong (Mishra and Ramakrishnan, 1981). The study sites at villages (25.34°N and 91.56°E) around Shillong in the Khasi hills of Meghalaya are at an altitude of about 1,500 m. The soil is acidic with pH ranging from 4 to 6. The climate is monsoonic with about 80% of the annual rainfall of 3,070 mm occurring during May – September (Gangwar and Ramakrishnan, 1987a). The average maximum and minimum temperatures during this season are 23°C and 13°C, respectively. The winter is from November to February with average maximum and minimum temperature of 18°C and 7°C, respectively, and with occasional showers. Frosts occur during December–January. The mild dry summer during March–April has an average maximum and minimum temperature of 22°C and 11°C, respectively.

The vegetation consists of sparsely distributed pine trees (*Pinus kesiya*) with much undergrowth of shrubs and herbs. The pine trees are not felled, but the lower branches are slashed in December. The slash is arranged in parallel rows running down the slope and then allowed to dry (Plate 1.5). In the month of March, soil is placed on top of the slash, so as to form ridges alternating with furrows of compacted soil running down the slope. Consequently the burn of the slash is slow and controlled. A fire line of cleared vegetation around the plot helps to check its spread.

The crop mixtures differ from that of the low elevation jhum in that tuber crops such as *Solanum tuberosum*, *Ipomoea batatus* and *Colocacia antiquorum* are planted on the ridges (Mishra and Ramakrishnan, 1981).

Plate 1.4 A jhum plot at Lailad (296 m) in Meghalaya under mixed cropping. *Zea mays* and *Colocasia antiquorum* are clearly visible. In this plot under a 10-year jhum cycle 10–13 crop species are grown together

These are planted soon after the burn and before the onset of the monsoon. Sowing of *Zea mays*, *Phaseolus vulgaris* and a few cucurbits are done just after the onset of the monsoon. Along each ridge, potato and *Zea mays* are sown together in three distinct rows. *Colocasia antiquorum* is generally confined to the top and bottom part of each ridge and the cucurbits are sown at random. *Phaseolus vulgaris* is sown around pine trees for their support.

After the harvest of the tuber crops in July and August, (Plate 1.6) a winter crop of potato is often sown along the ridges. Harvesting of *Zea mays* and the legume *Phaseolus vulgaris* is done in September and October, after which *Brassica oleracea* seedlings are planted along with the winter crop of *Solanum tuberosum.* The second potato crop is harvested during November, and then the field is left uncultivated between December and March. If a second year of cultivation is done, the same procedures are followed; otherwise, the land is fallowed for natural regrowth of vegetation.

The mixture of crops used would vary depending upon the jhum cycle (see Table 1.10). Under a long jhum cycle of 15 years, cropping is done for 1 year only, and no fertilizer is used. Under a 10-year cycle, organic manure in the form of pig dung and vegetable manure is applied at the rate of

Plate 1.5 A Khasi lady is laying down small branches and twigs of trees and undergrowth for drying. In this modified version of jhum at Shillong (1,500 m) in Meghalaya, the dried slash is subsequently topped over by a thin soil layer and subjected to controlled burning. These ridges would alternate with compacted furrows; the crops are sown on the ridges

600 $kg\,ha^{-1}yr^{-1}$ (oven dry weight). Under a 5-year jhum cycle, cropping is carried out for 2–3 years continuously after slash-and-burn, but the crops grown are only *Solanum tuberosum*, *Zea mays*, and *Brassica oleracea*. Variations in crop mixtures do occur (Gangwar and Ramakrishnan, 1987a); occasionally a monocrop of potato may be raised under shorter cycles. During cropping, both organic (pig dung and vegetable manure) and inorganic (nitrogen : phosphorus : potassium in the ratio of 1:1:1) fertilizers are applied at the rate of 1,000 $kg\,ha^{-1}$ and 10 $kg\,ha^{-1}$, respectively, for the first and second cropping in the 1st year of cultivation, and 1,850 $kg\,ha^{-1}$ and 20 $kg\,ha^{-1}$, respectively for the first and second cropping, in the 2nd year.

Herbaceous weeds grow from seeds, rhizomes or root sprouts. *Eupatorium adenophorum*, a noxious weed, may arise through seeds or root sprouts. *Imperata cylidrica* comes up chiefly through fire resistant underground rhizomes and is common in the low elevation jhum too. *Pteridium aquilinum* and *Dicranopteris linearis* are other important weeds that, along with root or stem sprouts of trees and tree seedlings, are kept under control through frequent slashing. Hand-hoeing along the ridges may be done 2–3 times during the season.

ECONOMIC YIELD PATTERN UNDER JHUM

A view point commonly expressed, more often not based on precise data, is that jhum as an agricultural system is unviable on economic considerations compared to settled farming like terracing of the hill slopes. Where data are given, these studies do not define the parameters under which the observations were made. In any study of this nature, one ought to specify the jhum cycle, the crop mixtures, the comparability of plots in terms of ecological factors, the various inputs that go into the system as manure and labour, and the conversion factors used while expressing mixed crop yield as yield of rice or as monetary return. This confusion regarding crop yield also exists for agroecosystems in north-eastern India. In fact, the rice yield under jhum was shown to be as low as 190 kg ha^{-1} (Borthakur *et al.*, 1978) to a high value of 1,200 kg ha^{-1} (Misra, 1976). The Agro-Economic Research Centre, Jorhat, Assam conducted surveys on jhum rice yield and concluded that the average yield of 800–900 kg ha^{-1} in the Garo hills, Mizoram and Arunachal Pradesh is comparable to the average yield of 1,145 kg ha^{-1} for the country as a whole for 1971–1972. A study of rice yields at Burnihat in the lower elevation of Meghalaya (Sahu, 1978) gave yearly outputs of 853 kg ha^{-1} compared to 3,428 kg ha^{-1} under valley cultivation and 738 kg ha^{-1} under terrace cultivation. A report by the Indian Council of Agricultural Research (Borthakur *et al.*, 1978) related rice yield under terracing (1,860 kg ha^{-1}) compared to a very low 190 kg ha^{-1} under jhum.

Unfortunately, none of these studies specify the fertilizer input under different land uses nor do they indicate the jhum pattern, the cycle of which determines the yield. None of the studies also specify whether yields from other crops are included in the final figures and how the conversions to rice yield was arrived at for the mixed cropping jhum system. We shall evaluate some of these aspects in the following pages considering actual yield, and using money and energy as other currencies for evaluation. We shall also consider the ecological implications of the results obtained for better management of the jhum system.

Yield pattern under jhum (typical version) of varied cycles at lower elevations

In a comparative analysis of jhum done by the Garos, under 30-, 10- and 5-year cycles, differences were observed, between them at Burnihat, at lower elevations in Meghalaya (Toky and Ramakrishnan, 1981a). Under the long jhum cycle of 30 years, cereals constitute the major component of the crop mixture. However, under a 5-year cycle the emphasis shifts to seasamum,

Table 1.3 Biomass of crops under different jhum cycles of the Garos at Burnihat in Meghalaya (from Toky and Ramakrishnan, 1981a)

	Biomass (kg ha^{-1} yr^{-1})					
	30-yr		*10-yr*		*5-yr*	
Crop species	*Root*	*Shoot*	*Root*	*Shoot*	*Root*	*Shoot*
Seed grain						
Oryza sativa	1,213	7,614	268	2,440	53	381
Sesamum indicum	163	1,519	103	1,637	8	150
Zea mays	243	3,342	113	1,623	9	136
Setaria italica	69	716	13	121	5	41
Phaseolus mungo	1	38	–	–	–	–
Ricinus communis	131	996	–	–	–	–
	(1,820)	(14,225)	(497)	(5,821)	(75)	(708)
Leaf and fruit vegetables						
Hibiscus sabdariffa	23	301	72	1,177	90	1,232
Hibiscus esculentus	–	–	8	89	–	–
Capsicum frutesence	–	–	2	4	–	–
Lagenaria leucantha	3	189	2	131	–	–
Cucurbita maxima	3	88	–	–	–	–
Cucumis sativa	1	49	–	–	–	–
Momordica charantia	–	–	2	17	–	–
Musa sapientum	–	–	825	3,350	3,325	11,475
	(30)	(627)	(911)	(4,768)	(3,415)	(12707)
Tuber and rhizome						
Manihot esculenta	339	313	1352	960	690	620
Colocasia antiquorum	334	43	373	27	221	25
Zingiber officinalis	10	4	–	–	–	–
	(683)	(360)	(1,725)	(987)	(911)	(645)

Values in parentheses are totals of each category

leafy vegetables and tuber crops, although rice is an important crop (Table 1.1).

Such a shift in pattern due to shortening of the cycle is fairly common in the region as was also shown for the Nishis of Arunachal Pradesh (Maikhuri and Ramakrishnan, 1990b), and this is significant. Under short cycles where the soil is relatively infertile compared to that under longer cycles, the farmer shifts his crop species towards those which have a better nutrient-use efficiency. Such a shift, with emphasis on perennial crops also gives better protective cover to the soil, checking water-related erosive losses more effectively, once the cover is established (Toky and Ramakrishnan, 1981b).

The total biomass (root and shoot) obtained from grain and seed yielding crops under a 30-year jhum cycle was 2.5 and 20.5 times more than that under 10-and 5-year cycles, respectively (Table 1.3). The total biomass from leafy and fruit vegetable crops was maximum under a 5-year jhum cycle and it was 2.8 and 24.5 times more than under 10- and 30-year cycles,

Table 1.4 Comparison of mean economic yield of crops under different jhum cycles of the Garos at Burnihat in Meghalaya (Values in parentheses are totals of each category, – indicates absence) (from Toky and Ramakrishnan, 1981a)

	Total yield (kg ha^{-1} yr^{-1})			*Moisture content* %
	30-yr	*10-yr*	*5-yr*	
Grain and seed				
Oryza sativa	1,161	378	66	20
Sesamum indicum	446	541	25	7
Zea mays	770	397	30	15
Setaria italica	193	23	9	17
Phaseolus mungo	10	–	–	25
Ricinus communis	5	–	–	12
	(2,585)	(1,339)	(130)	
Leaf and fruit vegetables				
Hibiscus sabdariffa	44	139	96	75
Hibiscus esculentus	–	50	–	65
Capsicum frutesence	–	1	–	80
Lagenaria leucantha	140	81	–	93
Cucurbita maxima	62	–	–	91
Cucumis sativa	16	–	–	92
Momordica charantia	–	5	–	82
Musa sapientum	–	105	488	78
	(262)	(381)	(584)	
Tuber and rhizome				
Manihot esculenta	339	1,352	690	59
Colocasia antiquorum	260	294	180	79
Zingiber officinalis	10	–	–	52
	(609)	(1,646)	(870)	
Silk worm				
Cocoon (silk)	4	–	–	2
Pupae (without cocoon)	0.2	–	–	90

respectively. The biomass obtained from tuberous and rhizomatous crops under a 10-year cycle was almost twice the output of the other two cycles.

Maximum economic yield per hectare for rice, maize and *Setaria italica* was obtained under a 30-year cycle; the reduction in the economic yield of grain and seed yielding crops was 48% and 98% under 10-and 5-year cycles, respectively (Table 1.4). However, the total economic yield of leafy and fruit vegetables, and tuberous and rhizomatous crops was higher under 5-and 10-year jhum cycles than under a 30-year cycle. The high yields of seasamum, tapioca, banana and leafy vegetables under a short jhum cycle may arise because of rapid initial growth giving quick cover which also shades out weeds. Further, deep and extensively branched root systems ensure efficient use of available nutrients (Toky and Ramakrishnan, 1981a).

Table 1.5 Monetary input–output (Rupees ha^{-1} yr^{-1}) under different jhum cycles at Burnihat in Meghalaya (from Toky and Ramakrishnan, 1981a)

	Jhum		
	30-yr	*10-yr*	*5-yr*
Input	2,616	1,830	896
Output	5,586	3,354	1,690
Net gain/loss	2,970	1,524	794
Output/input	2.13	1.83	1.88

Table 1.5 compares the monetary output/input analysis of the jhum under different cycles. The reduction in overall economic return is very obvious, when cost is calculated on the basis of prevailing market rates. The decrease in economic return under short jhum cycles is related to reduced soil fertility (Ramakrishnan and Toky, 1981) and increased weed potential of the site (Saxena and Ramakrishnan, 1984b). Similar results are shown through studies done elsewhere in Venezuela (Watters, 1971). According to this study, maize yield decreased from 803 kg ha^{-1} to 640 kg ha^{-1} and yucca and bean production fell 60% and 100%, respectively under short fallow systems. Labour which is the only input into the jhum system, used for slash-and-burn of the fallow and for weeding, is in a sense free to the farmer, because it is provided by the farmer's family.

Comparative analysis of jhum (typical version) by three tribes coexisting at lower elevations of Meghalaya

The study area at Lailad, about 70 km north of Shillong, (25.45°N, 91.45°E) is at an altitude of 296 m, and is located in the Khasi hills of Meghalaya, with four communities, namely the Garos, the Khasis, the Mikirs and the Nepalis living together. The area receives an annual average rainfall of 1,435 mm (Maikhuri and Ramakrishnan, 1990a). The rainy season extends from May to October. This is a warm period with high humidity. The average maximum temperature during the season is 35°C and the minimum is 27°C. The mild winter, with an average maximum temperature of 28°C and an average minimum of 17°C extends from November to mid-February. A short dry windy summer from mid-February to May has a maximum temperature of 34°C and a minimum of 23°C.

Jhum as practised by the Garos, the Khasis and the Mikirs living together in the same area at Lailad differed in terms of crop mixtures and the cultivation procedures (Table 1.6). Amongst the communities considered here, the Nepalis, being non-tribal, do not have jhum as a land use. Though mixed cropping is an essential feature of this land use system (Nye and Greenland, 1960; Mutsaers *et al.*, 1981; Ramakrishnan, 1984a), the number

Table 1.6 Major differences in cropping between different communities at Lailad in Meghalaya (from Maikhuri and Ramakrishnan, 1990a)

	Tribal		*Non-tribal*	
Agriculture system	*Garos*	*Khasis*	*Mikirs*	*Nepalis*
Slash-and-burn agriculture	Community labour	Community labour	Family members: stems of larger trees undisturbed	–
Labour for slashing	Stems of larger trees undisturbed	Total slashing		
Crop mixture	10–13	10–13	4–7	–
Sowing time	April	April	February	–
Organic manure input	–	Under 10 and 5 year cycles only	–	–
Weeding	3–4 times	Once	–	–
Harvesting	Sequential (June–December)	Sequential (June–October)	Simultaneous (August–September)	– –
By-products	Left in the plot	Left in the plot	Removed as cattle feed	–
Valley				
Yearly croppings	2	–	1	1
Labour input	Human and animal	–	Human	Human and animal
Organic manure input	Yes	–	Yes	Yes
Weeding	–	–	Once	3 times
Harvesting of crop	July first crop (December for second crop)	–	September	September
By-products	Left *in situ*	–	Removed as cattle feed	Removed as cattle feed

– indicates land use not practised

of crop species are minimal for the Mikirs who are tribes from the plains, and this is suggestive of their affinity to the plains people of the Assam valley who only monocrop rice. The hill tribes, like the Khasis and the Garos, do sequential harvesting of crops but not the Mikirs. Again, the Khasis and the Garos recycle their organic residue into the jhum plots which is characteristic of most hill tribes and unlike the Mikirs (Mishra and Ramakrishnan, 1982). Such differences in the jhum procedures based on socio-cultural affinities of the tribals are significant.

The monetary yield under different jhum cycles essentially followed a similar pattern as the one discussed above. Within a given cycle, however, the Garos obtained more yield from grain and seed crops, and from tuber and rhizome crops compared with the Khasis; the yield was least for the Mikirs (Table 1.7). The net economic return from jhum systems was generally higher for the Garos, which may be largely linked to better

Table 1.7 Mean ($\pm$ SE) economic yield (kg ha^{-1}) under different jhum cycles of different tribal communities at Lailad in Meghalaya (from Maikhuri and Ramakrishnan, 1990a)

Jhum cycle	*Crop category*			*Grand total*
	Grains and seeds	*Leaves and fruit*	*Tubers and rhizomes*	
20-year				
Garos	2,188 ± 174	370 ± 27	872 ± 68	3,430 ± 238
Khasis	1,920 ± 168	290 ± 22	452 ± 40	2,662 ± 189
Mikirs	1,244 ± 104	74 ± 6	468 ± 37	1,786 ± 154
10-year				
Garos	1,625 ± 131	349 ± 31	988 ± 81	2,962 ± 211
Khasis	1,007 ± 96	401 ± 29	859 ± 56	2,267 ± 169
Mikirs	807 ± 67	141 ± 11	411 ± 33	1,359 ± 133
5-year				
Garos	953 ± 71	374 ± 22	1307 ± 97	2,634 ± 178
Khasis	642 ± 43	495 ± 37	453 ± 32	1,590 ± 128
Mikirs	583 ± 32	403 ± 34	–	986 ± 76

– indicates no data

organization of the crop mixtures by these traditional jhum farmers, with emphasis on grain and seed crops and tuber and rhizome crops.

Comparative analysis of jhum (typical version) of the Nishis and the Hill Miris of Arunachal Pradesh

The study site at Balijan, about 110 km north of Itanagar in Arunachal Pradesh (26.56°N, 93.32°E) is located at an elevation of 200 m. (Maikhuri and Ramakrishnan, 1990b). The climate is typically monsoonic with an annual rainfall of 730 mm, 85% of which occurs during May to September. The average maximum temperature during the summer is 32.6°C and the minimum is 26.2°C; it is 26.5°C and 15.7°C, respectively during the winter.

The Nishis of Balijan in Arunachal Pradesh (200 m altitude) do jhum, more commonly under 5-year cycles, though cycle lengths of up to 60 years may be found in more remote areas (Maikhuri and Ramakrishnan, 1990b). Twenty-five to 33 crop species are grown together. The yield pattern under different jhum cycles remained essentially the same as discussed earlier. Although the return under jhum generally declined with the shortening of the jhum cycle, the net return under a 10-year cycle was higher than under all others (Table 1.8). This is because of the reduced labour costs involved in slash-and-burn operations under a 10-year jhum cycle than under longer cycles on the one hand, and the poor crop yield due to reduced soil fertility under a too short 5-year cycle (Ramakrishnan and Toky, 1981; Gangwar and Ramakrishnan, 1987b; Maikhuri and Ramakrishnan, 1990b). This

Table 1.8 Energy (MJ) and economic (Rupees $ha^{-1} yr^{-1}$) analysis of jhum done by the Nishis in Balijan in Arunachal Pradesh. Monetary values are also presented. (from Maikhuri and Ramakrishnan, 1990b)

	Jhum cycle (yr)									
	60		*30*		*20*		*10*		*5*	
Production measure	*Energy*	*Monetary*	*Energy*	*Monetary*	*Energy*	*Monetary*	*Energy*	*Monetary*	*Energy*	*Monetary*
Inputs										
Labour	2,788	4,624	2,231	3,650	1,543	2,980	1,155	2,740	819	2,186
Seed	67	32	63	38	56	36	39	26	34	29
Total	2,855	4,658	2,294	3,888	1,599	3,016	1,194	2,766	853	2,215
Output										
Grain and seed	51,639	5,532	47,425	5,058	38,874	4,137	33,075	4,377	9,585	1,110
Leaf and fruit vegetables	3,555	338	4,171	396	4,772	454	6,984	560	6,889	563
Tuber and rhizome	11,977	1,560	9,243	1,203	7,281	1,043	10,643	1,527	18,581	2,663
Total	67,171	7,430	60,839	6,657	50,927	5,634	50,702	6,464	35,055	4,336
Output/input ratio	24	1.6	27	1.7	32	1.9	43	2.3	41	2.0

essentially implies that a 10-year cycle is the cut-off length from the point of economic efficiency as shown in this study (Maikhuri and Ramakrishnan, 1990b) and as observed in our earlier studies (Toky and Ramakrishnan, 1981a).

The hill Miris of Arunachal Pradesh living at Raga (27.44°N, 93.51°E) at an altitude of about 1,000 m also practise the typical version of jhum under 7- to 10-year jhum cycles (Kumar and Ramakrishnan, 1989a). The output total is somewhat lower than the jhum under comparable cycles done by the Nishis.

Jhum (typical version) practised by the Khasis at mid-elevations of Meghalaya

The study site at Nayabunglow (25.45°N and 91.54°E) at an altitude of 910 m in the Khasi hills of Meghalaya is about 30 km north of Shillong. The climate is typically monsoonic with over 80% of the total annual rainfall of 1,800 mm occurring during May–September (Patnaik and Ramakrishnan, 1989). The monsoon is followed by winter during October–February. March and April represent a brief warm dry period. The mean monthly maximum and minimum temperatures during the monsoon are 28.6°C and 17.1°C, respectively and for the winter, they were 21.3 and 3.9°C, respectively.

At Nayabunglow, the Khasis practise jhum under cycle lengths of 5–10 years (Patnaik and Ramakrishnan, 1989), the emphasis here being exclusively on tuber, root and vegetable crop species, such as ginger, *Colocasia antiquorum*, *Ipomoea batatus*, *Capsicum frutesence* and cucurbits. This shift in emphasis towards tuber crops is related to the sharp reduction in soil fertility under short cycles at this elevation where soil fertility recovery is slower under the cooler climate. The Khasis obtain higher economic returns (Table 1.9) through this shift towards more nutrient-use efficient tuber/root crops, as also seen from many other studies (Toky and Ramakrishnan, 1981a; Ramakrishnan, 1984a). Indeed, the monetary output total under 10-year jhum cycles here is six times higher than that under the jhum practised by the Garos at lower elevations of Meghalaya (cf Table 1.4).

Modified version of jhum of the Khasis at higher elevations of Meghalaya

In the mixed cropping system used on the slopes prepared into ridges and furrows, the major crop grown by the Khasis at Shillong was rice, until a few decades ago. But when the British introduced potato as a crop into this region towards the turn of this century, it became a popular crop very rapidly, although rice is still the staple diet of the local tribe, and is frequently grown.

Table 1.9 Monetary (Rupees $ha^{-1} yr^{-1}$) analysis of jhum of the Khasis at Nayabunglow in Meghalaya (from Patnaik and Ramakrishnan, 1989)

	Jhum cycle (yr)		
	5	*7*	*10*
Input total	7,431	9,693	10,548
Human labour	4,800	6,154	6,902
Bullock power	–	–	–
Seed	2,631	3,539	3,646
Inorganic fertilizer	–	–	–
Output total	7,520	14,194	18,370
Output/input ratio	1.01	1.46	1.74

At higher elevations, where the pine trees are sparsely distributed, slash-and-burn with clear-felling of the forest is not feasible because of slower forest regeneration capabilty of the site under the sub-temperate climate. (Mishra and Ramakrishnan, 1983a). The preparation of the site into alternate ridges and compacted furrows running down the slope with the compacted furrows acting as water channels would help in conserving nutrient losses through water (Mishra and Ramakrishnan 1983c), which could, otherwise, be higher. This is particularly important because the soil fertility recovery is slower at these elevations (Mishra and Ramakrishnan, 1983b,d) and the soil under pine forests is highly acidic, further aggravating the nutrient availability (Ramakrishnan and Das, 1983; Das and Ramakrishnan, 1985). Burning the limited slash that is available after stacking in parallel rows, and slow burning of the dried slash, after topping it over with a thin soil layer to form the ridges, is an efficient way of resource management. The nutrient enriched furrows alone then are used to support crop growth. In such a situation of generally reduced soil fertility level that becomes further aggravated by the shortened jhum cycle of 4–5 years, the shift towards tuber crops that can give better economic yield in nutrient poor soils (Ramakrishnan, 1984a), is understandable. No wonder that when potato was introduced a few decades ago, it rapidly became popular.

A comparative analysis of the modified version of jhum carried out by the Khasis under varied cycle lengths shows that the economic returns are very high, although the yield declines with shortening of the jhum cycle (Table 1.10). The monetary return under a 10-year jhum cycle (Table 1.11) is about 5-times more than that under a similar cycle length done at lower elevation (cf Table 1.5).

This high net monetary return and economic efficiency (output/input ratio) is in spite of the high input required for land preparation in the high elevation modified system. Potato which has higher monetary value is largely produced for export from the village ecosystem.

Table 1.10 Economic yield of crops (kg ha^{-1} yr^{-1} oven dry wt) (± SE) under different jhum cycles of the Khasis at Shillong in Meghalaya (from Mishra and Ramakrishnan, 1981)

	Jhum			
Crop species	*15-yr cycle*	*10-yr cycle*	*1-yr crop*	*11-yr crop*
Root and tuber crops				
Solanum tuberosum	1,579.95 ± 23.75	1099.10 ± 5.55	748.76 ± 2.29	449.25 ± 1.86
Ipomoea batatus	444.78 ± 7.85	148.26 ± 2.10	–	–
Colocasia antiquorum	4.87 ± 0.27	5.57 ± 1.91	–	–
Cereals				
Zea mays	42.35 ± 1.07	59.29 ± 1.92	61.62 ± 1.39	8.57 ± 0.31
Oryza sativa	–	–	–	–
Legume				
Phaseolus vulgaris	6.66 ± 0.27	3.99 ± 0.26	–	–
Fruit vegetables				
Cucurbita maxima	272.43 ± 3.67	257.55 ± 1.95	–	–
Cucumis sativus	9.08 ± 0.58	6.31 ± 0.35	–	–
Leafy vegetables				
Brassica oleracea	76.48 ± 2.11	27.41 ± 1.29	163.90 ± 1.42	50.25 ± 1.88
Colocasia antiquorum	2.38 ± 0.13	2.47 ± 0.15	–	–
Cucurbita maxima	3.09 ± 0.24	2.79 ± 0.27	–	–
Ipomoea batatus	0.98 ± 0.07	1.13 ± 0.15	–	–
Other vegetables				
Colocasia antiquorum (stem)	0.44 ± 0.02	0.41 ± 0.02	–	–
Cucurbita maxima (flower)	0.41 ± 0.02	0.47 ± 0.03	–	–
By-products				
Oryza sativa (straw)	–	–	–	–

– indicates absence

Table 1.11 Cost–benefit analysis (Rupees ha^{-1} yr^{-1}) for different jhum systems of the Khasis at Shillong in Meghalaya (Mishra and Ramakrishnan, 1981)

	Jhum cultivation			
			5-yr cycle	
Production measures	*15-yr cycle*	*10-yr cycle*	*1st yr crop*	*2nd yr crop*
Inputs				
Labour	2,220	2,049	1,613	1,604
Organic manure	–	175	287	537
Fertilizer	–	–	19	39
Seed	1,061	1,206	1,235	1,254
Interest on working capital (12%)	394	412	378	412
Output				
Root and tuber crops	14,711	9,875	6,463	3,914
Cereal grains	65	90	94	13
Legumes	16	9	–	–
Fruit and other vegetables	4,220	3,908	–	–
Leafy vegetables	778	289	1,631	500
By-products	–	–	–	–
Cost of production (a)	3,675	3,842	3,532	3,846
Gross return (b)	19,790	14,171	8,188	4,427
Net return (b – a)	16,115	10,329	4,656	581
Return per rupee	5.4	3.7	2.3	1.2

Conclusions

Cropping and yield patterns under jhum vary considerably, as seen from the preceding pages. Partly this is related to the ecological, socio-economic and socio-cultural background where the system operates. Thus for example, the Garos being hill tribesmen are better organized jhum farmers than the Mikirs at Lailad, because the latter are basically plains tribesmen of the Assam valley.

The length of the jhum cycle has been drastically shortened during the recent past with a 4–5-year cycle length being more common, though a 60-year cycle could be rarely found as for the Nishis of Arunachal Pradesh. It is interesting to find a shift towards tuber/rhizomatous crops under shorter cycles; this was seen for the Garos at Burnihat in Meghalaya and for the Nishis of Arunachal Pradesh. Such distantly placed tribes showing a similarity in shift of the cropping pattern under shortening jhum cycles suggests that the farmers, based upon empirical knowledge collected over a period of time, adapt the crop mixture in a manner to maximize production. They find, for example, that certain crop species are more appropriate under reduced fertility status in the soil – in this case tuber and rhizome crops under short jhum cycles (Ramakrishnan, 1984a). Such a shift to tuber crops by the Khasis at higher elevations is also a response to the ecology of the area.

With a highly heterogeneous social environment, with many tribes coexisting in the same area, or tribes changing over short distances, wide variations occur for the yield pattern under jhum, which is related to crop mixtures used and cropping pattern itself. It follows from this that often by mere manipulation of crop mixture, one could improve the yield from the jhum system, as seen from the discussion here. However, it is important to keep a minimum cycle length of 10 years if the system, as it functions now, is to be economically viable.

ENERGY BUDGET UNDER JHUM

Apart from economic yield of crops and monetary cost–benefit analysis, energy budgets and efficiences provide one of the many ways to evaluate agroecosystem function. Fossil energy (fertilizer, fuel, etc.) land and labour are the three major resources for modern agriculture. These three are related to one another and could compensate, at least in part, one for the other.

In the jhum system under consideration here, fossil fuel energy is not normally used except where the cycle is short and soil fertility very low. Land is a limiting factor to the extent that with reduced land area and increased population pressure, the farmer resorts to shortened jhum cycles (Ramakrishnan, *et al.*, 1981a,b). Labour input is from within the family, except for labour intense activities where labour is provided collectively, a few families joining together. Soil fertility is maintained through natural processes of forest fallow regrowth during secondary succession (Ramakrishnan and Toky, 1981; Mishra and Ramakrishnan, 1983d). The present study analyses the energy budget under different jhum systems with varied inputs in terms of labour, land availability that determines jhum cycle and organic fertilizer input. The energy efficiency (output/input ratio) of the different jhum systems in relation to these components and the crop mixtures used is evaluated.

Energy budget under different cycles of typical jhum

In a comparative analysis of the jhum system of the Garos under 30-, 10-, and 5-year cycles at lower elevations of Meghalaya, we considered in detail the implications of energy budget under shortening jhum cycles (Toky and Ramakrishnan, 1982a). A major conclusion arising from this comparative analysis is that jhum in the hill areas of north-eastern India has survived all these years chiefly because of the high energy efficiency of the system associated with longer cycles where the only energy input is in the form of

Table 1.12 Energy inputs (MJ ha^{-1} yr^{-1}) for agriculture under different jhum cycles of the Garos in Burnihat in Meghalaya (from Toky and Ramakrishnan, 1982a)

	Energy input (MJ ha^{-1} yr^{-1}) *under jhum*		
Agricultural operation	*30-yr cycle*	*10-yr cycle*	*5-yr cycle*
Labour input (total)	1,641	1,176	494
Clearing understorey vegetation	274	331	78
Felling trees and bamboos	583	319	–
Collection of debris and burning	51	35	5
Terrace preparation	–	–	–
Dibbling broadcasting and transplanting	40	43	43
Hut construction	23	24	24
Weeding	63	76	87
Guarding the field from wild animals	71	70	64
Rearing caterpillars	123	–	–
Harvest	187	198	173
Transportation	21	13	8
Threshing	77	25	5
Shelling	128	42	7
Seed input	24	15	16
Fertiliser	–	–	–
Total energy input into the system	1,665	1,191	510

manual labour which is provided by the farmer, a major fraction of which goes for slash-and-burn operation (Table 1.12). Further, this labour input is uniformly distributed throughout the year as shown by us in one of the studies on the Khasis working under a 5-year jhum cycle (Fig 1.1) (Patnaik and Ramakrishnan, 1989). According to Rappaport (1971), the 'Tsembaga' people of the New Guinea Highlands obtained an average of 16 units of food energy for each unit of human energy employed during farming; this may increase to 20 under more favourable conditions. Others have reported equally high, or even higher, efficiency values with output/input ratios of up to 54 (Lewis, 1951; Norman, 1978; Uhl and Murphy, 1981).

Most of these studies, however, do not mention the length of the jhum cycle and its relationship to energy efficiency. Our study of three jhum cycles of 30, 10, and 5 years give different energy input and output patterns, as well as overall efficiency values for the system (Table 1.13). This is understandable in view of the fact that the labour energy input into the system varies for the following reasons: Firstly, the bamboo/dicotyledonous tree dominated fallow slashed under long cycles involves greater expenditure of energy compared with the weedy herbaceous growth slashed under a 5-year cycle, secondly, the energy expended on weeding increased with

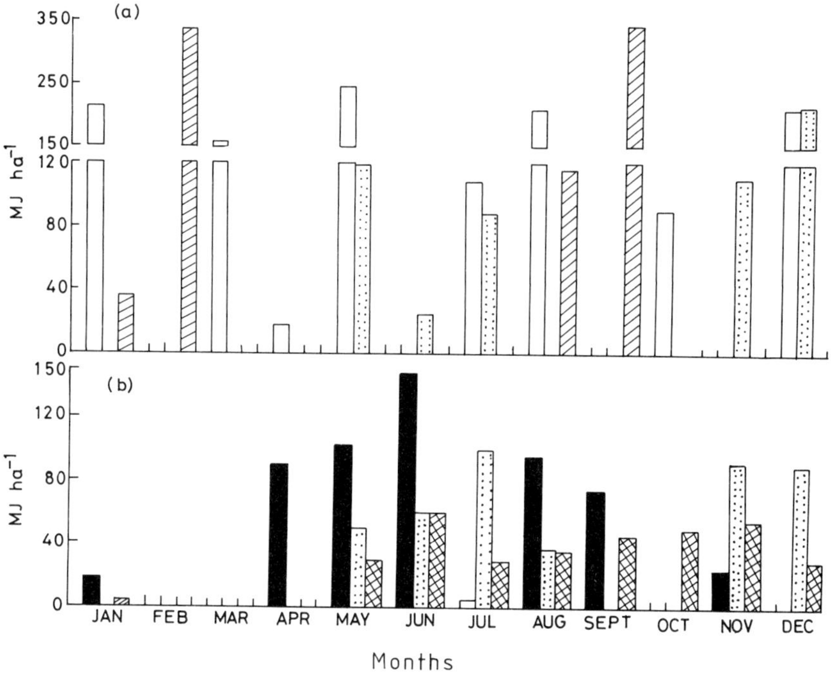

Figure 1.1 Distribution of labour over a year under jhum of the Khasis compared with other land use systems of the Khasis (a) another Nepalis (b) at Nayabunglow in Meghalaya. □, 5-year jhum cycle, ▨, valley cultivation; ▨, pineapple plantation; ▩, kitchen garden; ■, fodder grass cultivation (from Patnaik and Ramakrishnan, 1989)

Table 1.13 Energy ratio for agriculture under diffferent jhum cycles of the Garos in Burnihat in Meghalaya (Toky and Ramakrishnan, 1982a)

Agriculture system	*Energy* (MJ ha^{-1} yr^{-1}) *Input*	*Ouput*	*Output/input ratio*
Jhum			
30-year cycle	1,665	56,766	34.1
10-year cycle	1,191	56,601	47.5
5-year cycle	510	23,858	46.7

shortening of the jhum cycle as the weed problem is aggravated here (Saxena and Ramakrishnan, 1984b) so much so that the secondary succession is often arrested at the weed stage under a 5-year cycle (Toky and Ramakrishnan, 1983a), and finally, considerable reduction in yield occurred under a 5-year cycle, partly related to poor nutrient recovery during the short fallow phase (Ramakrishnan and Toky, 1981) and more severe weed problem as mentioned above.

Table 1.14 Energy input and output (MJ $ha^{-1} yr^{-1}$) under different jhum cycles of different tribal communities at Lailad in Meghalaya (from Maikhuri and Ramakrishnan, 1990a)

	Production measures					
		Output				
Jhum cycle	*Input*	*Grains and seeds*	*Leaves and fruit*	*Tubers and rhizomes*	*Total*	*Output/ input ratio*
20-year						
Garos	1,688	41,089	5,846	13,342	60,277	35.7
Khasis	1,352	37,487	4,582	6,916	48,985	36.2
Mikirs	1,043	24,649	1,169	7,160	32,978	31.6
10-year						
Garos	1,200	31,509	5,517	15,116	52,142	43.4
Khasis	1,659	18,065	6,336	13,143	37,544	22.6
Mikirs	794	14,642	2,228	6,288	23,158	29.0
5-year						
Garos	810	18,852	5,909	19,997	44,758	55.2
Khasis	1,470	11,934	7,821	6,931	26,686	18.1
Mikirs	546	9,462	6,367	–	15,829	29.0

– indicates no data

In another study carried out on the Nishis of Balijan at Arunachal Pradesh (Maikhuri and Ramakrishnan, 1990b), we compared energy efficiencies of jhum cycles of 60, 30, 20, 10 and 5 years, and found that a too long jhum cycle such as a 60-year one is less efficient (Table 1.8). The major disadvantage of a 60- or even 30-year cycle lies in the high labour energy input needed for slash-and-burn of a denser secondary forest. This would explain the more favourable output/input ratio of the jhum system, under a 10-year cycle compared to all others, including the 5-year cycle where the low output reduced the efficiency.

Comparative energy analysis of jhum systems of different tribes

A comparative analysis of energy efficiency (output/input ratio under the typical version of the jhum as practised by different tribes living is the same area in the north-eastern region showed wide variation (Table 1.14). Between the Garos, the Khasis and the Mikirs living in the same area, at Lailad at lower elevations (296 m) of Meghalaya the Garos have higher efficiency values (Maikhuri and Ramakrishnan, 1990a). However the Khasis, operating at Nayabunglow at mid-elevations of Meghalaya (960 m), with less variety in crop mixture, growing exclusively tuber and rhizomatous crops, obtain very low energy efficiencies for their system (Patnaik and Ramakrishnan, 1989). The energy efficiency for the Nishis of Arunachal Pradesh (Maikhuri and Ramakrishnan, 1990b) is generally comparable to that of the Garos at lower elevations of Meghalaya. Such wide variation between tribes, even under comparable jhum cycles, could often be largely linked with variation in the emphasis on the crop mixture, particularly when the communities live in the same area. If they are from different areas, labour expended in land preparation and differences in the ecology of the area could also be factors.

Energy budget under modified jhum at higher elevations of Meghalaya

The broad generalizations about the typical version of jhum discussed above also apply to the modified version of the high elevation system (Mishra and Ramakrishnan, 1981) except for a major difference in the energy budget. The output/input ratios for jhum here are generally lower with values of 26, 10, and 5 for 15-, 10-, and 5-year jhum cycles, respectively. This is because of a heavier labour input for land preparation into ridges and furrows and also because of organic (pig dung and vegetable compost) and inorganic (NPK, 1:1:1) fertilizer inputs which could be at the rate of 1,000 kg ha^{-1} and 10 kg ha^{-1}, respectively during the first year of cropping under a 5-year jhum cycle. During the second year of cropping, which is usually done, these quantities almost double. Under a 10-year jhum cycle, these quantities may be lower than that under a 5-year cycle. Such a high input of energy for cropping is necessitated by poor soil fertility at higher elevations, because of lower temperature that slows down soil fertility recovery processes (Mishra and Ramakrishnan, 1983d), higher soil acidity under pine, and slowly decomposing pine litter. Indeed, in a study of the modified version of the jhum system near Shillong involving potato as the major crop but with variations in the mixture of other crop species under 15-, 8-, and 5-year jhum cycles, the energy ratio (output/input) varied between 3.6 and 9.8, which is very low indeed (Gangwar and Ramakrishnan, 1987a).

In this jhum system, as also in all others discussed earlier, women play an important role in jhum operations. Only heavy labour such as that for slash-and-burn is shared by men. Figure 1.2 shows the allocation pattern of labour between males and females for a 5-year jhum cycle of the modified version as done by the Khasis, which was generally the same for longer cycles too. More than 65% of the work was done by the female members of the family. However, the work done by both male and female members for the initial operation of slashing of understorey vegetation was more or less the same. While preparation of land into ridges and furrows along with burning and seed sowing operations was an exclusively female function, weeding and maintenance of the field was almost equally shared between male and female members. The implications of these results are obvious. In any consideration of the jhum system, the viewpoints of the female

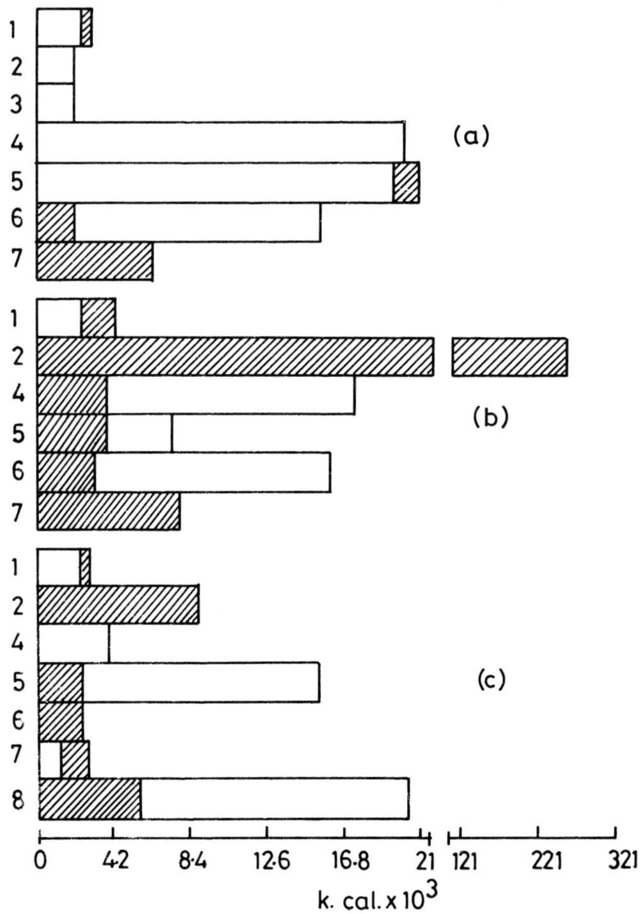

Figure 1.2 Allocation pattern of labour between male and female members of the Khasi family under jhum compared with other land use systems at Shillong in Meghalaya. a, 5-year jhum cycle; b, terrace agroecosystem; c, valley agroecosystem; hatched column, male labour; open column, female labour; 1, clearing the vegetation; 2, land preparation; 3, burning; 4, seed sowing; 5, weeding and field maintenance; 6, harvesting; 7, transportation; 8, threshing and milling (from Mishra and Ramakrishnan, 1981)

members of the society should find adequate consideration. We shall consider this further in our subsequent discussions.

Conclusions

Energy in relation to land under jhum

The comparative energy efficiency of the jhum cycles cannot be considered in isolation but needs to be discussed in relation to the land use pattern or else energy efficiency values *per se* could lead to a distorted comparison for different jhum cycles (e.g., 5-year cycle versus a 10- or 30-year cycle). The studies of Toky and Ramakrishnan (1982a) for example, indicate the trade-off between land area and energy input. If land is not a limiting factor, then the greater input to a larger area of the jhum system with a longer cycle could be used to offset imported energy and this would ensure harmony of the long cycle with the environment, at the same time ensuring rational returns to the farmer. However, with increase in population pressure and decrease in acreage available for jhum, due to environmental degradation with permanent weedy communities of arrested succession or even extreme desertification (Ramakrishnan, 1985d; Ramakrishnan *et al.*, 1981a,b), land is in short supply, resulting in very short jhum cycles of 4–5 years. In such a situation, the longer the jhum cycle, the larger the area of land that is needed, although effective output per hectare would decrease due to a correction factor of $^1/_{30}$, $^1/_{10}$ or $^1/_5$ for jhum cycles of 30, 10 and 5 years so that an effective output of only 1,892, 5,660 or 4,772 MJ per hectare is obtained (cf Table 1.13). On this basis too, a 10-year jhum cycle is the most efficient in terms of energy ratio and land use. Similar conclusions emerge from our other studies of jhum systems in the north-eastern region considering in one instance cycle lengths ranging between 5 and 60 years (Maikhuri and Ramakrishnan, 1990b).

Energy efficiency and sustainability of jhum

With relatively high energy efficiency, jhum under a 10-year cycle is sustainable. However, wide variations exist in the energy efficiency of jhum practised by different tribes in the same area. Efficiency in jhum operations such as slash-and-burn, sowing practices, weed management and crop organization, all contribute to overall energy efficiency. The variations that are available could be exploited for improving this land use, for the benefit of the farmer. Indeed certain crops such as *Flemingia vestita* used by jhum farmers at higher elevations of Meghalaya (Gangwar and Ramakrishnan, 1989b) could even replace the input of fertilizer required under shorter cycles. This is because of the nitrogen fixed by this legume when included in the crop mixture.

IMPLICATIONS OF MIXED CROPPING UNDER JHUM

A high species diversity is characteristic of many tropical agroecosystems (Whittaker, 1972, 1975). Such a high species diversity, as under jhum (Plate 1.6) contributes to agroecosystem stability. With a multilayered canopy and with a high leaf area index for efficient light capture and a layered root mass distribution in the soil profile for optimal nutrient use, crop

Plate 1.6 A jhum plot at Shillong (1,500 m) in Meghalaya under mixed cropping. *Zea mays* is clearly visible on the ridges. Potato has already been harvested by the Khasi farmer. This mixed cropping system with species grown on the ridges has compacted furrows running down the slope acting as water channels

productivity is usually high (Table 1.15). The values for the primary productivity under jhum in north-eastern India (Toky and Ramakrishnan, 1981a) compare with those of the secondary successional fallows up to 20 years in the same area (Toky and Ramakrishnan, 1983a).

A characteristic feature of jhum is the high rate of biomass accumulation in relation to economic output. The high rates of organic matter production measured under jhum (16–22 tons ha^{-1}) are close to values obtained from natural plant communities (14.8 tons ha^{-1}) for 20-year-old forest fallows in the same area of north-eastern India. With higher crop diversity, it should be possible to combine the need for increased harvestable food production with the need for maintaining high organic biomass content in the system as a whole (Trenbath, 1974). Without this high organic matter production it would become necessary to constantly import costly inorganic fertilizers which are hard to come by and whose effectiveness in the face of high temperature and heavy rainfall is questionable (Gliessman, 1980; Gliessman *et al.*, 1981; Ramakrishnan *et al.*, 1981a,b), particularly when the soil is extremely porous contributing to heavy infiltration losses (Toky and

Table 1.15 Economic yield and community traits for agriculture under different cycles for the Garos at Burnihat in Meghalaya (from Toky and Ramakrishnan, 1981a)

	Fallow period (yr)		
	5	*10*	*30*
Economic yields (tons ha^{-1} yr^{-1})			
Seeds	0.107	1.153	2.180
Leaf fruit	0.129	0.074	0.024
Tubers	0.320	0.613	0.192
Total	0.556	1.840	2.396
ANP	14.060	11.576	15.213
Growth rate (g m^{-2} day^{-1})	3.8	3.2	4.2
NPP	18.461	14.709	17.746
Growth rate (g m^{-2} day^{-1})	5.1	4.0	4.9
Number of cultivars	8	12	14
LAI	0.59	1.49	3.20
HI	0.030	0.125	0.135
HI (grain + seed)	0.184	0.230	0.182
Labour (days ha^{-1} yr^{-1})	149	305	436

ANP = annual net production
NPP = net primary productivity
LAI = leaf area index

HI = Crop yield/NPP because seeds, leaves and tubers are harvested
HI (grain + seed) = above ground crop yield/ANP

Ramakrishnan, 1981b; Mishra and Ramakrishnan, 1983c). Being a no-tillage agroecosystem, substantial organic matter (Mishra and Ramakrishnan, 1982) is conserved and is comparable to old-field natural plant communities (Ramakrishnan, 1988a; Toky and Ramakrishnan, 1983a, Stinner *et al.*, 1983).

Sequential harvesting of crops is an effective way of managing up to 35–40 crop species over both space and time. Thus after harvesting early maturing species such as maize and *Setaria italica*, rice gets more space at the peak of its growth period. Successive harvests of cereals create additional space for the remaining perennial crops, which also receive humus and nutrients (Ramakrishnan, 1984a). Such an efficient recycling of organic crop residues may be illustrated for an important nutrient such as nitrogen. The amount of nitrogen ploughed back as by-products into the system during one cropping season amounts to 0.6–0.8 kg ha^{-1} depending upon the jhum cycle, at higher elevations of Meghalaya (Mishra and Ramakrishnan, 1984). Under a 15-year jhum cycle, this amounts to $^{1}/_{6}$th of the total of 4.1 kg ha^{-1} yr^{-1} of nitrogen through non-edible biomass which goes into the jhum system either directly or through the manure pit in the village (Mishra and Ramakrishnan, 1982).

Mixed cropping as practised in this traditional agroecosystem is receiving considerable attention from modern agricultural scientists as a biological pest suppressant (Litsinger and Moody, 1976). The use of native varieties would probably ensure that a high degree of natural chemical defences are maintained (Janzen, 1973). Further, under mixed cropping, it is unlikely that any one of the pest populations of insects, fungi, bacteria or nematodes would reach epidemic levels due to high genetic diversity.

On a steep slope of 30–40° angle, the soil nutrient environment is very transient. One of the important objectives of the jhum farmer, therefore, is to capitalize upon the highly transient resource base as quickly and effectively as possible through a mixed cropping system. Maximizing economic yield in such a transient environment is critical. The jhum farmer ensures effective use of the nutrient gradient on the steep hill slope by emphasizing upon species that have a high nutrient-use efficiency along the nutrient-poor top of the slope and by organizing less use-efficient species along the nutrient-rich base of the slope (Ramakrishnan, 1984a). By this he is able to achieve a high leaf area index for optimizing photosynthesis and maximizing economic yield.

Erosion from plots being a major cause for loss of sediment and nutrients could be effectively countered by good soil cover afforded by the crop vegetation. Under longer jhum cycles, the losses of nutrients become minimized (Toky and Ramakrishnan, 1981b; Mishra and Ramakrishnan, 1983c) with a consequent higher soil fertility status (Ramakrishnan and Toky, 1981; Mishra and Ramakrishnan, 1983d). This is partly because there is more than a doubling of soil cover by the crop from 10-to 30-year jhum cycles and over a five-fold increase from 5- to 30-year cycles (Table 1.15). In the ultimate analysis, mixed cropping maximizes production, minimizes losses, provides a wide food resource base for the tribal society by providing cereals, legumes, vegetables and even fibre, and at the same time ensures leisure by effectively spreading out labour all the year round.

GENERAL CONSIDERATIONS

One of the major problems with jhum in north-eastern India, as is also true elsewhere, is the drastic shortening of the jhum cycle to 4–5 years during recent years. This has adversely affected economic yield with gradual decline in yield over a period of time when short cycles are imposed (Fig 1.3).

Jhum as practised in the north-eastern hill region of India is a highly complex system, with wide variations based on cropping and yield pattern. These variants, based on ecological, social and cultural variations in the tribal societies, are highly insulated because of topographical and language barriers. Yet these jhum systems may markedly differ from each other in

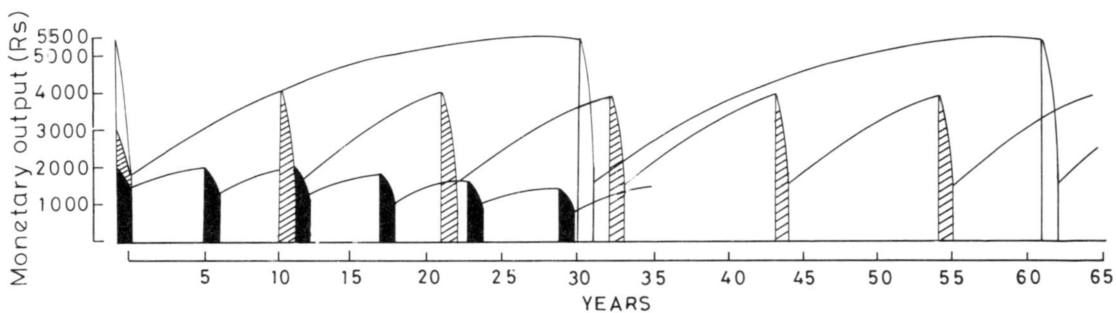

Figure 1.3 Economic yield pattern under different jhum cycles of 30 (□), 10 (▨), and 5 (■) years, over a period of time, at Burnihat at lower elevations of Meghalaya (from Ramakrishnan, 1985c)

terms of their economic and energy efficiencies. It is in this context that transfer of traditional technology from one tribal community to another in terms of crop organization and crop management becomes important. Often, by mere manipulation of the crop mixture, the economic returns to the farmer could be increased many-fold. This approach offers an important channel for redeveloping jhum, without any additional input from outside technology to strengthen it, provided a reasonable cycle length could be ensured.

On the basis of our economic and energy analysis of the jhum system, a 10-year jhum cycle is the minimum cycle length, if jhum in the present form is to be sustained. Therefore, wherever a minimum cycle length of 10 years is feasible mere manipulation of the crop mixture discussed above may prove to be adequate. It is when the cycle length comes down to a more common and very short 4–5 years, as at present, that distortions set in and the system starts breaking down. In such a situation, one is left with two options: one, take the pressure off the land for jhum by strengthening alternate land use systems discussed in the next chapter, or alternatively, and two, to stabilize jhum under a 5-year cycle by strengthening the agroforestry inputs as discussed elsewhere.

CHAPTER 2

ECOLOGICAL AND ECONOMIC EFFICIENCIES OF OTHER LAND USE SYSTEMS

INTRODUCTION

Although jhum is the predominant land use system in the north-eastern hill region of India (Ramakrishnan, 1985b), tribals here have other important traditional agroecosystem types. Monocropping of rice on valley lands, and multiple cropping of a complex nature through home gardens are sedentary systems with much variation in cropping and yield patterns (Patnaik and Ramakrishnan, 1989; Maikhuri and Ramakrishnan, 1990a). Traditionally, the tribal farmer has extensively cultivated pineapple (Patnaik and Ramakrishnan, 1989) and ginger (Kumar and Ramakrishnan, 1989b) as cash crops, although other cash crop systems such as coffee and tea are recent introductions of an experimental nature (Kumar and Ramakrishnan, 1989b). Terrace cultivation introduced as an alternative to jhum is a sedentary system that is not very common. Terracing of land promoted through the efforts of governmental agencies has not found acceptance from the local communities (Toky and Ramakrishnan, 1981a; Mishra and Ramakrishnan, 1981). This land use, which involves heavy monetary input for terracing and its maintenance, and needs heavy fertilizer application during cropping, is taken up temporarily, as long as subsidies are available. However, the terraces are abandoned once the subsidies are withdrawn and the farmer returns to his jhum system. An understanding of these different land uses as practised by the different tribes is important in order to work out appropriate strategies for land use development.

VALLEY LAND AGROECOSYSTEM

General description

Unlike jhum, valley agriculture is practised throughout the hill terrain, both at low and high elevations. It is a sedentary form of wet rice (*Oryza*

Plate 2.1 The Apatanis have an extremely efficient wet rice cultivation linked with pisciculture. This system has high economic returns and an energy output/input ratio of about 80. The Apatani men and women at Balijan, Arunachal Pradesh are seen harvesting rice

sativa) cultivation (cf Plate 1.1) and is a complementary system to jhum. It is done wherever the terrain permits, on flat lands between hill slopes (Toky and Ramakrishnan, 1981a). Obviously, this cropping system is restricted due to topography. The soil in valley lands is fertile due to nutrient wash-out from the hill slopes and therefore does not need added fertilizers. The main advantage is that the land gives sustained yield year after year.

At lower elevations of Meghalaya, often two crops are planted annually by the Garos on the same land (Toky and Ramakrishnan, 1981a; Maikhuri and Ramakrishnan, 1990a), although others such as the Mikirs and the immigrant Nepalis raise only one crop. Seedlings are generally raised in nursery beds in the month of March, transplanted at the beginning of April and the crop harvested by the end of July or early August. If a second crop is raised soon after the first crop, the land is again prepared and harvesting completed by October–November. The land is subsequently fallowed between November and March.

At mid-and high elevations of Meghalaya, the land is well prepared before sowing the seeds by broadcasting after the first few showers in June (Mishra and Ramakrishnan, 1981; Patnaik and Ramakrishnan, 1989). Harvesting is done in November, after which the land is fallowed until the following May or June. Only one crop is raised per year.

Table 2.1 Energy (MJ $ha^{-1} yr^{-1}$) and economic (Rupees $ha^{-1} yr^{-1}$) output/input ratios of valley land agroecosystem of the Garos at Burnihat in Meghalaya (from Toky and Ramakrishnan, 1981a, 1982a)

Valley land agroecosystem	*Input*	*Output*	*Output/Input ratio*
1st crop	1,502 (2,602)	25,639 (2,820)	17.1 (1.1)
2nd crop	1,341 (2,241)	24,957 (2,745)	18.6 (1.2)
Both crops	2,843 (4,843)	50,596 (5,565)	17.8 (1.1)

In parentheses are monetary values

In all these areas, the operations of weeding, transplanting, harvesting and threshing are done manually by both male and female members of the family. Only male members prepare the field. The average size of land under valley cultivation for a family of five members consisting of two adults and three children is 0.5–1 ha (Toky and Ramakrishnan, 1981a); larger families may have land area up to 2 ha (Maikhuri and Ramakrishnan, 1990a). Not more than 10–15% of the cultivated area would be under this land use.

Analysis of low elevation valley land agroecosystem of the Garos at Burnihat in Meghalaya

In a study of the valley land agroecosystem of the Garos at Burnihat at lower elevations of Meghalaya (Toky and Ramakrishnan, 1981a), 3,710 kg $ha^{-1} yr^{-1}$ of rice yield was shown as output per hectare, with an overall energy efficiency (output/input ratio) of about 18 (Table 2.1). This is double the reported average efficiency value of nine, for traditional rice cultivation in the Indian plains (Mitchell, 1979). These traditional systems, therefore, are better than most western agricultural systems, which yield only one or two units of food energy per unit of energy input (Spedding, 1975; Leach, 1976; Spedding and Walsingham, 1976; Pimental and Pimental, 1979). The energy efficiency of 18 for the valley system is much lower than that for jhum under a 10-year cycle in the same area (cf Table 1.13). This is partly due to lower energy output from a monoculture of rice compared with that obtained through a mixture of crops under jhum. The higher labour cost for land preparation and weeding in the valley system is also partly responsible for the lower energy efficiency (Table 2.2).

The monetary output from the valley system at Burnihat is comparable to one mixed cropping under a 30-year jhum cycle at this site (Table 2.1; also cf Table 1.5). However, the net gain here was a quarter of that under jhum with a 30-year cycle, for reasons cited above.

Table 2.2 Energy inputs (MJ ha^{-1} yr^{-1}) in raising two crops under valley land agroecosystem of the Garos at Burnihat in Meghalaya (from Toky and Ramakrishnan, 1982a)

Agricultural operation	*Energy input* (MJ ha^{-1} yr^{-1})
Labour input (total)	2,813
Field preparation	1,564
Nursery preparation	62
Seedling transplantation	49
Irrigation	64
Weeding	537
Harvesting	137
Transportation	53
Threshing	134
Shelling	213
Seed input	30
Total input into the system	2,843

Comparative analysis of low elevation valley land agroecosystems of different tribes coexisting at Lailad in Meghalaya

In a comparative analysis of a cluster of villages at Lailad (296 m altitude) in Meghalaya, valley land agroecosystems as practised by the Garos, the Mikirs and the Nepalis were analysed for their ecological and economic efficiencies (Maikhuri and Ramakrishnan, 1990a). In these valley systems, unlike the one discussed above for the Garos of Burnihat, some organic manure (386–1,137 kg ha^{-1}) is applied by the different tribes. At lower elevations of Meghalaya, the Nepalis and the Mikirs take only one crop during the monsoon, whereas the Garos make more effective use of the land by taking two crops, one during the monsoon and another during the winter. Nepalis in this region, take only one crop in a year, perhaps because of their preoccupation with a well organized cattle husbandry system for milking. It may be noted here that traditionally tribal communities of north-eastern India do not maintain cattle and this animal husbandry practice wherever found is a recent introduction into the region by the immigrant communities.

The Garos with two croppings in a year had a higher energy input for each cropping under valley agriculture compared with other tribes; the Nepalis expended least energy (Table 2.3). With the Nepalis obtaining higher yield from one cropping compared with the total from two croppings of the Garos, the former had higher energy and economic efficiencies (output/input ratio) for their system. The Nepalis who traditionally practise sedentary agriculture, are not only better organized but the returns are equal to that from jhum under a 20-year cycle of the Garos in the region (Maikhuri and Ramakrishnan, 1990a).

Table 2.3 Energy input and output (MJ $ha^{-1} yr^{-1}$) under valley cultivation among different tribal and non-tribal communities at Lailad in Meghalaya. Values are also given for monetary input/output pattern (Rupees $ha^{-1} yr^{-1}$) ± SE (from Maikhuri and Ramakrishnan, 1990a)

Production measures	*Garos*			*Mikirs*	*Nepalis*
	1st crop	*2nd crop*	*Total*		
Input total					
Energy	6,382	5,219	11,601	4,313	3,479
Monetary	1,400 ± 68	988 ± 36	2,388 ± 123	1,333 ± 62	1,316 ± 58
Output total					
Energy	27,656	14,282	41,938	34,895	43,172
Monetary	2,556 ± 52	1,320 ± 18	3,876 ± 72	3,440 ± 96	4,460 ± 122
Output/input ratio					
Energy	4.3	2.7	3.6	8.0	12.4
Monetary	1.8	1.3	1.6	2.6	3.4

Comparative analysis of low elevation valley land agroecosystems of different tribes coexisting at Balijan in Arunachal Pradesh

Wet rice cultivation

At Balijan in Arunachal Pradesh, the local Nishi tribe coexists with the immigrant tribals of the nearby plains of Assam and Bangladesh, such as the Karbis, the Kacharis and the Chackmas. The Nishis, the Chackmas and the Karbis alone practise valley cultivation, the former raising one crop of rice annually, the latter raising two crops (Maikhuri and Ramakrishnan, 1990b). Organic manure use is minimal (123–170 kg ha^{-1}). The first cropping of rice is done by transplanting seedlings between July and November. The second cropping done by the Karbis alone is a mixture of rice and maize raised between February and June, but the seeds are broadcast. Bullock power is used for ploughing.

The Nishis and the Chackmas had similar yield patterns for their valley agroecosystem ((therefore, the values for valley cropping as carried out by the Nishis alone are given here) Table 2.4). With labour as the chief input, the energy yield under the valley system was equal to the yield under a 5-year jhum cycle as carried out by the Nishis, but the total of two croppings executed by the Karbis exceeded the yield under a 10-year jhum cycle of the Nishis (Maikhuri and Ramakrishnan, 1990b).

From a monetary view point, one cropping under valley as carried out by the Nishis and the Chackmas was comparable to the output under a 10-year jhum cycle of the Nishis, whereas two croppings of the Karbis yielded more than that under a 60-year jhum cycle (Maikhuri and Ramakrishnan, 1990b). Thus, wide variations in yield patterns are evident from these comparative studies.

Table 2.4 Energy output/input (MJ $ha^{-1} yr^{-1}$) under valley compared with mustard cultivation done by the Nishis and the Karbis of Balijan in Arunachal Pradesh. Values are given also for monetary output/input pattern (Rupees $ha^{-1} yr^{-1}$) ± SE (from Maikhuri and Ramakrishnan, 1990b)

	Valley cultivation		*Mustard cultivation*	
Production measure	*1st crop (Nishis and Karbis)*	*2nd crop (Karbis)*	*Flatland (Nishis and Karbis)*	*Kitchen garden (Karbis)*
Input total				
Energy	2,286	2,458	720	456
Monetary	1,583 ± 72	1,154 ± 52	1,070 ± 40	558 ± 21
Output total				
Energy	37,491	12,791	10,237	4,535
Monetary	4,940 ± 169	2,351 ± 101	2,017 ± 40	1,042 ± 48
Output/input ratio				
Monetary	16.4	5.2	14.2	9.9
Energy	3.1	2.0	1.9	1.9

Nishis take only one crop from valley land

Mustard cultivation

Mustard (*Brassica campestris*) is cultivated by the Nishis and the Karbis between October and February on flatlands, as a monoculture. Initially, the weedy growth is slashed and ploughed-in and the seeds are broadcast. Labour is the only input. Mustard seeds are used for oil extraction, for cooking; otherwise the economic yield from the system is minimal (Table 2.4), but with a reasonably high energy efficiency. Karbis often raise mustard on much smaller plots, as part of the kitchen garden, but the returns are lower than that under larger flatland.

Comparative analysis of valley land agroecosystems of the Khasis and the Nepalis at mid-elevations at Nayabunglow in Meghalaya

The local tribe, the Khasis and the immigrant Nepalis living at Nayabunglow practise valley cultivation of wet rice (Patnaik and Ramakrishnan, 1989). The Napalis are basically cattle farmers. As these immigrants are not permitted to own land in rural areas for agriculture, they hire the valley land from the Khasi landlords of the near-by Shillong town.

Both the Khasis and the Nepalis take only one crop of rice annually. Some inorganic fertilizer at the rate of 102 kg ha^{-1} is added to the soil by both communities. Ploughing and land preparation is carried out during the months of May and June using bullocks. The Khasis hire bullocks for the land preparation. Other operations such as transplanting, weeding, harvesting, threshing, etc. are done manually by both male and female

Table 2.5 Energy (MJ ha^{-1} yr^{-1}) and monetary (Rupees ha^{-1} yr^{-1}) efficiency of valley land agroecosystem of the Khasis and the Nepalis at Nayabunglow in Meghalaya (from Patnaik and Ramakrishnan, 1989)

	Khasis		*Nepalis*	
	Energy	*Monetary*	*Energy*	*Monetary*
Input total	4,903	4,257	4,623	2,890
Human labour	678	3,150	427	1,897
Bullock power	491	533	307	333
Seed	193	300	171	300
Inorganic fertilizer	3,541	274	3,718	360
Output total	47,795	6,334[a]	50,825	6,460[b]
Output/input ratio	9.8	1.5	11	2.2

[a] 5,501 Rupees were from grains and 833 Rupees from rice straw
[b] 5,460 Rupees were from grains and 1000 Rupees from rice straw

members of the family. Both the Khasis and the Nepalis give land revenue as rent to the Khasi owner of the land in the city. There is considerable interaction between the Khasis and the Nepalis in managing their valley land agroecosystem. Consequently the output from valley cultivation of the Nepalis is close to that of the Khasis, but with a slightly higher efficiency for the former because of reduced labour input (Table 2.5).

Wet rice cultivation in valley lands by the Apatanis of Arunachal Pradesh

Small farmers in the wet tropics have always concentrated upon rice cultivation to obtain their basic needs, if the land surface is flat. This is because rice, like maize, has a particularly high calorie content (Finck, 1970). These traditional systems based on technology developed over many generations (Schultz, 1964) are often energy efficient. The Apatanis with a highly developed valley cultivation of rice perfected over centuries (Plate 2.1) has often been suggested to be one of the relatively advanced tribal societies in the north-eastern region of India (Furer-Haimendorf, 1962, 1985). Therefore, an understanding of the valley land agroecosystem function of the Apatanis becomes significant.

Valley land agroecosystems of the Apatanis, like those of other tribes discussed earlier whose chief land use is jhum (Ramakrishnan, 1984a), is also dependent upon nutrient wash-out from the hill slopes. With crop harvest, considerable quantities of nutrients are lost through economic yield (Rosswall and Panstian, 1984; Swamy and Ramakrishnan, 1988c) and must be replaced for sustainable land use. Recycling crop residues (Jones, 1976) and use of organic wastes of the village for sustaining soil fertility, as carried out by the Apatanis, is an effective way of restoring soil fertility.

Table 2.6 Soil nutrient status (mean ± SE) in (0.7 cm depth) rice agroecosystems of Apatanis at Hari in Arunachal Pradesh (Kumar and Ramakrishnan, 1990)

Elements	*Within the village*	*Outside the village*
C (%)	2.3 ± 0.2	1.4 ± 0.09
N (%)	0.3 ± 0.02	0.2 ± 0.02
P (mg 100 g^{-1})	0.1 ± 0.01	0.1 ± 0.01
K (mg 100 g^{-1})	25.0 ± 1.2	15.0 ± 0.8
Ca (mg 100 g^{-1})	112.5 ± 9.4	67.5 ± 3.2
Mg (mg 100 g^{-1})	32.5 ± 1.0	27.5 ± 0.8

Irrigation farming such as wet cultivation of rice requires communal work to maintain it, and to improve the water delivery system (Ruthenberg, 1971). In the absence of a disciplined schedule and scale of water distribution among the beneficiaries, very often economic returns could decline drastically. The Apatanis with co-operative effort under the overall supervision of the village headman, have optimized water use in their rice fields. In this activity, both men and women participate.

Study area

Hari is an Apatani village in Arunachal Pradesh (27.36° N, 93.49° E) at an altitude of 1,570 m, and has 418 families and a total population of 2,021. The average size of a family is about five, living in a large bamboo hut of 15 × 4 m, floor area. Wet rice cultivation is their only agricultural land use.

The climate is monsoonic with over 75% of the total rainfall of 1,758 mm (Kumar and Ramakrishnan, 1990) occurring during the monsoon period of May to September. The average maximum and minimum temperatures during the summer were 24.1°C and 12.7°C and during the winter were 17°C and 2.6°C, respectively.

Cropping and yielding patterns

Apatanis, at Hari, make effective use of their irrigated land by planting early and late ripening varieties of rice. The early variety is sown further away from the village where disturbance by animals and poorer irrigation facilities could be major constraints (Kumar and Ramakrishnan, 1990). Lower nutrient status on plots further away from the village is another consideration. Thus rice plots closer to the village are nutritionally richer than those further away (Table 2.6). Closer to the village, where conditions are more favourable, the late variety is preferred. Further, rice is supplemented with *Eleusine coracana* cultivated on elevated partition bunds between the rice plots.

Table 2.7 Density, basal area and mean economic yield (dry weight ± SE) of rice agroecosystem of Apatanis at Hari in Arunachal Pradesh (from Kumar and Ramakrishnan, 1990)

	Agroecosystem[1]	
	Rice (late variety)[2]	*Rice (early variety)*
Density (plant m^{-2})	32 (112)	48 (104)
Basal area ($cm^2\,m^{-2}$)	400 (25.7)	216 (23.9)
Mean economic yield		
g $plant^{-1}$	12.7 ± 0.4	7.2 ± 0.3
	(2.8 ± 0.1)	(2.4 ± 0.1)
kg ha^{-1}	4,064 ± 141	3,456 ± 163
	(102.5 ± 3.0)	(129.5 ± 5.0)

Values in parentheses are for *Eleusine coracana*
[1] Average bund area ha^{-1} rice plot was 519 m^2 and 327 m^2 for early and late varieties, respectively
[2] Fish production was 48 kg ha^{-1}

The early variety of rice had higher density but with reduced basal area compared to the late variety (Table 2.7). Economic yield per plant and per unit area of the early variety was significantly lower when compared to the late variety. The yield per hectare of *Eleusine coracana* grown on partition bunds of rice plots was higher in plots with the early variety than in those with the late variety. Fish production along with rice was substantial.

With human labour as the major input and with very little organic manure use, the Apatanis obtain a high energy output (Table 2.8). Labour input for rice/rice plus millet where the early variety of rice is grown was higher than for the late variety of rice. The Apatanis do not use draught animal power. The output from the system and the net return to the farmer was higher under the late variety grown alone or with millet/fish, than under the early variety of rice. The economic and energy efficiencies, and output per unit labour hour were also higher under the agroecosystem with the late variety of rice.

The exceptionally high energy efficiency of this valley land agroecosystem (60–80 units per unit energy input) is markedly different from the values discussed earlier in this chapter, for other tribes in this region, with an output/input ratio of 4–18 (Mishra and Ramakrishnan, 1981; Toky and Ramakrishnan, 1982a; Maikhuri and Ramakrishnan, 1990a). This high energy efficiency also contrasts with that recorded for jhum in north-east India, which does not exceed 50 units per unit energy input (cf Chapter 1). It is far superior to the traditional wet cultivation of rice of the Indian plains with an efficiency value of nearly 9 units per unit energy input (Mitchell, 1979) and that elsewhere in the Philippines (Nguu and Palis, 1977). With 27–35 MJ units of energy output per labour hour, this system compares favourably with similar systems of China (32.8 MJ) (Dazhong and Pimentel, 1984) and more modern agriculture of industrialized societies

Table 2.8 Energy (MJ ha^{-1} yr^{-1}) and monetary (Rupees ha^{-1} yr^{-1}) (in parentheses) input–output patterns of rice agroecosystem of Apatanis at Hari in Arunachal Pradesh (from Kumar and Ramakrishnan, 1990)

	Rice agroecosystem	
Production measures	*Late variety*	*Early variety*
Input total		
Rice	846.5 (2,481)	904.5 (2,677)
Rice + millet	870.2 (2,551)	946.0 (2,798)
Rice + millet + fish	906.6 (2,753)	–
Labour		
Rice	713 (2,163)	769 (2,337)
Millet	23 (67)	40 (116)
Fish	36 (102)	–
Organic manure	125 (250)	125 (250)
Seed		
Rice	8.5 (68)	10.5 (90)
Millet	0.7 (3)	1.5 (5)
Fish	0.4 (100)	–
Output total		
Rice	66,284 (8,941)	56,367 (7,603)
Rice + millet	67,956 (9,102)	58,480 (7,817)
Rice + millet + fish	68,182 (10,062)	–
Energy efficiency		
Output/input ratio		
Rice	78.3 (3.60)	62.3 (2.84)
Rice + millet	78.1 (3.57)	61.8 (2.79)
Rice + millet + fish	75.2 (3.65)	–
Energy output/labour hour :		
Rice	35.6	27.6
Rice + millet	35.4	27.4
Rice + millet + fish	34.1	–

(Leach, 1976), such as the United Kingdom, with an output of 40 MJ units. In the present context, it is significant that human labour is a free input being largely obtained from within the family itself, and for specific tasks alone through co-operative efforts.

With a highly developed wet rice cultivation, the Apatani economy is largely based upon agriculture alone (Kumar and Ramakrishnan, 1990). The net per capita monetary return through agriculture is high. As much as 40% of the rice produced is sold to the economically weaker neighbouring tribes such as the Nishis and the hill Miris. However, the agroecosystem of the Apatanis could be improved through appropriate crop rotation and productive utilization of the land during the winter season. In spite of these possibilities, the Apatani village ecosystem is a good example of economic self-sufficiency of a traditional agricultural society that practises ecologically sound sedentary agriculture in the north-eastern hill region of India.

Conclusions

The variety of valley land agroecosystem patterns that one finds here is indicative of the range that exists in the north-eastern India. Many of them represent under-utilization of land and are inefficient. Traditional wet land cultivators of rice such as the local Apatanis of Arunachal Pradesh and the immigrant Nepalis provide examples of a high degree of ecological and economic efficiency for this system. However, valley land cultivation is constrained by the availability of water for cultivation during the relatively dry winter months. Where tribes like the Garos at Lailad take two crops in a year, they essentially restrict their activity to the monsoon period.

A few possibilities exist for redeveloping this system so that the farmer is able to obtain better returns : (1) mere transfer of technology from one tribe to another could be helpful. The Apatani system of farming is one of the most evolved and highly organized systems of wet cultivation of rice; (2) introduction of early maturing and improved varieties of rice would help in obtaining two or even three harvests in a year; (3) water is a major constraint outside the monsoon period. Rain water harvesting and storage in tanks offers immense possibilities for irrigation. The run-off water from hill slopes could be collected in 70–80 m^3 tanks lined with heavy duty polythene sheets. They are cheap, easy to maintain and have a life span of 20–25 years (K.G. Saxena and P.S. Ramakrishnan, unpublished). A programme is established with the author to put up such tanks for hill irrigation (cf Plate 10.1)

HOME GARDENS

Shifting agriculture (jhum) is essentially an agroforestry system, with the tradition extending into antiquity. Through this and other agroforestry practices, the farmer has linked his family to the forest ecosystem and also effectively incorporated animal husbandry (Mishra and Ramakrishnan, 1982; Patnaik and Ramakrishnan, 1989; Kumar and Ramakrishnan, 1989a, 1990; Maikhuri and Ramakrishnan, 1990a,b). Thus, agriculture, animal husbandry and domestic sub-systems of the village are all closely linked with the forest ecosystem, providing food, fodder, fuelwood and timber. These and other agroforestry systems have helped to conserve and protect the often limited natural resources of many societies (Nair, 1984). The ultimate objective here is to optimize the beneficial effects of interactions of the woody species with the crop and/or the animal components to

Table 2.9 Ecological traits of a 40-year-old home garden in Latin America (from Ewel *et al.*, 1982)

Parameters	
(a) Stratification	0 – > 14 m
(b) Leaf area index	3.9
(c) Percentage cover	100.0
(d) Leaf biomass ($g\,m^{-2}$)	307.0
(e) Root biomass up to 25 cm depth ($g\,m^{-2}$)	307.2
(f) Small root (< cm diam.) surface area ($m^2\,m^{-2}$) ground surface to 25 cm depth	2.1

diversify production, reduce external inputs and make the farming system sustainable (Gliessman *et al.*, 1981). Of all these agroforestry systems, home gardens, variously termed as kitchen gardens or forest gardens are one of the most complex and highly diversified systems, an interesting agroecosystem from the point of view of resource management for sustainable agriculture (Gliessman, 1989a).

It would be appropriate at this stage to briefly consider the conclusions arising from some of the studies on tropical home gardens, as the conclusions arising from these are generally applicable to the north-eastern Indian situation too. These home gardens, as in north-eastern India, have rich plant species diversity, dominated by woody perennials and are stratified (Alcorn, 1984; Gangwar and Ramakrishnan, 1989a; Maikhuri and Ramakrishnan, 1990a). With a mixture of annuals and perennials forming a multi-storeyed structure, they resemble a natural forest. From a plot of 0.5–2 ha, located close to the habitation, the farmer obtains food products, firewood, medicinal plants, spices and ornamentals, all the year round (Gonzalez, 1985; Gangwar and Ramakrishnan, 1989a; Maikhuri and Ramakrishnan, 1990a).

In a comparative analysis of a variety of ecological features for nine plant communities of tropical Latin America, a 40-year-old home garden was found to be an ecological efficient system by Ewel *et al.* (1982). The traits analysed by him (Table 2.9) are suggestive of efficiencies in relation to light capture, nutrient turnover and storage in the biomass, and soil conservation (Gliessman, 1989a). Stratified community organization may increase habitat diversity which in turn may contribute to better biological control in the system (Altieri and Letourneau, 1982).

Home gardens meet a variety of requirements of the farmer, all the year round, as shown through a study in Costa Rica (Table 2.10) (Gliessman, 1989a). Apart from meeting the varied needs of the farmer, the produce from home gardens may provide cash income to the farmer as shown in the Costa Rican study, and our own studies in north-eastern India (Gangwar and Ramakrishnan, 1989a; Maikhuri and Ramakrishnan, 1990a). Indeed, they even perform social or aesthetic functions, serving as an indicator of

Table 2.10 Comparison of plant species in a home garden over 2 years in Canas, Guancaste, Costa Rica (from Gliessman, 1989a)

Species/individual (numbers)	*1985*	*1986*
Total species	71	83
Tree species	17	16
Food species	21	18
Ornamental species	23	31
Medicinal species	7	9
Firewood species	3	5
Spice species	0	4
Total Individuals	940	1,873

Garden size = 1,240 m^2

social status of the owner or improving environmental quality (Gliessman, 1989a).

Home gardens of the Mikirs at lower elevations at Lailad in Meghalaya

At Lailad, in Meghalaya, cash crops such as arecanut (betel nut) (*Areca catechu*), betel leaf (*Piper betel*), black pepper (*Piper nigrum*) and banana (*Musa* spp.) are cultivated by the Mikirs, in plots 0.5–1.5 ha per family (Plate 2.2). Arecanut and black pepper are harvested during October–November. The climbing perennials (betel and black pepper) are sown around a legume tree (*Erythrina stricta*) which provides support. Betel leaf and the fruits of banana are harvested throughout the year.

With greater emphasis on arecanut and banana, in the already established home garden of the Mikirs, the energy input is only in the form of human labour for maintenance of the garden; this is minimal. With high energy and economic returns, compared to the maintenance cost, the efficiencies (output/input ratio) are very high (Table 2.11). This relatively simple home garden of the Mikirs, with effective utilization of space gives 3-to 10 -fold economic return compared to the jhum system at Lailad as carried out by the Mikirs, Garos and the Khasis (cf Table 1.7).

Home garden of the Khasis of the desertified Cherrapunji area in Meghalaya

Study area

The village of Tynriang (25°N 92°E) is located 80 km southeast of Shillong, at an altitude of 1,150 m. The area is highly desertified (Khiewtam, 1986; Ramakrishnan and Ram, 1988), except for small patches of sacred grove

Plate 2.2 The home garden of the Mikir tribe at Lailad (290 m altitude) in Meghalaya and that of others in the region, with betel nut (*Areca catechu*) as an important tree component. A variety of grain, tuber, and leafy and fruit vegetable crops are grown at the ground level

Table 2.11 Mean plant density, economic yield, monetary (± SE) and energy input and output pattern under home garden of the Mikirs at Lailad in Meghalaya (from Maikhuri and Ramakrishnan, 1990a)

Production measures/ crop species	*Density* (ha^{-1})	*Yield* ($kg\,ha^{-1}$)	*Money* ($Rs\,ha^{-1}\,yr^{-1}$)	*Energy* ($MJ\,ha^{-1}\,yr^{-1}$)
Input total	–	–	1,650 ± 110	1,140
Output total	–	–	14,667 ± 685	26,794
Arecanut	1,250	1,112 ± 98	10,489	16,791
Betel leaf	450	81 ± 6	3,240	996
Banana	800	568 ± 36	568	8,463
Black pepper	90	37 ± 2	370	544
Net return	–	–	13,014	–
Output/input ratio	–	–	8.8	23.5

– indicates no data
RS = rupees

forest that are protected by the local people for religious reasons (Boojh and Ramakrishnan, 1983b). The annual average rainfall here is 1,073 cm which may go up to 2,456 cm in an exceptional year as in 1974. The climate is typically monsoonic with about 96% of the annual rainfall occurring between May and October (Gangwar and Ramakrishnan, 1989a). The mean monthly maximum and minimum temperatures during this period are 24°C and 14°C, respectively. The mean monthly maximum and minimum temperatures during the winter are 17°C and 6°C, respectively. The brief summer of March–April has a mean monthly maximum and minimum temperatures of 20°C and 12°C, respectively.

Desertification and land use

Although jhum is the traditional land use of the Khasis of Meghalaya (Mishra and Ramakrishnan, 1981, 1982), and their staple diet is rice and fish, the people living in Tynriang close to Cherrapunji have abandoned this land use system because of large-scale desertification that has occurred (Ramakrishnan *et al.*, 1981; Ramakrishnan, 1985b). In many areas in the Khasi hills of Meghalaya, where forests have almost disappeared and the land is being desertified because of the reduction in jhum cycle length to less than 5 years, the Khasis have adopted a fallow system of agriculture (FAO/SIDA, 1974) in which the burning operation of jhum is eliminated (Gangwar and Ramakrishnan, 1987a). This new system may cause even more rapid desertification of the landscape (Ramakrishnan, 1985b, 1988c) due to rapid depletion of soil fertility such that the fallow/sedentary agricultural systems often become uneconomic. This is because, under the fallow system of agriculture, the fallow regrowth is only herbaceous weeds that are ploughed back into the system and this alone is inadequate to sustain soil fertility. Tree crops have been suggested as an alternative to the slash-and-burn agriculture system elsewhere (Ruthenberg, 1971; Andreae, 1980; Watson, 1983; Akachukwu, 1985) and this has implications for north-eastern India (Ramakrishnan, 1985a,c) too. It is in this context, the home garden of the Khasis evolved, as a response to large-scale desertification in the region becomes significant (Gangwar and Ramakrishnan, 1989a).

Cropping and yield patterns

The home gardens of the Khasis at Tynriang are 1.5–2 ha in size and are more complex than the one discussed earlier for the Mikirs of Lailad. Nurseries of bay leaf (*Cinnamomum obtusifolium*, betel nut (*Areca catechu*) orange (*Citrus sinensis*) and jack fruit (*Artocarpus heterophyllus*) trees are raised separately by seed. Four-year-old saplings are transplanted into the home garden plots in May, before the onset of monsoon, on land prepared earlier. Betel leaf (*Piper betel*) and black pepper (*Piper nigrum*) vines are introduced directly around the tree saplings, 2 years after transplantation. During the first 6 years (before the tree plantations mature), banana (*Musa sapientum*), pineapple (*Ananas comosus*) and sweet potato (*Ipomoea batatus*) are cultivated (Gangwar and Ramakrishnan, 1989a). Weeding is done twice

Table 2.12 Cost-benefit analysis (± SE) of the home garden of the Khasis during the early 6 years of establishment, at Tynriang, Meghalaya (from Gangwar and Ramakrishnan, 1989a)

Production measure	Rupees ha^{-1} yr^{-1}
Input total	7,120 ± 550
Seed	270
Planting	340
Weeding	5,250
Harvesting	480
Transportation	780
Output total	5,178 ± 445
Ananas comosus	2,050
Ipomoea batatas	1,048
Musa sapientum	2,080
Net return	−1,941
Output/input ratio	0.7

a year until the plantation is established. The trees are pruned during the winter. Harvest of bay leaf, betel nut and orange starts 6 years after transplantation, whereas black pepper and betel leaf are harvested 2 years after planting. Since black pepper and betel leaf yield only for a period of about 7–15 years, at such times old vines are removed and replaced by young vines. Harvesting of black pepper occurs in April, jack fruit in May–June, orange in November–December and betel nut and bay leaf in December–February. Betel leaves are harvested three times a year in May, September and December.

The first 6 years of raising the home garden is expensive. Land preparation, raising of tree saplings in the nursery and their transplantation resulted in an initial expenditure of 11,481 rupees per hectare. During the first 6 years, labour input for weeding in particular was high (Table 2.12). During this period, banana, pineapple and sweet potato together gave a net return which was less than the yearly cost for maintenance.

From the 7th year onwards when the tree plantation becomes established, the economic output per year increased sharply (Table 2.13). Amongst trees, betel nut had the highest density ($11.1/100\,m^2$) followed by bay leaf ($6/100\,m^2$) and orange ($2.3/100\,m^2$). Weeding cost declined in older plantations. From an economic output/input ratio of 0.7 (exclusive of initial cost) during the first 6 years, the efficiency value increased to 1.8, for the subsequent period. In older plantations betel nut gave the highest monetary output (Plate 2.3).

Labour is the chief input into the plantation system. In older plantations, after 6 years, the energy efficiency ratio is high (output/input ratio of 22.3)

Table 2.13 Cost-benefit analysis ($\pm$ SE) of an established home garden of the Khasis, at Tynriang in Meghalaya (from Gangwar and Ramakrishnan, 1989a)

Production measure	Rupees ha^{-1} yr^{-1}
Input total	13,093 $\pm$ 972
Pruning/girdling	380
Weeding	3,500
Harvesting	6,443
Transportation	2,150
Storage	620
Output total	23,155 $\pm$ 1,463
Areca catechu	12,750
Artocarpus heterophyllus	1,238
Citrus sinensis	1,617
Piper betel	1,350
Cinnamomum obtusifolium	4,680
Piper betel	1,520
Net return	10,062
Output/input ratio	1.8

Plate 2.3 The Khasis of Tynriang (1,150 m altitude) in the Cherrapunji region of Meghalaya grow betel nut (*Areca catechu*) as a cash crop for export

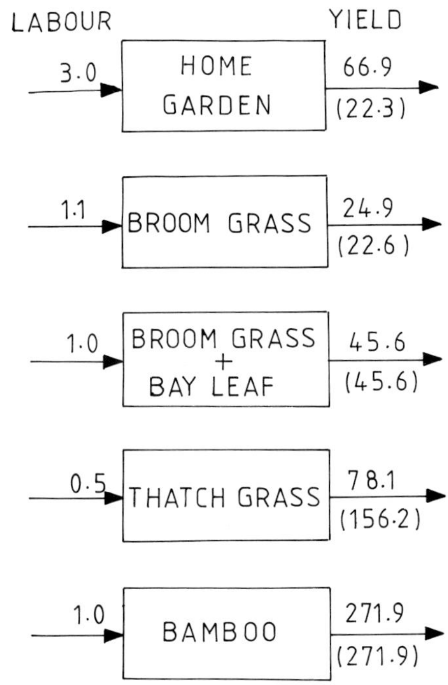

Figure 2.1 Energy (MJ × 10^3 ha^{-1} yr^{-1}) output/input pattern of home garden and cash crop systems of the Khasis at Tynriang in Meghalaya. Efficiency ratios are in parentheses (from Gangwar and Ramakrishnan, 1989a)

(Fig. 2.1). Here, males do more labour intensive chores such as girdling, harvesting and transportation (Fig. 2.2). Females play a major role in weeding. A major limitation, however, is the need to import labour from outside the village boundary. Home gardens are environmentally compatible for desertified areas, as an effective alternative to jhum. However, such an economy can be sustained only through import of food from adjoining areas or from outside the region. This is possible because of the cash income generated through substantial export of produce from the home garden.

Conclusions

With wide variations in socio-economic and socio-cultural backgrounds of the tribal communities in north-eastern India and with high heterogeneity in ecological setting, wide variations in the structure and organization of the home gardens are to be expected. Some of them, such as the small kitchen gardens maintained by the Nepalis of Nayabunglow represent a low level in their organization and function in that the Nepalis grow only

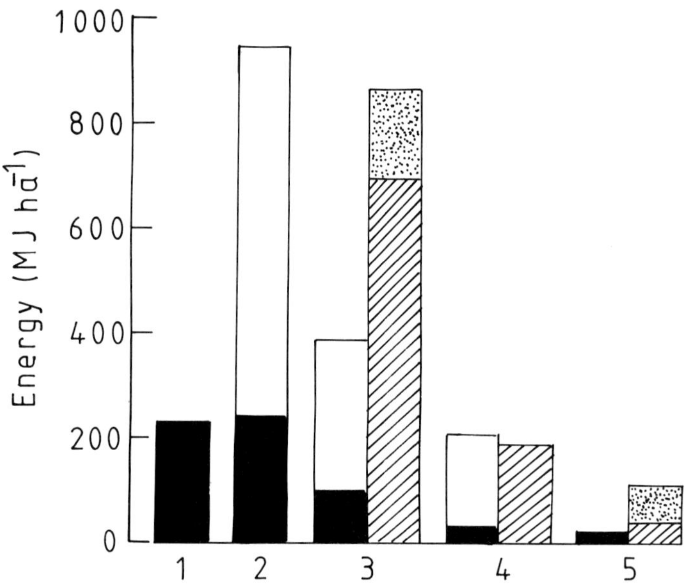

Figure 2.2 Distribution of labour between male and female members of the family and imported from outside, for different activities in the home garden of the Khasis at Tynriang in Meghalaya. 1, pruning and girdling; 2, weeding; 3, harvesting; 4, transportation; 5, storage; ■, male from within; □, female from within; ▒, male imported; ▨, female imported (from Gangwar and Ramakrishnan, 1989a)

grain and legume species along with leafy and fruit vegetables (Patnaik and Ramakrishnan, 1989). Others such as the home gardens of the Mikirs of Lailad and the Khasis of Tynriang in Meghalaya are extremely complex. The home gardens are not only efficient ecologically and economically, but also meet the diverse needs of the society. As is happening in other parts of the world, here in the north-east too, these highly sustainable systems are on the decline because of rapid deforestation and untenable developmental activities. We need to know more about these agroforestry systems, so that we are able to choose the best from the already available traditional variants and even build upon them in order to make them more productive.

With a traditional land use such as jhum becoming more and more untenable because of the rapid shortening of the jhum cycle, the home garden concept offers immense possibilities to ensure sustainable land use development. Indeed, it is interesting to find that the traditional jhum farmer of the Tynriang village and others in the Cherrapunji region of Meghalaya have, on their own, shifted towards home gardens of a complex organization and function, under the threat of large-scale desertification.

Plate 2.4 The black pepper (*Piper nigrum*) plantation being maintained by the Nishis of Arunachal Pradesh and other tribes in the region

CASH CROP ECOSYSTEMS

With rapid shortening of the jhum cycle in the north-eastern hill region (Ramakrishnan, 1984a,b, 1985c), the traditional jhum system has become non-viable. A shift towards plantation and cash crops (Plate 2.4) is one of the possibilities (Ramakrishnan, 1985c). If approached correctly, taking into consideration ecological and social angles, this land use system could be complementary to a redeveloped jhum system, if not a replacement. The tribals of the north-east have been, over the years, cultivating many traditional cash crops, such as pineapple, arecanut and ginger. A few non-traditional cash crops such as coffee and tea have been tried on an experimental scale, for possible introduction into the region. A brief consideration of these land use systems is appropriate at this stage, so as to assess their possible value to tribal economy.

Cash crop systems of the Khasis in the desertified Cherrapunji area

Broom grass cultivation

In the village of Tynriang (1,150 m altitude) in the Cherrapunji area, near Shillong in Meghalaya, the Khasis grow broom grass (*Thysanolaena*

Plate 2.5 Broom grass (*Thysanolaena maxima*), an early successional weed harvested by the Khasis of Tynriang (1,150 m altitude) in the Cherrapunji region of Meghalaya, as a cash crop

maxima) in wastelands either by itself or as an undergrowth with sparsely distributed bay leaf (*Cinnamomum obtusifolium*) trees (Gangwar and Ramakrishnan, 1989a). Used for making brooms, the grass is planted during May–June, after the plot is burnt. Some sprouts may arise also from underground rhizomes present at the site, since this is an early successional weed species coming up after slash-and-burn operations (Saxena and Ramakrishnan, 1983a). Harvesting is done at yearly intervals for a period of about 7 years (Plate 2.5). Subsequently, the grass is burnt and replanted. Weeding is done once or twice a year.

Broom grass yield increased with plantation age up to 3 years and stabilized during the subsequent 3–7 years (Table 2.14). Although the yield of the grass in plantations of *Cinnamomum obtusifolium* was slightly lower, the bay leaf was an additional item harvested and, therefore, the total economic returns per hectare were higher. The input for various operations under pure and mixed cultivation of broom grass decreased with the age of the grass plantation, but the monetary output increased. Monetary returns were also higher under the mixed system than under pure grass. The mixed system was also more energy efficient. The lower energy efficiency of the broom grass compared to the thatch grass discussed below is because, in the former only the upper portion of the shoot with the mature inflorescence is harvested and not the entire above-ground biomass.

Table 2.14 Monetary output/input ratio (Rupees $ha^{-1} yr^{-1}$) ($\pm$ SE) of a broom grass pure and broom grass mixed with bay leaf plantation of different age categories, maintained by the Khasis, at Tynriang in Meghalaya. The economic yield of broom grass is given in dry weight (kg $ha^{-1} yr^{-1} \pm$ SE) (Gangwar and Ramakrishnan, 1989a)

	Pure			*Mixed*[a]		
	1 yr	*2 yr*	*3–7 yr*	*1 yr*	*2 yr*	*3–7 yr*
Input total	4,670 ± 275	3,930 ± 220	3,870 ± 245	4,995 ± 325	3,886 ± 255	3,303 ± 292
Clearing and burning	340	–	–	340	–	–
Transplantation[b]	1,350	–	–	1,040	–	–
Weeding and maintenance	1,740	1,740	870	1,652	1,652	826
Harvesting and transportation	1,240	2,190	3,000	1,963	2,234	2,477
Output total	3,348 ± 275	5,913 ± 524	8,100 ± 745	6,474 ± 481	7,953 ± 621	9,250 ± 756
Economic yield of broom grass	620 ± 51	1,095 ± 97	1,500 ± 138	390 ± 22	660 ± 48	904 ± 73
Output/input ratio	0.72	1.50	2.09	1.30	2.04	2.80

[a] Bay leaf yield worth 4,368 Rupees (1,820 ± 151 kg) was obtained
[b] The cost of bay leaf plantation has not been included

Bamboo and thatch grass cultivation

Bamboo and thatch grass are also early successional species (Toky and Ramakrishnan, 1983a) cultivated in the village (Gangwar and Ramakrishnan, 1989a). Bamboo (*Dendrocalamus hamiltonii*) used for house construction, basket making and for export (Plate 2.6) is planted in May, through rhizomes harvested selectively first after 6 years and subsequently at yearly intervals, for about 25 years. Thatch grass (*Imperata cylindrica*) used in hut construction grows wild as a secondary successional weed. Therefore, transplantation is done only in free spaces to cover the ground.

Monetary return from bamboo plantation is higher compared to thatch grass harvest (Plate 2.7) per hectare (Table 2.15). However, the monetary efficiency of bamboo is lower than that of thatch grass, because of higher labour input for the former.

With labour as the only input, the energy efficiencies are high (Fig. 2.1). With a fast growth rate for bamboo (Rao and Ramakrishnan, 1988a) and a large biomass, the energy efficiency of this plantation is the highest. Thatch grass has a higher energy efficiency than the broom grass discussed earlier.

Plate 2.6 Bamboo (*Dendrocalamus hamiltonii*), an early successional species, harvested by the Khasis of Tynriang (1,150 m altitude) in the Cherrapunji region of Meghalaya, as a cash crop, is ready for export. Inset is a house constructed from bamboo

Plate 2.7 Thatch grass (*Imperata cylindrica*), an early secondary weed, harvested by the Khasis of Tynriang (1,150 m altitude) in the Cherrapunji region of Meghalaya, as a cash crop

Table 2.15 Monetary output/input ratio (Rupees $ha^{-1} yr^{-1}$ ± SE) of a thatch grass and bamboo plantation of the Khasis, at Tynriang in Meghalaya (Gangwar and Ramakrishnan, 1989a)

	Thatch grass	*Bamboo*
Input total	1,716 ± 125	3,695 ± 242
Maintenance	180	270
Harvesting	720	2,055
Transportation	816	1,370
Output total	3,827 ± 260 (4,133 ± 280)	6,850 ± 408 (14,385 ± 850)
Net return	2,121	3,155
Output/input ratio	2.23	1.85

Values in parentheses are economic yield (dry weight) (kg $ha^{-1} yr^{-1}$)

Cash crop ecosystems of the Khasis at Nayabunglow in Meghalaya

Description of the agroecosystem

At Nayabunglow in Meghalaya, ginger and pineapple are two traditional cash crop systems, whilst coffee and tea are introduced into the region on an experimental scale (Kumar and Ramakrishnan, 1989b).

A coffee (*Coffea arabica*) plantation was established on terraces in the early 1960s by government agencies. The plantation was raised with a spacing of 3 × 2 m (Kumar and Ramakrishnan, 1989b). *Schima wallichii* and *Bauhinia purpurea* are two important shade tree species used. Average plant height of coffee, after about 20 years growth is 1.5 m. Seed picking is done in December–January. Pruning of the plant is done soon after. Weeding is done once, followed by application of pesticides in September–October. Inorganic fertilizer (N and P, 5:3) is applied twice a year at the rate of 155 kg ha^{-1}.

Transplantation of tea (*Camellia sinensis*) was executed on slopes (60 × 90 cm intervals) during 1978 using 1-year-old saplings (Kumar and Ramakrishnan, 1989b). *Albizzia odoratissima* is the important shade tree species along with *Albizzia chinensis*, planted at distances of 3 × 5 m. Trimming of the bush at 75 cm height is done in February. Plucking of tea leaves is done during April–October, at 10-day intervals. Weeding is done three times during the rainy season. Apart from weedicide/pesticide applications, inorganic fertilizer (NPK, 2:1:2) is used twice in a year at the rate of 555 kg ha^{-1}. The freshly plucked tea leaves are sold at the rate of 1.50 rupees kg^{-1}.

Ginger (*Zingiber officinale*) is traditionally cultivated on terraces (Kumar and Ramakrishnan, 1989b). After preparation of the land into ridges and

Table 2.16 Energy (MJ ha^{-1} yr^{-1}) analysis of plantation/cash crop ecosystems compared with jhum under 10-year cycle at Nayabunglow, Meghalaya (from Kumar and Ramakrishnan, 1989b)

	Coffee	*Tea*	*Pineapple mixed cropping*	*Ginger*	*Jhum 10-yr cycle*[a]
Input	8,855	19,425	973	1,3854	1,302
	(2,754)	(14,314)	(3,096)	(18,845)	(1,830)
Output	8,450	18,1310	17,085	33,392	56,655
	(4,560)	(37,125)	(12,090)[b]	(42,435)	(3,354)
Output/input	0.95	9.33	17.56	2.41	43.51
ratio	(1.66)	(2.59)	(3.90)	(2.25)	(1.83)

[a] Toky and Ramakrishnan, 1981a
[b] 9,095 Rupees for pineapple and 2,997 Rupees for rhizome and tuber crops

furrows and after repairing the old terraces, sowing is done in April at a 20 × 30 cm interval, following an application of organic manure at the rate of 720 kg ha^{-1}, before crop sowing. Inorganic fertilizer (NPK, 1:1:1) is applied twice a year at the rate of 500 kg ha^{-1}. Harvesting of rhizome is done in December–January.

Pineapple (*Ananas comosus*) in one study was being cultivated on terraces for the last 6 years (Kumar and Ramakrishnan, 1989b). This along with rhizome and tuber crops (*Colocasia antiquorum, Curcuma longa* and *Manihot esculentus*) are sown in March and harvested during the following December. Two harvests of pineapple are done from the same field, once in July–August (monsoon variety) and another in December–January (winter variety). Weeding is done twice in June and December. Weed biomass and old pineapple biomass are put back into the plot as organic manure.

Economic and energy analysis

Energy input for tea was higher followed by ginger; pineapple mixed cropping system had minimum energy input. The major energy input for coffee, tea and ginger was inorganic fertilizer, followed by labour. Of all the inputs labour cost was the highest for tea followed by ginger. Since field preparation for tea, coffee and pineapple mixed cropping systems are done only when cultivation starts, this cost is not recurring. While tea is raised on slopes, pineapple cultivation is done on terraces. Except for the pineapple mixed cropping system, inorganic fertilizer was applied and this was minimal for coffee. Yearly seed input was required for rhizome and tuber species grown with pineapple and for ginger cultivation.

Although the economic output from ginger was higher than that of other crop systems, and this was followed by tea, the input for these two crops was high (Table 2.16). With low input into pineapple mixed cropping system and a relatively higher return from it, its economic efficiency is higher. A

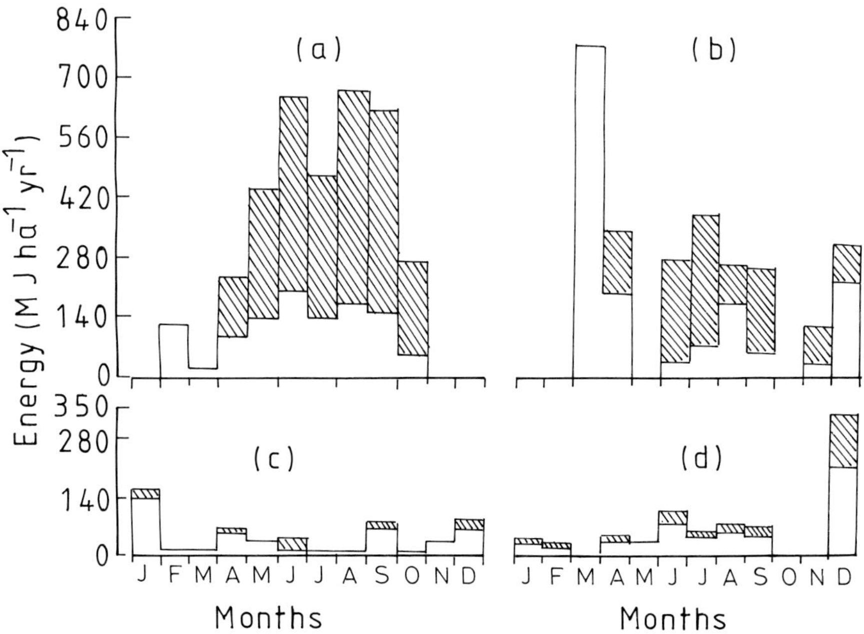

Figure 2.3 Monthly labour energy input distribution pattern for plantation/cash crop ecosystems of the Khasis at Nayabunglow in Meghalaya a, tea; b, ginger; c, coffee; d, pineapple mixed cropping; ▧,female; □, male

comparison of the output/input ratio of these plantation/cash crop systems with jhum under a 10-year cycle (Toky and Ramakrishnan, 1981a, 1982a) suggests that, except for coffee, the efficiency values are higher for the former.

Labour requirements for tea and ginger are high throughout the year; female labour was higher than male labour input unlike that shown for coffee and pineapple systems (Fig 2.3). These differences are related to the agricultural operations involved at different times of the year. From this point of view, the pineapple system seems to be attractive. Indeed, apart from other considerations of high yield, low labour input for the pineapple mixed cropping system may account for its wider acceptance as a land use system in the region.

The energy efficiency in terms of output/input ratio was maximum for the pineapple system, amongst all cash crops (Table 2.16). Although this efficiency for jhum under a 10-year cycle is high, this cannot be considered in isolation but has to be discussed in relation to land use pattern, or else energy efficiency value *per se* would lead to distorted comparisons. In other words, the apparent high efficiency of the jhum system becomes real only if land for jhum under 10-year cycle is not a limiting factor. Otherwise,

Table 2.17 Energy ($MJ\,ha^{-1}\,yr^{-1}$) and monetary ($Rs\,ha^{-1}\,yr^{-1}$) efficiency of pineapple plantation of the Khasis at Nayabunglow in Meghalaya (from Patnaik and Ramakrishnan, 1989)

	Energy	*Monetary*
Input total (human labour)	820	3,461
Output total	46,279	77,000
Output/input ratio	56.4	22.3

with a correction factor of a tenth to make comparisons valid with other systems considered here, the efficiency for jhum would decline drastically.

Although the energy output from ginger is low, the economic returns are high, suggesting that energy efficiency has to be reconciled with economic efficiency for an objective evaluation of the system. With substantial inorganic fertilizer input going into these cash crop systems, except pineapple, the efficiency of these sedentary systems are generally low. However, efficient recycling of resources with emphasis on organic manure would contribute to better ecological efficiency of these cash crop systems.

Pure plantations of pineapple (*Ananas comosus*) are cultivated on slopes of 30–40° angle, by the Khasis of Nayabunglow, taking two harvests, once in July–August (monsoon variety) and another in December–January (winter variety) (Patnaik and Ramakrishnan, 1989). With little labour input for land preparation and weeding, the energy and monetary efficiencies are very high (Table 2.17), in fact much higher than pineapple mixed cropping system on terraces discussed above.

Conclusions

The major problem with a plantation crop such as tea is the need for heavy labour input, often involving importation of labour from outside the village boundary, indeed even outside the region. This is often costly. Under the given socio-economic and socio-political framework in which traditional societies operate at present, it may be desirable to limit plantations into smaller family-size units that can be handled internally, labour coming exclusively from within the family. It is in this context, a mixed land use pattern with a redeveloped and stabilized jhum system and/or valley cultivation to ensure self-sufficiency of food crops, and with cash crop economy organized on a co-operative basis (Ramakrishnan, 1984b) becomes relevant. For maintenance of soil fertility in these cash crop systems, it is also important to consider lesser dependence upon inorganic fertilizers. Emphasis on recycling organic residues is important, as this is an ecological concept on which traditional societies are based.

Table 2.18 Cropping patterns at higher elevations of Meghalaya around Shillong in Meghalaya (from Gangwar and Ramakrishnan, 1987a)

Jhum system	*Fallow system*	*Sedentary system*
15-yr cycle	3-yr cycle	
(1) *Solanum tuberosum*	(1) *Solanum tuberosum*	(1) *Solanum tuberosum*
Zea mays	2-yr cycle	*Zea mays*
Perilla ocimoides	(2) *Flemingia vestita*	*Perilla ocimoides*
Phaseolus mungo	*Brassica oleracea* var.	*Phaseolus mungo*
Colocasia antiquorum	*capitata*	*Colocasia antiquorum*
Cucurbita maxima	(3) *Raphanus sativus*	*Cucurbita maxima*
8-yr cycle	(4) *Solanum tuberosum*	(2) *Solanum tuberosum*
(2) *Solanum tuberosum*	1-yr cycle	*Zea mays*
Zea mays	(5) *Brassica oleracea* var.	*Perilla ocimoides*
Perilla ocimoides	*botrytis*	(3) *Brassica oleracea* var.
(3) *Solanum tuberosum*	*Perilla ocimoides*	*botrytis*
3-yr cycle	(6) *Solanum tuberosum*	*Perilla ocimoides*
(4) *Solanum tuberosum*	*Oryza sativa*	(4) *Brassica oleracea* var.
Zea mays	(7) *Raphanus sativus*	*capitata*
Perilla ocimoides	(8) *Solanum tuberosum*	*Perilla ocimoides*
Phaseolus mungo	(9) *Brassica oleracea* var.	(5) *Solanum tuberosum*
Colocasia antiquorum	*botrytis*	*Digitaria cruciata* var.
Cucurbita maxima	(10) *Ipomoea batatas*	*esculenta*
		(6) *Raphanus stivus*
		(7) *Solanum tuberosum*
		(8) *Brassica oleracea* var.
		botrytis
		(9) *Brassica oleracea* var.
		capitata
		(10) *Flemingia vestita*
		(11) *Fagopyrum esculentum*
		(12) *Eleusine coracana*
		(13) *Oryza sativa*
		(14) *Digitaria cruciata* var.
		esculenta

FALLOW AND SEDENTARY SYSTEMS

During the recent past, as a result of increasing population pressure, there has been a shift from more extensive to more intensive systems of land use in many parts of the under-developed world (Boserup, 1965; Okafor, 1987). Longer fallow systems have been replaced by shorter bush fallow systems of agriculture (FAO/SIDA, 1974) and ultimately by permanent or sedentary agriculture. The fallow system is variously termed also as 'semi-permanent cultivation' (Nye and Greenland, 1960; Allan, 1965) and 'stationary cultivation with fallowing' (Faucher, 1949). Emphasis on the crops grown would vary considerably (Table 2.18) (Gangwar and Ramakrishnan, 1987a).

Around city centres such as Shillong in Meghalaya, because of intense population pressure, the modified version of the jhum (Mishra and

Plate 2.8 A general view of the fallow/sedentary systems of agriculture around Shillong (1,500 m altitude) in Meghalaya. In the background on the hill slopes where forest cover is available, the Khasis of this area do the modified version of jhum. All these land uses involve elaborate preparation of land into ridges and furrows

Ramakrishnan, 1981) has often been replaced by the fallow and sedentary systems of agriculture (Plate 2.8). These land use systems take over when the jhum cycle comes down below 5 years, as has happened amongst village communities closer to the Shillong township. The modified version of jhum found at sites further away are similar to that discussed in Chapter 1 (Mishra and Ramakrishnan 1981; Gangwar and Ramakrishnan, 1987a) and therefore are not considered here.

Fallow system

In the fallow system of agriculture (FAO/SIDA, 1974), the weed biomass of the fallow phase is slashed in January and organized in parallel rows covered over by a thin layer of soil and allowed to decompose. The crop is sown on these ridges in March. The rest of the land is compacted into alternating furrows, running down the slope, as in the modified jhum system (Mishra and Ramakrishnan, 1981).

Two croppings are done in a year, one between March and June and another between August and November. During the first cropping, crop species are raised singly or in mixtures of two with emphasis on vegetables (Table 2.18). The second crop raised in the year is always a monoculture

Plate 2.9 *Flemingia vestita*, a lesser-known plant of food value, is a much branched trailing herb (branches up to 100 cm long) grown for its tuber under fallow/sedentary system of agriculture around Shillong (1,500 m) in Meghalaya. The tuber is edible. Inset is a general view of its cultivation

of potato. Since *Flemingia vestita* raised in March (Plate 2.9) cannot be harvested until October, this species is raised along with another crop species, but with only one cropping in a year. For seed purposes, only potato tubers are purchased from the market whilst all others are raised from within. Three hundred and twelve to 5,400 kg ha^{-1} yr^{-1} (dry weight) of organic manure are applied before crop sowing. Two hundred to 1,275 kg ha^{-1} yr^{-1} of inorganic NPK (1:1:1) fertilizer are used, with two to three applications during cropping. The quantities used varied depending upon the crop mixture and the location. Of all the species cultivated under the fallow system, *Flemingia vestita* and *Perilla ocimoides* (Plate 2.10) are two lesser known crops of food value of the Khasis (Gangwar and Ramakrishnan, 1989b). They are of food value during the leaner months of the year when traditional food crops are in short supply.

Economic and energy efficiencies varied considerably depending upon the cropping pattern (Table 2.19). Potato is an important crop from a monetary view point. This, as a second crop, along with mustard gave maximum monetary return to the farmer. However, energy efficiency was maximized where *Flemingia vestita*, a legume crop was involved. This may be related to the nitrogen accretion in the soil due to its inclusion in the crop mixture (Gangwar and Ramakrishnan, 1989b). Better performance of

Plate 2.10 *Perilla ocimoides*, a lesser-known plant of food value, is an aromatic bushy annual growing to a height of up to 150 cm, grown under fallow/sedentary systems of agriculture around Shillong in Meghalaya. The grains are edible

tuber crops under reduced soil fertility is understandable because of better nutrient-use efficiency (Ramakrishnan 1984a) under short fallow systems.

Sedentary systems

The economic and energy returns varied significantly depending upon the crop species involved (Table 2.20). In general, vegetable (Plate 2.11) and tuber crops gave higher economic returns compared to seed crops (items 11 to 14 of Table 2.18). As discussed earlier, the energy efficiency was high where *Flemingia vestita* was involved, because of the ability of this species to improve soil nitrogen fertility.

Table 2.19 Monetary (Rupees ha^{-1} yr^{-1}) and energy (MJ ha^{-1}) output/input during first cropping under different fallow systems around Shillong in Meghalaya (from Gangwar and Ramakrishnan 1987a)

	Input		*Output*		*Output/Input ratio*	
Fallow	*Money*	*Energy*	*Money*	*Energy*	*Money*	*Energy*
(1)	7,977(17,439)	5,292(19,993)	10,452(24,202)	35,446(82,084)	1.31(1.39)	6.70(4.11)
(2)	5,393	1,038	21,404	20,643	3.97	19.89
(3)	5,728(14,518)	5,718(16,568)	11,461(34,670)	14,502(54,343)	2.00(2.39)	2.54(3.28)
(4)	8,986(17,775)	9,054(19,904)	12,096(23,844)	41,030(80,871)	1.35(1.34)	4.53(4.06)
(5)	7,244(14,591)	15,449(23,637)	26,990(37,587)	38,089(74,024)	3.73(2.58)	2.47(3.13)
(6)	10,750(18,097)	17,415(25,603)	18,240(28,837)	62,544(98,479)	1.70(1.59)	3.59(3.85)
(7)	7,582(14,929)	8,656(16,844)	16,829(27,426)	18,098(54,033)	2.22(1.84)	2.09(3.21)
(8)	9,914(17,261)	11,131(19,319)	15,742(26,339)	53,397(89,332)	1.59(1.53)	4.80(4.62)
(9)	6,039(13,386)	10,988(19,176)	25,780(36,797)	25,408(61,343)	4.27(2.75)	2.31(3.20)
(10)	4,036(11,383)	7,367(15,555)	10,824(21,421)	41,254(77,189)	2.68(1.88)	5.60(4.96)

Values in parentheses include the second crop grown in the year except in the case of item 2 where only one crop is grown
The serial number corresponds to those given in Table 2.18

Table 2.20 Monetary (Rupees $ha^{-1} yr^{-1}$) and energy ($MJ\,ha^{-1}$) output/input pattern during the first cropping under different sedentary agriculture systems around Shillong in Meghalaya (from Gangwar and Ramakrishnan, 1987a)

Sedentary systems	*Input*		*Output*		*Output/Input ratio*	
	Money	*Energy*	*Money*	*Energy*	*Money*	*Energy*
(1)	10,271(19,663)	18,404(32,981)	14,777(26,141)	55,647(94,186)	1.44(1.33)	3.02(2.86)
(2)	12,001(21,393)	21,363(35,940)	18,897(30,261)	70,769(109,308)	1.57(1.41)	3.31(3.04)
(3)	6,214(15,606)	11,018(25,595)	19,040(30,404)	26,995(65,534)	3.06(1.95)	2.45(2.56)
(4)	8,610(18,002)	14,172(28,749)	23,050(34,414)	17,301(55,840)	2.68(1.91)	1.22(1.94)
(5)	6,710(16,102)	12,424(27,001)	15,049(26,413)	44,008(82,547)	2.24(1.64)	3.54(3.06)
(6)	5,748(15,140)	4,697(19,274)	13,158(24,522)	19,034(57,573)	2.29(1.62)	4.05(2.99)
(7)	10,075(19,467)	11,108(25,685)	13,748(25,112)	46,633(85,172)	1.36(1.29)	4.20(3.32)
(8)	6,814(16,206)	15,365(29,942)	18,340(29,704)	36,291(74,830)	2.69(1.83)	2.36(2.50)
(9)	8,475(17,867)	14,216(28,793)	21,820(33,164)	15,414(53,953)	2.57(1.86)	1.08(1.87)
(10)	4,891	984	11,016	21,812	2.25	22.17
(11)	1,931(11,323)	4,436(19,013)	4,533(15,897)	21,022(59,561)	2.35(1.40)	4.74(3.13)
(12)	1,663(11,055)	2,588(17,165)	3,705(15,069)	26,770(65,309)	2.23(1.36)	10.34(3.80)
(13)	1,066(10,458)	1,703(16,280)	1,969(13,333)	7,405(45,944)	1.85(1.27)	4.35(2.82)
(14)	856(10,248)	1,114(15,691)	3,890(15,254)	8,977(47,516)	4.54(1.49)	8.06(3.03)

Values in parentheses include the second crop grown in the year
The serial numbers correspond to those given in the Table 2.18

Plate 2.11 A closer view of vegetable (cauliflower) being grown on ridges around Shillong (1,500 m) in Meghalaya, as a component of the sedentary system of agriculture

Conclusions

With rapid shortening of the jhum system around Shillong, the farmer went in for a variety of croppings under fallow and sedentary systems of agriculture. Under these systems, the energy input had a substantial organic/inorganic fertilizer component, apart from human labour. Potato, which gives better returns to the farmer under reduced soil fertility, as compared with cereals, is an important species in these systems.

Flemingia vestita along with *Perilla ocimoides* form two important lesser known crops of food value. Of these two, the former, when grown along with other traditional crops, improves the ecological efficiency of the system. Indeed after a few years of cropping, the farmer would often raise a pure crop of this legume because of its nitrogen fixing properties through root nodules. The nitrogen fixed by this species was about 250 $kg\,ha^{-1}\,yr^{-1}$ (Gangwar and Ramakrishnan, 1989b). With a range of values between 65 and 224 $kg\,ha^{-1}\,yr^{-1}$ for many traditional leguminous crop species (Nutman, 1976), which occasionally may go up to 500 $kg\,ha^{-1}\,yr^{-1}$ (National Research Council, 1979), the nitrogen fixed by *Flemingia vestita* compared favourably. Indeed, with a matty growth form and dense crop cover, this species would also check losses of nutrients through run-off water during the monsoon (Mishra and Ramakrishnan, 1983c). A few years of vegetable/tuber cropping involving traditional species with or without

Flemingia vestita followed by a year of cropping under this legume alone, therefore, makes a good rotation practice.

With wide variations in energy/economic output/input patterns over a small area, these systems provide opportunities to obtain higher returns for the farmer, through mere manipulation of the crop mixture, with minimal technology input from outside. However, continued cropping under fallow/sedentary systems over a long time period in these high rainfall areas has often resulted in rapid deterioration in site quality leading to site desertification (Ramakrishnan, 1985b,c). Traditional technologies such as the use of *Flemingia vestita* for site quality maintenance should be combined with the appropriate soil conservation measures and organic manure usage, if these land use systems are to be sustained in the region.

TERRACE AGROECOSYSTEM

Terrace cultivation is not a traditional land use in the north-eastern region. However, in order to discourage farmers from jhum cultivation, bench-terraced land prepared by the local Soil Conservation Department was provided free to the local farmers, from time to time. This land use suggested as an alternative to jhum has not found acceptance from the local people. As long as monetary incentives were provided by the governmental agencies, this land use system was adopted, and as the incentives were withdrawn, the farmer reverted back to jhum. In order to evaluate this land use system, we shall look at how this land use system is organized.

Terrace agroecosystem at lower elevations at Burnihat in Meghalaya

Terraced farming carried out by the Garos at Burnihat at lower elevations of Meghalaya is similar to the jhum system (Toky and Ramakrishnan, 1981a; cf Chapter 1) in the organization of mixed cropping of 14 crop species raised and harvested sequentially (cf Table 1.2). A major difference from the jhum system concerns the heavy input of fertilizers, initially provided free by the local Soil Conservation Department, but with the cost subsequently borne by the farmers.

Apart from labour input for a variety of operations, inorganic fertilizer is a major input into the system (Table 2.21). During the 1st year of terrace cultivation, the labour input for land preparation is particularly high (2,478 MJ ha^{-1}) and this has to be imported from outside the family unit. Maintenance of the terrace, year after year, also involves considerable labour energy input (984 MJ ha^{-1}). Inorganic fertilizer is a major input into the system and acts as a major disincentive for terrace cultivation. In this case, the farmer used as much as 60 kg of N, 30 kg of P_2O_5 and 30 kg of K_2O per hectare per year, costing about 800 rupees.

Table 2.21 Economic and energy efficiencies (output/input) of terrace agroecosystem at Burnihat (from Toky and Ramakrishnan, 1981a, 1982)

	Money (Rupees $ha^{-1} yr^{-1}$)	*Energy* (MJ $ha^{-1} yr^{-1}$)
Input total	2,542(4,544)	6,509(8,003)
Output total	3,658	43,602
Output/input ratio	1.4(0.8)	6.7(5.4)

Values in parentheses are for the first year
Of the total input, fertilizer was 5,323 MJ of energy per year

Table 2.22 Economic and energy efficiencies of terrace agroecosystem at Shillong in Meghalaya (after Mishra and Ramakrishnan, 1981)

	Money (Rupees $ha^{-1} yr^{-1}$)	*Energy* (k cal $ha^{-1} yr^{-1}$)
Input total	6,004(12,243)	3,080,952(3,341,549)
Output total	12,561	5,236,548
Output/input ratio	2.1(1.02)	1.7(1.6)

Energy output under terrace cultivation through grain and seed crops was almost a third of that obtained under a 30-year jhum cycle here, although energy output through leafy and fruit vegetables, and tuber and rhizomatous crops, was higher than that under a 30-year jhum cycle (Toky and Ramakrishnan, 1982a).

With substantial labour input for terracing, particularly during the 1st year, along with a heavy input of fertilizers, this land use is inefficient from energy and economic view points (Table 2.21). This is in spite of the fact that the output here is only a little less than that under a 10-year jhum cycle when land use reuse pattern is not considered, but with an effective output of 43,602 MJ compared with 5,660 MJ under a 10-year jhum cycle (cf Table 1.13), when related to land use.

Terrace agroecosystem of the Khasis around Shillong at higher elevations of Meghalaya

Terrace cultivation at higher elevations emphasizes on *Solanum tuberosum*, along with *Zea mays* and *Brassica oleracea* (Mishra and Ramakrishnan, 1981), with two croppings in a year as under jhum here; the second winter crop is potato alone (cf Chapter 1). The fertilizer input is high involving 3,000 kg $ha^{-1} yr^{-1}$ of organic manure (pig dung and vegetable compost) and about 740 kg of inorganic NPK (1:1:1) fertilizer. With substantial input of labour and high fertilizer input, the energy and economic efficiencies were low (Table 2.22). In fact the energy efficiency was much lower than

Table 2.23 Weed biomass (kg ha^{-1}) under terrace agroecosystem compared with jhum under a 7-year cycle at Nayabunglow, Meghalaya (from Patnaik, 1988)

	Terrace age (yr)		*Jhum*
	4	*12*	*7-year cycle*
Total weed biomass	2,050(1,038)	2,759	2,650
Weed recycled	507(1,503)	1,009	870

Values in parentheses are for 2nd cropping; weed recycled here comes from intervening fallow period

that noted for the terrace system at lower elevations discussed above. This is in spite of the high gross energy and monetary returns due to potato, the major crop emphasized here (Mishra and Ramakrishnan, 1981).

Terrace agroecosystem of the Nepalis at mid-elevations at Nayabunglow in Meghalaya

At Nayabunglow, 30 km north of Shillong, at an altitude of 960 m, terrace cultivation is done only by the immigrant Nepalis; the local Khasi tribe prefers to do jhum, as they find it difficult to sustain terrace cultivation that demands high fertilizer energy input. After the land is terraced, the Nepali farmer executes two croppings in a year on younger terraces. On older terraces, however, he does only one mixed cropping in a year.

Terrace cultivation resembles jhum practised in the area (Patnaik and Ramakrishnan, 1989) in that the slash burning is followed by mixed cropping during the monsoon season. However, the slash material here is only the weed biomass of the previous December–March period and the crop residue from the previous cropping season. Where two croppings are done, a mixed cropping between April and August is followed by a monocropping of *Eleusine coracana.* A heavy dose of organic manure (2,550 kg ha^{-1} yr^{-1}) is applied before crop sowing in April. *Zea mays, Vigra sinensis, Phaseolus vulgaris* and *Cucurbita maxima* are sown simultaneously, followed by sequential harvesting as mature crops, between August and December. Weeds are removed by hand-hoeing.

Although the emphasis is on *Zea mays,* tuber crops alone are put under 12-year-old terraces. Such an emphasis on tuber crops, with the elimination of second cropping of *Eleusine coracana* and indeed curtailing the number of species in the crop mixture to just five species, as was seen under a 15-year-old terrace, (Patnaik, 1988) may all be linked with reduced soil fertility in the older terraces.

Weeds pose a serious problem. Weed potential of the site, when the first cropping is compared, increased in the older terrace (Table 2.23), this being comparable to the jhum system in this area under a 7-year cycle (Patnaik

Table 2.24 Monetary (Rupees $ha^{-1} yr^{-1}$) and energy (MJ $ha^{-1} yr^{-1}$) analysis of terrace agroecosystem at Nayabunglow in Meghalaya (from Patnaik, 1988)

	Money		*Energy*	
	4-yr-old terrace	*12-yr-old terrace*	*4-yr-old terrace*	*12-yr-old terrace*
Input total	6,217 ± 365 (2,296 ± 180)	3,642 ± 87	4,846 ± 112 (465 ± 44)	3,820 ± 185
Labour	4,139 (2,116)	1,869	943 (441)	534
Organic manure	1,726	1,627	3,836	3,254
Seed	352 (180)	146	67 (24)	32
Output total	11,791 ± 251 (4,968 ± 66)	5,963 ± 111	101,525 ± 1,516 (53,952 ± 23)	50,815 ± 1,073
Output/input ratio	1.9 (2.2)	1.6	21 (116)	13

Table 2.25 Labour energy (MJ $ha^{-1} yr^{-1}$) input pattern for terrace agroecosystem compared with jhum under a 7-year cycle at Nayabunglow in Meghalaya (from Patnaik, 1988)

	Family input	*Co-operative input*	*Hired input*
4-yr-old terrace	803	51	90
7-yr-jhum cycle	1,224	361	0

and Ramakrishnan, 1989). Intensified weed potential could then be a major constraint for sustaining terraces in the region, as is also true for very short jhum cycles (Saxena and Ramakrishnan, 1984b). This is discussed in detail elsewhere (Chapter 3).

Under the 12-year-old terrace, the economic yield harvested is essentially from four crop species only, since *Ipomoea batatus* and *Colocasia antiquorum* were used only as a cover crop over the soil, although some leaves of the latter may be used as a leafy vegetable. A decline in crop yield in the older terrace is to be expected due to a decline in soil physical and chemical properties under continuous cropping (Patnaik 1988); this is also reflected in the single cropping done here. The energy output/input ratio also declined sharply in the older terrace (Table 2.24). A very high output/input ratio for energy during the second cropping is because it is largely dependent upon the inputs from the first cropping phase, except for some labour and seed. However, labour for terrace cropping, apart from that contributed from within the family and that coming on a co-operative basis from the village, was also hired (Table 2.25). This is in contrast to the jhum system under a 7-year cycle that is compared here.

Conclusions

Terracing suggested as an alternative to jhum has been rejected by the tribals because of a variety of reasons. Under a high rainfall situation, even if run-off losses of sediment and nutrient from steep slopes are checked, the infiltration losses are heavy and this would be accentuated on the terrace (Ramakrishnan and Toky, 1981; Mishra and Ramakrishnan, 1983c; Patnaik, 1988). Soil physical and chemical properties would deteriorate under continuous cropping on the terraces. We have already seen how weed potential would increase under continuous cropping, through the case study at Nayabunglow in Meghalaya. A heavy input of organic/inorganic fertilizers is required to sustain soil fertility. Inorganic fertilizers are difficult to obtain, they are expensive and their use efficiency is highly curtailed. This makes the system uneconomical. The terraces are washed out constantly along the sides due to heavy rain and therefore maintenance costs are high. Except in areas such as at Nayabunglow, where the soil is well developed, it becomes difficult to maintain terraces. Even at Nayabunglow, in spite of a deep soil profile that is favourable for maintaining terraces, the local Khasi tribe still practise jhum or valley cultivation, leaving terrace cultivation for the immigrant Nepalis. The immigrant Nepalis do terrace cultivation on land hired from the urban tribal landlords of Shillong, who hire out the land for a fixed price along with part of the produce from the land, as payment. The Nepalis being cattle farmers are able to put in organic manure in the form of cow dung, which the Khasis are unable to do. With a substantial amount of the produce going to the absentee landlord and with declining productivity of the land over a period of time, the Nepalis end up doing subsistence agriculture but with greater dependence upon their well organized cattle husbandry.

This land use and the partial terracing (Plate 2.12) suggested since 1975 by the Indian Council of Agricultural Research (ICAR) (Borthakur *et al.*, 1978) have not been able to make an impact because they are totally divorced from the socio-economic framework in which the tribal societies function (Ramakrishnan, 1985c). The ICAR model recommends bench terracing of the lower third of the slope, the middle third being for a hortipastoral system with half-moon terracing and the top third reserved for forestry. This rigid land use system conflicts with traditional land tenure and use patterns that otherwise provides considerable flexibility in land tenure and agricultural operations through labour management within the village community itself. The ICAR model, in terms of land use and cropping pattern, is based on experimental plot evaluation and is not linked to real life situations where social considerations often are critical.

Plate 2.12 Partial terracing of the hill slope as suggested by the Indian Council of Agricultural Research, which is a high fertilizer input system, not favoured by the tribals

GENERAL CONSIDERATIONS

The alternate land use options to jhum discussed here are significant in that they offer possible pathways for land use redevelopment. Sustainable development of land use in any region has to satisfy a few basic criteria:
(1) It should be sustainable without major input of energy from outside the region;
(2) It should be compatible with environmental quality on a short- and long-term basis;
(3) It should have the flexibility to meet not only present needs of the society, but also the long-term demands through adaptive innovations; and
(4) It should ensure equity and social justice.

Amongst the land use options considered here, terrace cultivation introduced as an alternative to jhum often fails in all these respects. It is energy intensive, environmentally non-compatible because of irreversible site degradation that it may often cause, and can be practised only by the upper strata of the society with adequate monetary resources. On the other hand, traditional land use systems such as valley agroecosystems and home gardens could respond to ecological, social and economic demands and adjust to changing situations. Therefore, these two land uses could be the basis for land use redevelopment.

Plantation/cash crop economy has often been suggested as a land use option for development of the region. For north-eastern India, this has long-term developmental implications. Horticulture and plantation crops along with forestry would ensure permanent plant cover on hill slopes. Studies from Sarawak, Malaysia suggest that cash cropping could be intensified and the Ibans of this region could grow enough rice with cash crops to meet their food needs (Chin, 1982). However, these plantations if carried out as a business enterprise over large areas of land acquired from the tribals, would not be able to conform to the sustainability criteria. It would only alienate the tribal societies from their land and could result in distortions, eventually leading to social tensions (Ramakrishnan, 1988c).

Therefore, the only way in which one could develop cash crop economy is to create small village level units. The home garden concept offers the most suitable approach to manage cash crop systems in the north-eastern region. Here, a family unit will be able to handle the cash crop system without too much dependence on imported labour. Such a system organized on a co-operative basis would fit into the social and the cultural milieu, as it is based on redeveloping a traditional technology that is already being practised very widely.

CHAPTER 3

VILLAGE ECOSYSTEM FUNCTION OF TRADITIONAL SOCIETIES

INTRODUCTION

The humid tropics throughout the world are characterized by dense rain forest ecosystem types with high species diversity but with extreme fragility (Whitmore, 1975; Raven, 1981; Ramakrishnan, 1985b). This ecosystem is subject to rapid conversions either due to timber extraction for industrial activities, fuelwood extraction or shifting agriculture (Myers, 1980; Ramakrishnan, 1985d). In north-eastern India, as elswhere in the world (Watters, 1971; Haswell, 1973; Dove, 1983), the local people depend upon forest resources for their varied economic activities, including shifting agriculture, as discussed in Chapter 1. Some of the tribes have other land use systems (cf Chapter 2) often linked to the forest such as the home gardens. These land use patterns vary depending upon the socio-economic and socio-cultural backgrounds of the tribal communities, as already discussed in earlier Chapters.

Since jhum, which is the chief land use for agriculture in north-eastern India, has in recent times fallen into disrepute because of shortening of the jhum cycle, this distorted land use has disrupted the traditional linkage between the tribal communities and the forest ecosystem on which they are dependent. Closely linked with jhum is the animal husbandry system largely involving swine husbandry and poultry (Mishra and Ramakrishnan, 1982). Goats and cattle have been introduced by the immigrant communities (e.g. the Nepalis) and the plain tribals of the Assam valley such as the Mikirs. Such introductions, if done on a large scale, could have adverse impacts on forest resources due to grazing and browsing effects. A holistic approach to an understanding of the village ecosystem function of traditional societies of the humid tropics, therefore, assumes significance. In fact, many of the studies that have looked at traditional societies have considered only a specific component of the village system, such as agriculture, animal husbandry or the domestic sector (Spedding, 1975; Chandra *et al.*, 1976) but few have considered the village as a functional unit (Reddy, 1981; Mishra and Ramakrishnan, 1982; Maikhuri and Ramakrishnan, 1990 a,b). Until recently, the indigenous agricultural village of Asia, Africa and South America functioned as nearly independent systems (Norman, 1978). In more recent times, however, this self-sufficiency has been more and more related to greater interdependence because of modern technology introductions (Pimental and Pimental, 1979) or due to the damage caused by environmental degradation (Mishra and Ramakrishnan, 1982; Gangwar and Ramakrishnan, 1989a). An understanding of the village ecosystem function of tribal societies of north-eastern India is important from this view point too, where limited natural resources often have to be shared by many communities.

The traditional societies function on the concept based upon the recycling of resources within the system (Spedding, 1979; Mishra and Ramakrishnan, 1982; Sundarraj and Mitchell, 1987) using human labour as an important energy input (Leach, 1976; Revelle, 1976; Mitchell, 1979). Such systems have the advantage of a certain degree of flexibility, being able to adjust to changes that may be imposed on them, unlike more industrialized societies. Apart from the fact that a solar-powered human ecosystem is intrinsically interesting, an understanding of its function is critical because of the frequent rejection of outside advice by the local communities, which is increasingly perceived as a central problem in developing countries.

COEXISTENCE OF TRIBAL SOCIETIES SHARING COMMON RESOURCES

With a large number of tribal communities of diverse social, cultural and linguistic backgrounds, north-eastern India is characterized by marked anthropogenic differences over very short distances. If ecological differences arising out of a hill topography are superimposed upon the anthropological diversity, then the heterogeneity becomes so complex that generalizations on village organization and function become difficult. However, these diverse communities often coexisting in the same area, have evolved ways in which they are able to do so, sharing common resources in a highly complementary manner (Patnaik and Ramakrishnan, 1989; Maikhuri and Ramakrishnan, 1990a,b). This section dealing with a few case studies, therefore considers the differences between different communities living in the same area, evaluating village ecosystem function using money and energy as currencies for economic and ecological efficiencies. This integrated and holistic approach to understanding the functioning of traditional societies living in village clusters looks at : (1) the traditional land management practices for agriculture and animal husbandry, (2) the link

Table 3.1 Structure of three tribal villages and a non-tribal village at Lailad in Meghalaya (from Maikhuri and Ramakrishnan, 1990a)

	Tribe/village			
	Garos/ Nongladoh	*Khasis/ Nongkindrih*	*Mikirs Umsophy*	*Nepalis/ Tasku*
Number of households	70	34	10	4
Total human population	624	205	86	27
Adult males	185	65	34	9
Adult females	178	54	30	8
Children 5–7 years old	58	18	10	3
Children 7–9 years old	66	22	6	2
Children 9–12 years old	137	46	6	5
Average family size	8.9	6.0	8.6	6.8
Total area under cultivation (ha)	94	30	45	4
20-year jhum cycle (ha)	–	30	–	–
10-year jhum cycle (ha)	58	–	–	–
5-year jhum cycle (ha)	–	–	20	–
Valley (ha)	36	–	22	4
Home garden	–	–	3	–
Total animal population	1,071	462	123	49
Bullocks and cows for slaughter	176	32	–	–
Milking cows	–	–	3	13
Bullocks for ploughing	46	–	–	4
Goats	–	39	12	6
Swine	78	92	24	–
Poultry	695	299	84	26
Ducks	76	–	–	–

– indicates no data

between the forests and other land management systems; and (3) the self-sufficiency of village communities as they now operate. The objective is to analyse how the coexisting societies share a common resource base.

Ecological analysis of a cluster of villages of different tribes at Lailad in Meghalaya

At the study area at Lailad (296 m altitude) the Garos, the Khasis, the Mikirs and the Nepalis coexist sharing common resources. These four communities belong to four different villages as indicated in the village structure (Table 3.1). The Garos, the Khasis and the Mikirs do jhum under different cycle lengths of 20, 10 or 5 years. Valley cultivation in this village cluster is done by the Garos, the Mikirs and the Nepalis. Home gardens are maintained by the Mikirs alone. Whereas the hill tribes represented by the Garos and the Khasis maintain cattle for meat, the plains tribe, represented by the Mikirs of the Assam valley and the immigrant Nepalis, use cattle for milking. Swine husbandry and poultry are the two important animal husbandry practices of all the tribes. Goats and cattle are introduced into the region by the immigrant Nepalis, and is therefore restricted.

Agriculture sub-system

Amongst the communities considered here, the Nepalis, being non-tribal, do not practise jhum. The jhum procedures of the different tribes differ based upon their socio-cultural affinities (cf Chapter 1; Table 1.6). The major conclusion here is that the jhum system of the Garos is better organized than that of others because of better organization of the crop mixture by this traditional community (Maikhuri and Ramakrishnan, 1990a). Similarly, the Nepalis who traditionally do sedentary farming have more productive wet rice cultivation on valley lands compared to the tribals who do jhum (cf Chapter 2; Table 2.3). The home garden with effective organization of the crop mixture and use of space through stratification in the plant community is a highly organized production system of the Mikirs (cf Chapter 2; Table 2.11), with 3-to 5-fold economic returns from it, compared to that under jhum done by the different tribes (cf Table 1.7).

Animal husbandry sub-system

The animal husbandry of the humid tropics is a low cost sub-system (Payne, 1985) and this contrasts sharply with that of the temperate region. Greater availability of natural resources in the humid tropics (Gomez-Pompa *et al.*, 1972; Raven, 1981) and relatively lesser population pressure as in the north-eastern hill region (Ramakrishnan, 1985c) are the reasons for this. However, the reduced cost of animal husbandry is reflected in the generally lower output from it (Table 3.2).

Of all the animal husbandry practices, swine husbandry (Plate 3.1) is one of the cheapest to maintain (Ramakrishnan, 1985c). Swine husbandry is based on efficient recycling of resources between it and the agricultural system. In a study of the output/input pattern for swine husbandry of the Khasis at Shillong (1,540 m altitude) in Meghalaya (Mishra and Ramakrishnan, 1982), it was found that crops accounted for 4.9% of the feed input into this sub-system while the crop residues plus browsing accounted for the rest of the feed input (Fig. 3.1). Besides juvenile animals, labour for rearing animals was another input into this sub-system. Apart from meat, dung was an important output from the system. The output/input ratio is low if by-products and grazing are included as inputs into the system but higher if they are excluded. The waste biomass from agriculture, including food items unfit for human consumption, are recycled through swine husbandry. This makes it very cheap to maintain. Therefore, it is reasonable to find that this detritus-based system is closely interlinked with

Table 3.2 Annual energy input/output pattern (MJ ± SE) for per fowl, duck, pig, goat and cow raised by different communities at Lailad in Meghalaya. Values are also given for monetary input/output pattern (Rupees yr^{-1} ± SE) (from Maikhuri and Ramakrishnan, 1990a)

	Garos[1]		*Khasis*[2]		*Mikirs*[3]		*Nepalis*[4]	
Production	*Energy*	*Monetary*	*Energy*	*Monetary*	*Energy*	*Monetary*	*Energy*	*Monetary*
Input								
Fowl	4,130 ± 364	24 ± 2	4,135 ± 335	8 ± 4	4,129 ± 330	20 ± 1.6	4,138 ± 340	19 ± 1
Duck	4,149 ± 326	40 ± 2	–	–	–	–	–	–
Pig	2,512 ± 226	38 ± 2	5,037 ± 460	92 ± 5	5,070 ± 467	164 ± 11	–	–
Goat	–	–	4,143 ± 335	72 ± 5	4,151 ± 340	37 ± 2	4,163 ± 357	65 ± 2
Cow	24,060 ± 2,022	160 ± 13	23,879 ± 1,888	64 ± 4	22,772 ± 1,781	85 ± 6	23,133 ± 1,856	157 ± 12
Output								
Fowl	91 ± 6	129 ± 10	74 ± 5	104 ± 8	78 ± 6	125 ± 9	65 ± 5	98 ± 5
Duck	91 ± 7	96 ± 6	–	–	–	–	–	–
Pig	294 ± 26	240 ± 18	690 ± 60	540 ± 33	282 ± 31	270 ± 16	–	–
Goat	–	–	252 ± 22	540 ± 47	260 ± 24	560 ± 47	211 ± 17	410 ± 28
Cow	8,852 ± 658	1,210 ± 18	9,134 ± 686	1,500 ± 105	6,791 ± 412	556 ± 30	8,073 ± 522	1,242 ± 110
Output/input								
Fowl	0.02	5.4	0.01	13.0	0.02	6.2	0.01	5.1
Duck	0.02	2.5	–	–	–	–	–	–
Pig	0.12	6.3	0.14	5.9	0.06	1.6	–	–
Goat	–	–	0.06	7.2	0.06	15.1	0.05	6.3
Cow	0.4	6.6	0.4	23.4	0.3	6.5	0.4	7.9

Slaughtering interval of pig: [1]6-month, [2]1-year, [3]3-year
Slaughtering interval of goat: [2,3]1-year, [4]2-year
Slaughtering interval of cattle: [1]3-year, [2]2-year
Cow milked after age of: [3]3-year, [4]4-year
– indicates no data

Plate 3.1 Swine husbandry is an important component of the village ecosystem function of all the tribes of north-eastern region. Pigs generally live along with the humans in huts constructed on stilts, but at the ground level

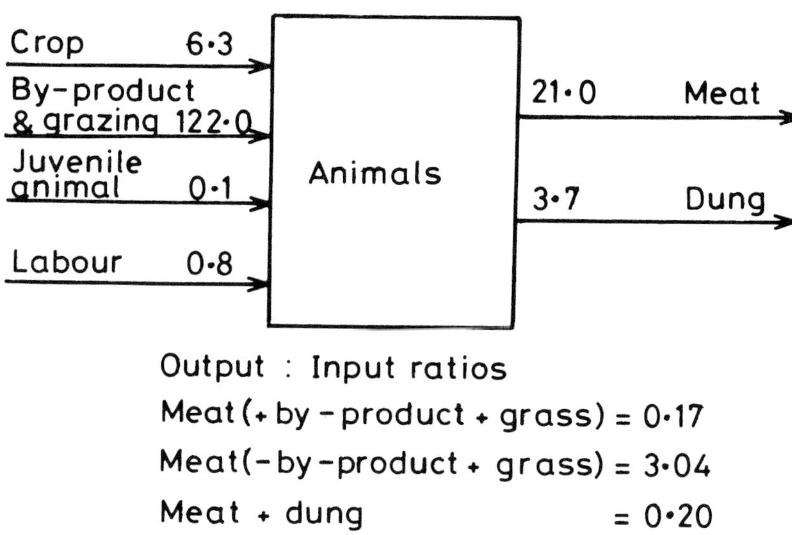

Figure 3.1 Energy input:output pattern and efficiency ratio for the animal husbandry subsystem at the village of Setthliew, near Shillong in Meghalaya. Unit = MJ $\times 10^3$ (from Mishra and Ramakrishnan, 1982)

shifting agriculture throughout the world (Rappaport, 1971; Pimental and Pimental, 1979; Mishra and Ramakrishnan, 1982).

At Lailad, marked differences exist in the input/output patterns for pigs and for other animals such as cattle and goats, partly because of differing frequencies for animal slaughter by different tribes. Thus, with slaughter at 6-monthly intervals, the annual return from swine husbandry was lower for the Garos than for the Khasis who slaughter at yearly intervals. However, both energy and economic efficiencies of swine husbandry of the Mikirs, who slaughter pigs at 3-yearly intervals, is markedly lower than that of the Garos and the Khasis. With an energy expenditure of 18.8×10^6 MJ over a 10-year period for raising a single pig that is slaughtered at festival time, under Tsembaga swine husbandry in the New Guinea highlands (Rappaport, 1971) and with only 1.5% return on the food energy feed to pig meat energy according to the calculations of Pimental and Pimental (1979), this system is not very efficient although it is a practical way of storing excess food energy. The efficiency of the system studied here is, however, much better because of the lighter demands on the farmers with dependence on waste recycling, browsing and poor quality tubers from agriculture that are unfit for human consumption. Slaughter of pigs every year as they mature, rather than at shorter or longer intervals keeps the cost of rearing to the minimum.

The Khasis do not feed their fowl any grain and therefore the economic efficiency was very high for this tribe compared to others. The energy efficiency for goat rearing was similar for all communities, but the economic efficiencies differed, with higher values obtained for the Mikirs; the labour cost of rearing goats was low for them. The Khasis graze cattle as a community effort between a few families so that the input cost is much reduced with a consequently higher economic efficiency.

Domestic sub-system

The hill tribes, the Garos and the Khasis depend upon the wild sources for food more than the non-tribal Nepalis. The food energy and protein consumption from agriculture was maximum for the Garos followed by the Khasis (Table 3.3). Although the energy consumption from animal husbandry was higher for the Nepalis, the animal protein consumption by them was only next to that of the Khasis. The Garos and the Khasis have a highly diversified food base, with items such as rats, snails, ants and termites collected from the forest. This is one of the reasons not only for the absence of malnutrition but for a protein-rich diet among the traditional societies, in spite of their subsistence farming. This is also true for other traditional societies in the region and those elsewhere in the humid tropics (National Research Council, 1982). Because of a less diversified food

Table 3.3 Mean food consumption pattern (capita^{-1} year^{-1}) by different communities at Lailad in Meghalaya. Values are also given for protein consumption (kg) (from Maikhuri and Ramakrishnan, 1990a)

	Energy (MJ)							
	Garos		*Khasis*		*Mikirs*		*Nepalis*	
Category	*Food*	*Protein*	*Food*	*Protein*	*Food*	*Protein*	*Food*	*Protein*
Agriculture sub-system	5,822	31.2	5,007	29.1	3,888	22.2	4,554	26.8
Animal husbandry sub-system	86	2.1	188	1.5	96	3.1	306	4.8
Forest sub-system	205	1.3	183	1.7	67	0.5	3	0.1
Imported	44	0.4	32	0.2	81	0.5	102	0.7
Total ± SE	6,157 ± 350	35.0	5,410 ± 300	36.0	4,132 ± 205	26.3	4,965 ± 290	32.4

Table 3.4 Monetary (Rupees $10^3 \pm$ SE) output/input pattern and efficiency ratio of the village ecosystem (family^{-1} year^{-1}) of different communities at Lailad in Meghalaya. Values are also given for annual energy export (MJ $\times 10^3$) (from Maikhuri and Ramakrishnan, 1990a)

Category	*Garos*	*Khasis*	*Mikirs*	*Nepalis*
Input				
Vegetable and legume	0.05	0.02	0.07	0.11
Mustard oil	0.12	0.06	0.24	0.20
Kerosene oil	0.03	0.05	0.04	0.06
Clothing	0.21	0.37	0.41	0.50
Medicine	0.04	0.05	0.02	0.10
Others	0.26	0.10	0.37	0.68
Total ± SE	0.71 ± 0.04	0.65 ± 0.02	1.15 ± 0.08	1.65 ± 0.10
Output				
Agriculture sub-system	1.72	2.98	1.24	2.43
Energy export	19.9	18.5	8.7	15.1
Home garden	–	–	3.78	–
Energy export	–	–	4.3	–
Animal husbandry sub-system	2.97	2.49	1.27	2.42
Energy export	4.73	2.94	0.61	2.31
Total ± SE	4.69 ± 0.30	5.47 ± 0.40	6.29 ± 0.33	4.85 ± 0.31
Energy export	24.63	21.44	13.61	17.41
Output/input ratio	6.6	8.4	5.5	2.9

– indicates no data

production base in the jhum system of the Mikirs (cf Table 1.6) or the valley system of the Nepalis and with less diversified animal protein consumption patterns of these two communities, their food import is higher.

Apart from food items such as grain, vegetables and cooking oil, kerosene is used for lighting. The energy cost for clothing, medicine and other miscellaneous items, such as tobacco, salt and matchboxes could not be included in our energy budget because of the difficulty in calculation, although they were considered for monetary analysis (Table 3.4). Mikirs alone exported cash crop produce such as arecanut, banana and pepper from the home gardens they maintained. However, with the agriculture and animal husbandry products exported, the monetary efficiency was maximum for the Khasis and minimum for the Nepalis.

The energy flow diagram (Fig. 3.2) showing interlinkages between animal husbandry, agriculture and domestic sectors suggest differences between the three tribes. Whilst per capita food consumption was highest for the Nepalis, the Khasis were better off than others. However, the Garos exported more food outside the village boundary, through the local market. With a high per capita output of organic manure from the animal husbandry of the Garos, the input into agriculture was higher than that by others.

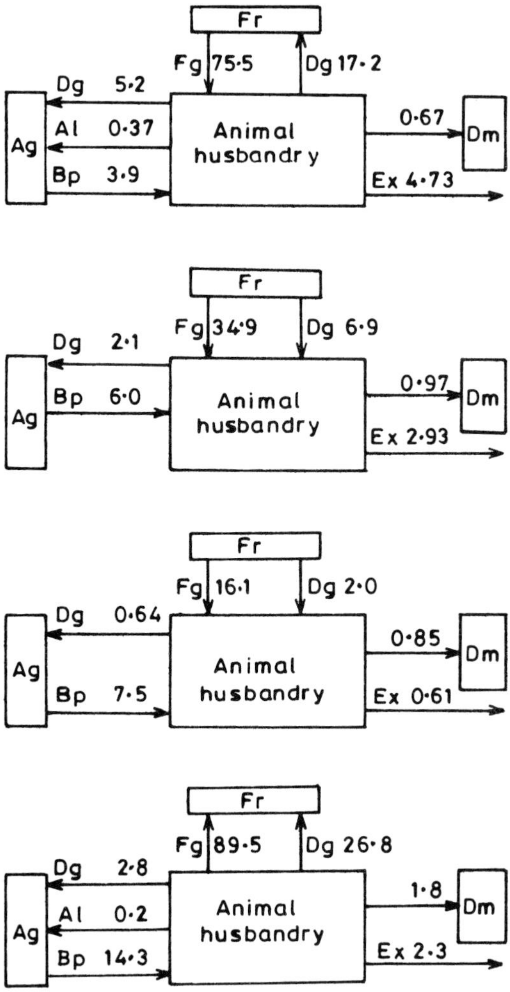

Figure 3.2 Energy flow through animal husbandry (family^{-1} year^{-1}) amongst the different communities at Lailad in Meghalaya. Unit = MJ × 10^3; (a) Garos, (b) Khasis, (c) Mikirs, (d) Nepalis. Fr, forest; Ag, agriculture; Dm, domestic; Dg, dung; Al, animal labour; Bp, crop by-products; Fg, fodder and grazing (from Maikhuri and Ramakrishnan, 1990a)

The Nepalis wasted a high proportion of the organic manure as the animals are grazed in the forest, making it difficult to collect dung.

The diagrammatic presentation of energy flow through the village ecosystem presented in Fig. 3.3 for the Garo community is typical of the linkages between agriculture, animal hubandry, forest and domestic sectors of the four communities. The dependence of these communities on forest for jhum, fodder, fuelwood and food from the wild collection is significant.

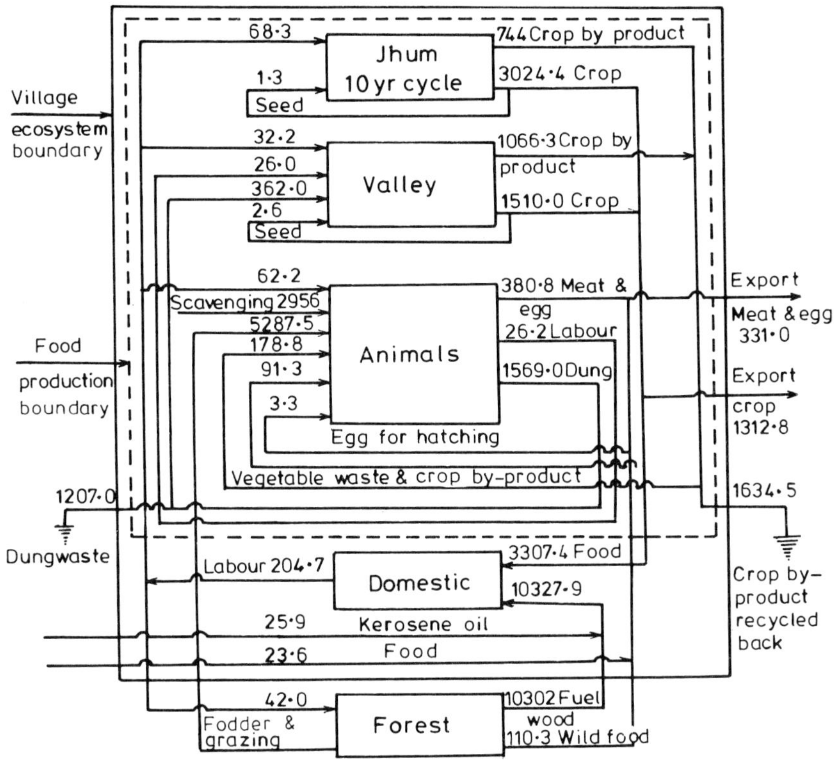

Figure 3.3 Energy flow diagram for the Garo village ecosystem at Lailad in Meghalaya. Unit = MJ × 10^3 (from Maikhuri and Ramakrishnan, 1990a)

The society is based essentially on recycling of natural resources within and between different sub-systems. With much of the dung produced by animals now being wasted by these communities, improved recycling of dung is an approach for increasing productivity of the agriculture sub-system. With abundant natural resources and relatively low population pressure, these village communities are self-sufficient and are able to coexist in the same area by a complementality in the pattern of sharing natural resources. Thus, even with respect to their land use pattern, whilst the Garos and the Khasis emphasize upon jhum and some valley land agriculture, the Nepalis depend exclusively on the valleys for agriculture and the Mikirs emphasize upon cash income through their home gardens.

Ecological analysis of a cluster of villages of different tribes at Balijan in Arunachal Pradesh

At the study area at Balijan (200 m altitude) in Arunachal Pradesh, four different tribes, the Nishis, the Karbis the Kacharis and the Chackmas live

Table 3.5 Structure of four tribal village communities at Balijan in Arunachal Pradesh (from Maikhuri and Ramakrishnan, 1990b)

	Tribes/Village			
	Nishi/ Tubung	*Karbi/ Balijan*	*Kachari/ Dhunabari*	*Chackma/ Mahogoni*
Number of households	30	28	19	22
Total human population	265	193	113	128
Adult male	82	66	38	42
Adult female	78	58	33	34
Children, 5–7 years old	22	15	10	20
Children, 7–9 years old	38	21	14	15
Children, 9–12 years old	45	33	18	17
Average family size	8.8	6.9	5.9	5.8
Total area under cultivation (ha yr^{-1})	74	39	–	25
Under jhum (ha yr^{-1})	32	–	–	–
Valley cultivation(ha yr^{-1})	12	18	–	25
Mustard cultivation (ha yr^{-1})	30	12	–	–
Kitchen garden (ha yr^{-1})	–	9	–	–
Total animal population	529	232	188	332
Bullock for ploughing	30	79	–	29
Cattle for slaughtering	125	–	–	–
Milking cow	–	18	–	13
Mithun	23	–	–	–
Goat	36	31	–	–
Swine	15	–	3	17
Poultry	300	104	185	273

in adjoining but separate villages (Table 3.5) sharing the common natural resources. The Nishis alone do jhum under 5-to 60-year jhum cycles depending upon the closeness of the jhum land to the village; the more remote sites had longer cycles. Wet cultivation of rice is done by the Nishis, Karbis and Chackmas, using bullocks for ploughing. Mustard (*Brassica campestris*) is cultivated on flat lands by the Nishis and the Karbis. Kacharis do not have any agriculture.

Cattle are maintained by the Nishis for meat whereas the Karbis and the Chackmas, the two plains tribals from the Assam valley and Bangladesh, respectively, use them for limited milk production. Swine husbandry and poultry are traditional animal husbandry practices. Mithun (*Bos frontalis*) is a traditional animal reared only by the Nishis. As for many tribes such as the Adis, the Sulungs and the Apatanis of Arunachal Pradesh, Mithun is of religious value being used for sacrifice, apart from its value for barter.

These semi-domesticated animals considered as status symbols of the family, require large forested grazing lands. Rapid deforestation has resulted in a decline in Mithun husbandry.

Agriculture sub-system

The jhum agroecosystem of the Nishis executed under 5- to 60-year cycles, had 25 to 35 species consisting of grain and seed, leaf and fruit vegetables, and tuber and rhizome species, the number of species in the mixed cropping system varying depending upon the length of the jhum cycle (Maikhuri and Ramakrishnan, 1990b). As is typical for the jhum system, the emphasis under long jhum cycles beyond 20 years was more on grain and seed crops; under shorter cycles the emphasis shifted towards leaf and vegetable crops (10-year cycle) or tuber and rhizome crops (5-year cycle). A major conclusion arising from the energy/monetary analysis (cf Table 1.8) was that a 10-year jhum cycle is the minimum cycle length for economic and ecological efficiency of the system.

While the Nishis and the Chackmas under-utilize their valley lands with one cropping of rice in a year, the Karbis have two croppings in a year, the second cropping being a mixture of rice and maize. Because of wide variations in cropping and yield patterns, possibilities exist for improving this land use for yield all the year round. Mustard cultivation is a minor land use activity and the seeds are for local consumption before or after oil extraction; leaves are used as vegetables.

Animal husbandry sub-system

The traditional swine husbandry is not done by the Karbis for religious reasons. Goat and cattle rearing introduced into the village cluster by the immigrant Chackmas of Bangladesh and the Karbis of the Assam valley provide meat and milk for the different tribes. Cattle are raised for meat by the Nishis, whereas Chackmas and the Karbis rear them for milk production.

The energy efficiency ratio generally followed a similar pattern for poultry of all the tribes except the Karbis, whereas the monetary output/input ratio was maximum for the Kacharis compared to other communities (Table 3.6). The energy and monetary efficiencies of swine husbandry of the Nishis was higher than that of others. The energy efficiency for goat husbandry was similar for the Nishis and the Karbis but the economic efficiency was higher for the Nishis. The energy output/input ratio for cattle was generally similar for all the tribes, whereas the monetary efficiency was higher for the Nishis. Mithun raised by the Nishis alone had an energy output/input

Table 3.6 Annual energy output/input pattern (MJ) for per fowl, goat, cow and mithun raised by different communities at Balijan in Arunachal Pradesh. Values are also given for monetary output/input pattern (Rupees yr^{-1}) (from Maikhuri and Ramakrishnan, 1990b)

	Nishis[1]		*Karbis*[2]		*Kacharis*[3]		*Chackmas*[4]	
Production	*Energy*	*Monetary*	*Energy*	*Monetary*	*Energy*	*Monetary*	*Energy*	*Monetary*
Input total								
Fowl	4,146 ± 341	36 ± 1.6	4,141 ± 321	52 ± 2	4,156 ± 350	31 ± 1.1	4,138 ± 346	44 ± 1.8
Pig	5,080 ± 460	105 ± 6.0	–	–	5,030 ± 436	155 ± 8	5,010 ± 445	225 ± 12
Goat	4,121 ± 320	122 ± 6	4,162 ± 328	155 ± 11	–	–	–	–
Cow	23,985 ± 1,932	246 ± 3.1	23,689 ± 1,902	450 ± 4.8	–	–	23,120 ± 1,892	580 ± 5
Mithun	36,800 ± 2,910	78 ± 0.5	–	–	–	–	–	–
Output total								
Fowl	86 ± 6	95 ± 4.0	60 ± 3	118 ± 5.7	72 ± 5	93 ± 3.5	95 ± 6	105 ± 5.8
Pig	500 ± 42	400 ± 25	–	–	400 ± 34	309 ± 18	350 ± 23	250 ± 12
Goat	215 ± 19	320 ± 20	200 ± 17	280 ± 14	–	–	–	–
Cow	9,400 ± 680	1,520 ± 88	8,100 ± 502	853 ± 50	–	–	8,400 ± 532	1,083 ± 60
Mithun	10,100 ± 700	1,200 ± 70	–	–	–	–	–	–
Output/input ratio								
Fowl	0.02	2.6	0.01	2.3	0.02	3.0	0.02	2.4
Pig	0.01	3.8	–	–	0.08	2.0	0.07	1.1
Goat	0.05	2.5	0.04	1.8	–	–	–	–
Cow	0.4	6.2	0.3	1.9	–	–	–	–
Mithun	0.3	15.4	–	–	–	–	0.4	1.9

Slaughtering interval of pig: [1] 1-year, [2] 3-year, [3] 5-year
Slaughtering interval of goat: [1] 1-year, [2] 3-year
Slaughtering interval of cattle: [1] 3-year
Slaughtering inteval of mithun: [1] 3-year
Cow milked after age of: [2] 4-year, [4] 4-year

pattern similar to that of the cattle but with higher economic efficiency. While these differences with respect to economic and energy efficiencies are related to free feed inputs that could be provided by different tribes through grazing/scavenging, the efficiencies improve with increased slaughtering frequency of the animals.

The Nishis generally had more efficient animal husbandry. The input into their animal husbandry both from forest and agriculture sectors was higher. They consumed more food from this sub-system and exported meat outside the village boundary. The Kacharis, with no agriculture for support and with little forest resource to exploit, had the weakest animal husbandry system. Although animal dung could be a very useful commodity, most of it is wasted, but with much variation in the utilization pattern.

Domestic sub-system

The dependence of the four tribes on the forest resources is significant. Apart from the use of forest for jhum and for animal fodder and fuelwood collection, these tribal communities gather food from the wild for their pigs and for themselves. The forest provides a wide variety of food items, particularly during the lean season when traditional food items could be supplemented. The Chackmas and the Nishis obtained more wild collections from the forest than others (Table 3.7). The food obtained from the wild are bamboo shoots, *Dioscorea bulbifera*, leafy vegetables, roots, tubers and mushrooms, whereas the animals from the wild are deer, frogs and snails. A major fraction of the food energy consumed by the different tribes, however, comes from agriculture (79–97% of the total). But the Kacharis obtained all their agriculture produce (99% of the total energy consumed) from the market. The per capita food energy consumption was minimal for the Kacharis compared to others. The Nishis obtained more protein of animal origin compared to all others. The Kacharis had the lowest protein in their diet compared to others.

Agriculture and animal husbandry products are exported by all the communities except the Kacharis. The Kacharis exported fuelwood in exchange for food. The monetary efficiency of the domestic sub-system of the Nishis was maximal and that of the Kacharis minimal.

The linkages within two contrasting village ecosystems, namely that of the Nishis (Fig. 3.4) and the Kacharis (Fig. 3.5) are shown through energy flow diagrams. The Karbis and the Chackmas are similar to the Nishis in their village organization and function. The dependence of the Nishis on the forest sector is for a variety of needs including jhum, whereas the Kacharis largely depend upon the forest for fuelwood collection for export. The Nishis with a variety of land use systems and well organized animal

Table 3.7 Mean food energy consumption (capita^{-1} yr^{-1}) by different communities at Balijan in Arunachal Pradesh. Values are given for protein consumption (kg) (from Maikhuri and Ramakrishnan, 1990b)

	Energy							
	Nishis		*Karbis*		*Kacharis*		*Chackmas*	
Category	*Food*	*Protein*	*Food*	*Protein*	*Food*	*Protein*	*Food*	*Protein*
Agriculture sub-system	4,132	24.2	4,968	29.6	–	–	4,965	28.4
Animal husbandry sub-system	1,054	13.0	73	1.3	27	2.4	211	3.8
Forest sub-system	40	1.1	25	0.4	25	0.7	56	2.2
Imported	18	0.2	56	0.4	3,524	14.3	68	0.6
Total ± SE	5,244 ± 290	38.5	5,122 ± 270	31.7	3,576 ± 160	17.4	5,300 ± 310	350.0

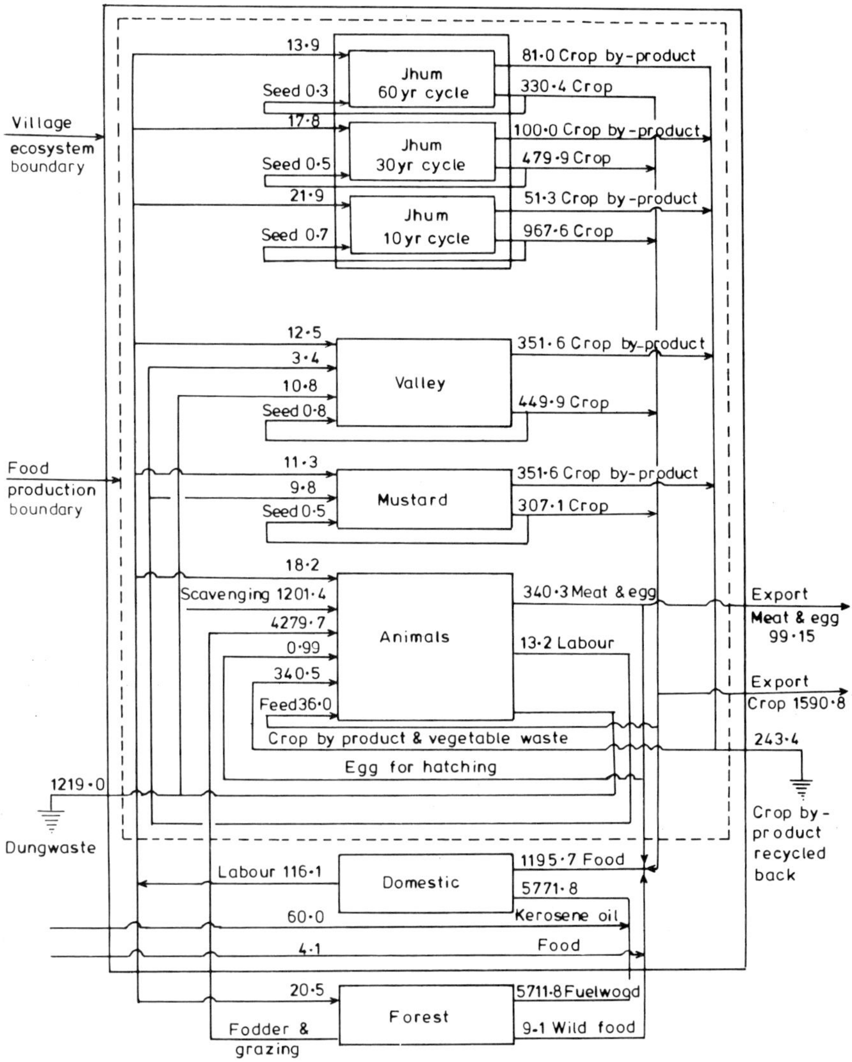

Figure 3.4 Energy flow diagram for the Nishi village ecosystem at Balijan in Arunachal Pradesh. Unit = MJ × 10^3 (from Maikhuri and Ramakrishnan, 1990b)

husbandry have an elaborate set of sub-systems and energy linkages within and between them. Consequently their village unit is self-sufficient. On the other hand, with no agricultural activity, the Kacharis have a distorted village ecosystem function and are largely dependent upon fuelwood collection from rapidly declining forests.

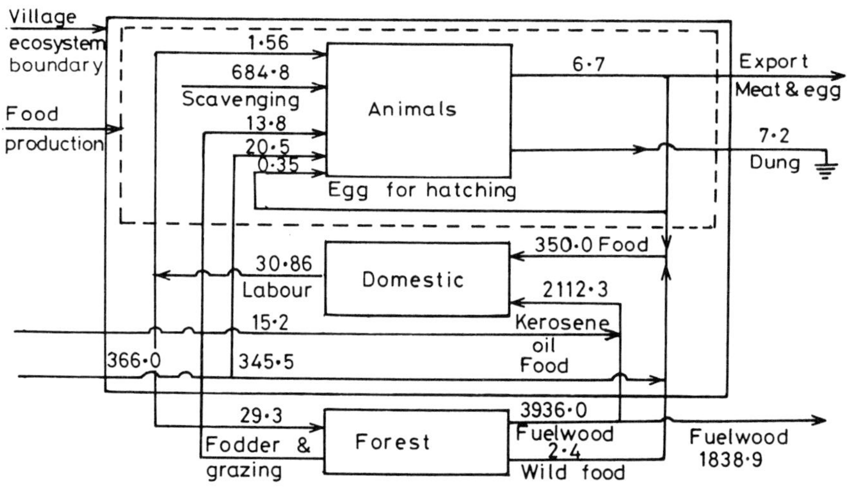

Figure 3.5 Energy flow diagram for the Kachari village ecosystem at Balijan in Arunachal Pradesh. Unit = MJ × 10^3 (from Maikhuri and Ramakrishnan, 1990b)

Comparative analysis of village ecosystems of two coexisting communities at Nayabunglow in Meghalaya

The study area at Nayabunglow (960 m altitude) in Meghalaya has the local Khasis coexisting with the immigrant Nepalis. The Khasis alone do jhum under 5-, 7- and 10-year cycles (Table 3.8). They also do some valley cultivation of rice; this land use is a highly specialized activity of the Nepalis who traditionally are sedentary farmers. While the Khasis grow pineapple on slopes of 30–40° angle, the Nepalis usually have a kitchen garden and also cultivate fodder grass (*Pennisetum purpurea*). Pigs and poultry are traditionally maintained by the Khasis; the latter alone is an activity of the Nepalis. The Nepalis have cattle and goats introduced into the region.

Agriculture sub-system

The Khasis do jhum at Nayabunglow under 5-, 7- and 10-year cycles. The cropping and yield patterns generally followed what is discussed in Chapter 1 for different jhum cycles, with a shift in the crop mixture towards tuber/root crops under shorter cycles. The economic efficiency (cf Table 1.9) and energy efficiency declined under shorter cycles (Patnaik and Ramakrishnan, 1989). The Khasis also grow pineapple (*Ananas comosus*) on steep slopes with an exceptionally high energy efficiency (56.4 as

Table 3.8 Ecosystem structure of typical Khasi and Nepali villages at Nayabunglow in Meghalaya (from Patnaik and Ramakrishnan, 1989)

	Community/Village	
	Khasis (Umtrou)	*Nepalis (Guntei)*
Number of households	40	42
Total human population	198	172
Adult male	64	59
Adult female	58	41
Children 5–7 years old	39	44
Children 7–9 years old	18	6
Children 9–12 years old	19	22
Total area under cultivation(ha)	14.3	27
Area under jhum	3	–
Area under valley	9	11.9
Area under pineapple plantation	2.3	–
Area under kitchen garden	–	13
Area under fodder grass cultivation	–	2
Total animal population	86	251
Pig	24	–
Milking cow	–	96
Bullock	–	55
Goat	–	53
Chicken	62	47

output/input ratio) and economic efficiency (22.3). This is because of a very low labour input with no fertilizer use (Patnaik and Ramakrishnan, 1989).

The Nepalis are basically cattle farmers who, being immigrant into the protected tribal area, are not permitted to own land in the rural areas for agriculture. Therefore they hire land from the Khasi landlords of the town and do valley cultivation of rice and/or have kitchen gardens on land owned by the local Khasis. Both these land uses provide yearly yields unlike jhum; the Nepalis traditionally being sedentary agriculturalists obtain a high efficiency from their valley lands (cf Chapter 2). These cattle farmers also cultivate the fodder grass (*Pennisetum purpurea*) in small plots of 0.1–0.3 ha, since they cannot obtain their fodder exclusively from the rapidly dwindling forests around.

Animal husbandry sub-system

With jhum as the major agricultural activity for the Khasis, swine husbandry and poultry-keeping are the only two animal husbandry activities of this

Table 3.9 Energy output/input ratio (MJ $year^{-1}$ $animal^{-1}$) for different animal husbandry systems of the Khasis and the Nepalis at Nayabunglow in Meghalaya (from Patnaik and Ramakrishnan, 1989)

	Khasis			*Nepalis*		
	Swine	*Poultry*	*Milking cow*	*Bullock*	*Goat*	*Poultry*
Input total	6,718	4,203	38,093	38,093	5,358	4,221
Labour	47	5	1,092	1,092	35	23
Feed	4,440	297	11,907	11,907	1,545	892
Firewood	1,597	–	7,985	7,985	1,198	–
Fodder	–	–	6,782	6,782	2,159	–
Grazing/browsing/ scavenging	634	3,901	10,327	10,327	421	3,306
Output total	750	196	6,722	902	237	176
Dung	106	161	583	583	146	147
Milk	–	–	6,139	–	–	–
Meat	644	6	–	–	91	5
Egg	–	29	–	–	–	24
Draught power	–	–	–	319	–	–
Output/input ratio	0.11	0.047	0.18	0.024	0.044	0.04

tribe, as for the Khasis elsewhere in the region (Mishra and Ramakrishnan, 1982; Maikhuri and Ramakrishnan, 1990a) and for other tribes in the region (Ramakrishnan, 1985c; Maikhuri and Ramakrishnan, 1990a,b). Such a close linkage between shifting agriculture and swine husbandry and even poultry-keeping is well recognized in the region and all the world over. The chief advantage of this linkage is that it is detritus-based and the cost for maintenance is minimal (Tables 3.9, 3.10).

Unlike the Khasis, the Nepalis being sedentary agriculture farmers and with greater dependence on cattle for milk, emphasize upon this, along with bullocks for draught power. They also keep some goats and poultry. Their own requirement for cow dung manure is high; a sizeable fraction of it is also sold in the market. The input into this system is substantial, unlike swine husbandry or poultry. The feed for cattle and goats involved fodder, stall-fed or foraged from the wild, as well as firewood for cooking the feed. Confinement feeding provides some advantages in that the energy (feeding) cost of foraging and disease recycling are eliminated; 'least cost' rationing allows optimum use of other foodstuffs (by-products and wastes (Dean *et al.*, 1972)). The output from cattle husbandry of the Nepalis is high so that the economic and energy efficiencies are comparable with swine husbandry or even better. From an economic view point, bullock maintenance was more expensive than cattle for milking (Table 3.10).

Table 3.10 Monetary output/input ratio (Rupees year^{-1} animal^{-1}) of different animal husbandry systems of the Khasis and the Nepalis at Nayabunglow in Meghalaya (from Patnaik and Ramakrishnan, 1989)

	Khasis			*Nepalis*		
	Swine	*Poultry*	*Milking*	*Bullock*	*Goat*	*Poultry*
Input total	660	87	2,576	2,576	354	102
Labour	228	32	915	915	152	20
Cattle	371	55	1,577	1,577	134	82
Firewood	61	–	84	84	68	–
Output total	1,100	113	8,960	916	415	87
Milk	–	–	8,760	–	–	–
Draught power	–	–	–	700	–	–
Dung	50	11	216	216	23	9
Meat	1,050	27	–	–	392	22
Egg	–	75	–	–	–	56
Output/input ratio	1.67	1.3	3.48	0.36	1.17	0.85

It is usually societies with large population densities and/or a small proportion of grazing land that concentrate on confinement feeding. Thus, swine and poultry require a much greater portion of grain in their diets than the ruminants and thus are more competitive with humans (although the tribals use largely grains of poorer quality). They have reasonably high feeding efficiency of over 30% of plant protein converted into animal protein with 80% of human edible returns (Loomis, 1984). The Nepalis depend upon forests to meet their fodder needs – fodder, fresh or dry which is grazed, browsed or gathered. However, with declining forest resources and reduced access to the forest resources for the immigrant community, they tend to depend largely upon *Pennisetum purpurea* raised in plots hired from the tribal Khasis.

Domestic sub-system

In many tribal societies in the region, the land is community-owned, the Khasis of Meghalaya also following such a land tenure system. The individual family holdings are regulated through the village headman, on payment of a nominal fee to him. The plot belongs to the farmer as long as he cultivates it. Therefore, the ownership traditionally is an incipient one, with all the produce from the jhum land obtained by the farmer being his own. However, in Nayabunglow, which is close to the city of Shillong, the land tenure pattern has undergone some changes because of the purchase of valley lands and a few hill slopes by the Khasis living in the town, who subsequently have leased out the same to the local Khasis and

Table 3.11 Comparative analysis of production and consumption patterns (Rupees $year^{-1}$) from agriculture of a Khasi and Nepali family at Nayabunglow in Meghalaya (from Patnaik and Ramakrishnan, 1989)

	Khasi	*Nepali*
Area (ha)	0.67	1.0
Input total	2,994	4,918
Human labour		
Inside	1,446	2,665
Outside	780	788
Animal labour		
Inside	–	310
Outside	320	–
Inorganic fertilizer	170	216
Organic manure	–	719
Seed	278	220
Output total	4,237	8,529
Consumption		
Human	1,509	2,689
Animal	47	1,948
Export	254	2,152
Revenue	1,850	1,805

Nepalis for cultivation. As a consequence, these land owners obtain revenue from the land which is 50% of the total produce from valley cultivation. Therefore, although valley cultivation is an economically viable system, the actual benefit to the local farmer is considerably reduced.

The agricultural produce is largely consumed within the family itself (Table 3.11). The input into the agriculture system per family and on a unit area basis was higher for the Nepalis than for the Khasis. Further, the Nepalis are able to export more of their agricultural produce after higher consumption within the family and after paying the land revenue.

The two communities export most of that produced from the animal husbandry (Table 3.12). The Nepalis consume 17% of the milk production. The Khasis export all the dung collected, whereas the Nepalis use 70% within the family, exporting the remainder. Some feed is purchased by both from the market, and is a major monetary input into animal husbandry (Table 3.13). Nepalis had much higher monetary output (about 15 times more) than the Khasis. The Nepalis also pay revenue to the Khasis for using the forest for grazing and pay the loan from the bank. The Nepalis collect fodder from the forest and both communities obtain firewood from

Table 3.12 Comparative analysis of production and consumption patterns (MJ year^{-1}) from animal husbandry of a Khasi and a Nepali family at Nayabunglow in Meghalaya (from Patnaik and Ramakrishnan, 1989)

	Khasi	*Nepali*
Number of animals		
Cattle	–	8
Goat	–	6
Pig	1	–
Poultry	4	1
Production		
Milk	–	12,278
Meat	668	551
Egg	29	24
Dung	750	5,687
Draught power	–	638
Consumption		
Egg	29	24
Milk	–	2,046
Export to market		
Milk	–	10,232
Egg	–	–
Meat	644	546
Dung	750	4,000
Import from market		
Feed	3,547	5,700
Import from forest		
Fodder	–	64,463
Firewood	3,194	71,068

the forest for cooking the feed for the animals. The Nepalis consume about 22 times more firewood per family than the Khasis, much of it for preparing the animal feed.

Although the food consumed by both the communities is of plant origin, the Khasis consume three times more meat than the Nepalis (Table 3.14). The Khasis also consume more fish. The energy and protein needs of both are met completely and, in fact, they obtain much more than their standard requirements (Ranjhan, 1977).

With a comparatively smaller family unit for the Nepalis, the input of energy into their domestic system is about the same as for the Khasis (Fig. 3.6). Energy output in the form of labour was much higher for the Nepalis than for the Khasis, the former expending the labour within their own domestic sector; the Khasis alone export labour. The energy efficiency of the domestic sector is very low. One of the important reasons for this is

Table 3.13 Comparative analysis of monetary production and consumption pattern from animal husbandry (Rupees year^{-1}) of a Khasi and a Nepali family at Nayabunglow in Meghalaya (from Patnaik and Ramakrishnan, 1989)

	Khasis	*Nepalis*
Input total	972	22,834
Human labour	320	8,252
Firewood	61	1,080
Cattle feed	591	13,502
Output total	1,532	23,225
Human consumption	300	2,998
Draught power	–	1,400
Export	1,232	16,958
Revenue	–	69
Loan	–	1,800

Table 3.14 Consumption pattern (capita^{-1}) by the Khasis and the Nepalis at Nayabunglow in Meghalaya. Values are given for the standard requirements (from Patnaik and Ramakrishnan, 1989)

	Khasis		*Nepalis*	
Category	*Energy* (MJ)	*Protein* (kg)	*Energy* (MJ)	*Protein* (kg)
Plant originated food				
Grains	4,580	19	4,758	22
Pulses	165	2	513	6
Green vegetables and tubers	105	0.1	145	0.13
Fruit	53	0.5	9	0.005
Mustard oil	1.2	–	242	–
Animal originated food				
Milk	–	–	369	3.0
Egg	4	0.2	3.4	0.14
Fish	53	0.5	13	0.2
Meat	7	0.7	2.3	0.13
Total	4,968	23	6,055	31.6
Standard requirement	4,211	20.08	4,211	20.0

high consumption of firewood energy. According to a calculation, the energy obtained through firewood is 7.6 times more in developing countries compared to the developed societies (Stout *et al.*, 1979). The lack of efficiency is because of energy-inefficient cooking stoves (Donovan, 1981; Fox, 1984). For the future, one ought to consider possibilities of energy-efficient stoves

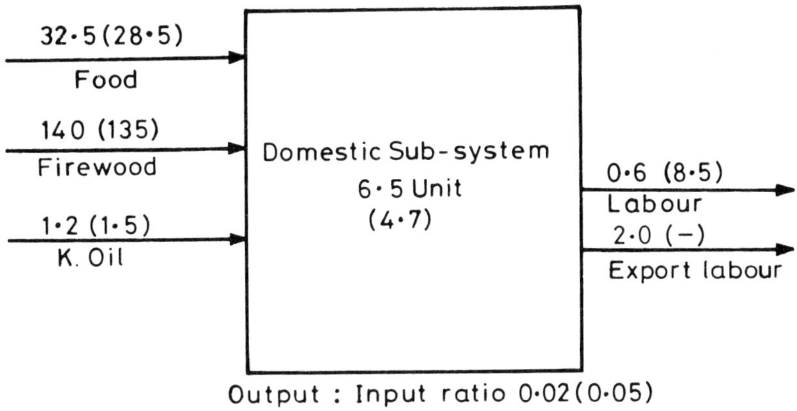

Figure 3.6 Comparative energy input and output analysis (MJ × 10^3 family^{-1} year^{-1}) of the domestic sub-system of the Khasis and the Nepalis (within parentheses) at Nayabunglow in Meghalaya (from Patnaik and Ramakrishnan, 1989)

to replace the inefficient open types now used. This would help to check to some extent the large-scale conversion of rain forests now occurring in the region (Ramakrishnan, 1985c,d).

A low land to man ratio, rapidly declining economic return from agriculture high revenue to the absentee landlord, rapidly declining forest resources, all adversely affect the economy of both the Khasis and the Nepalis. Khasis compensate for this to some extent through export of labour, and by migration to the city in search of work.

Conclusions

Agriculture is the major activity of all communities, the local tribals and the immigrants like the Nepalis. The tribals largely depend upon jhum, with limited valley cultivation and home gardens. The cropping patterns vary depending upon the tribe, which would determine the efficiencies of the system. Thus, at Lailad in Meghalaya, the Garos are more efficient jhum farmers than the Khasis or the Mikirs. Similarly, the Nishis at Balijan in Arunachal Pradesh, who are efficient jhum farmers, alone have access to jhum lands because of their dominance over the immigrant tribals like the Karbis, Chackmas and the Kacharis. Amongst all the communities considered here, the Nepalis have one of the most efficient wet rice cultivation in the valleys. Differential emphasis on land use by different communities ensures mutual avoidance of using the common natural resource base. The wide variation that exists in the land use provides the basis for designing redevelopment strategies for agriculture in the region, as also discussed elsewhere (cf Chapter 1 and 2).

Table 3.15 Per capita annual food consumption by the tribal communities of Arunachal Pradesh in north-east India with the percentage contribution of lesser-known plant and animal species (in parentheses) (from Gangwar and Ramakrishan, 1989b)

	Food source	
	Plant	*Animal*
Nishi		
Energy (MJ)	4,428(18)	133(23)
Protein (kg)	25.9(27)	1.4(21)
Hill Miri		
Energy (MJ)	4,603(17)	129(29)
Protein (kg)	27.4(25)	1.4(29)
Sulung		
Energy (MJ)	5,329(41)	96(78)
Protein (kg)	23.5(42)	1.1(73)

With increase in population pressure, the dependence of the tribals on the forest for jhum becomes less tenable. Therefore, they often abandon this land use and shift towards a cash crop economy as has happened in a Khasi village at Tynriang in the Cherrapunji area in Meghalaya (Gangwar and Ramakrishnan, 1989a).

In the context of increasing population pressure, many lesser known plants of food value cultivated in the agricultural plots by different tribes could form an important food source. With increased pressure on land, the Khasis of Meghalaya often go for fallow and sedentary agriculture systems (Gangwar and Ramakrishnan, 1987a, 1989b). As under jhum, such systems have lesser known plants in the crop mixture such as *Flemingia vestita*, a multipurpose species (that contributes also to nitrogen economy of the soil; cf Chapter 5) and *Perilla ocimoides* (Gangwar and Ramakrishnan, 1987a, 1989b). The Khasis of Meghalaya use these species as an important component of their diet. The Nishis, the Hill Miris and the Sulungs of Arunachal Pradesh cultivate a variety of lesser known plants of food value under jhum of 5- to 30-year cycles and they obtain energy through these crops to the extent of 10–20% of the total intake (Table 3.15). The consumption pattern during the lean season is more, when traditional crops are in short supply (Gangwar and Ramakrishnan, 1990). With improvement in these crop species, some of them could play an important role in the future.

Cattle and goats brought in by the immigrant communities from the plains make high demand on the forest for grazing and browsing. On the other hand, traditional swine husbandry and poultry are more easily sustained since they are detritus-based or use waste products from the

Table 3.16 Bamboo utilization pattern in two categories of villages in east Khasi Hills district of Meghalaya (from Rao and Ramakrishnan, 1987b)

	Category of village	
	Heavy consumption	*Low consumption*
Fencing (kg)	170	120
Repairing (kg)	490	460
Water pipes	90	63
Firewood (kg × 10^3)	48	19
Food (kg)	135	80
Fodder (kg × 10^3)	85	65
Total (kg × 10^3)	49	19.8
Export (kg × 10^3)	34	0
Grant total (kg × 10^3)	83.9	19.8
Industrial (kg × 10^3)	800	–
Number of villages	567	1,533
Average village size (number of families)	60	55

agriculture sub-system. The immigrant Nepalis compensate for lack of forest grazing land by cultivating fodder species as a family effort.

Improved recycling of dung, otherwise largely wasted, would improve productivity of the agriculture sub-system. Dung could be used directly or the slurry recycled after biogas generation. This would also ease the pressure on the declining forest and help in conserving this valuable resource. It is in this context that conserving an important forest resource such as bamboo, which is also used as firewood, becomes important. (Rao and Ramakrishnan, 1987b). In a study of 30 randomly selected villages in the Khasi hills of Meghalaya, most were classified as low consumption units because of severe local shortage of this resource (Table 3.16). Bamboo (*Dendrocalamus hamiltonii*) being an early successional species during forest succession, with its biomass peaking up in 20- to 25-year-old forest fallows, shortened jhum cycles of 4–5 years and site desertification (Ramakrishnan, 1985b, c) have adversely affected its availability to local communities (cf Chapter 6). Indeed, the demand for this species for industrial uses such as pulping is also putting additional strains on the availability of this commodity.

Redevelopment of agriculture, animal husbandry and forest resource base, therefore, should be based on local perception of needs. Such an approach would be environmentally viable and ensure people's participation.

HUNTING SOCIETY OF THE SULUNGS AND THE HIGHLY EVOLVED APATANIS : A COMPARATIVE ANALYSIS

Being a hunter–gatherer society of the remote wilderness areas of Arunachal Pradesh, the Sulungs are largely dependent upon the forest for meeting

Plate 3.2 The Sulungs, a hunter–gatherer society of Arunachal Pradesh at Tabumah village (1,080 m altitude)

their varied needs (Gangwar, 1987; Gangwar and Ramakrishnan, 1987b). On the other hand, the Apatanis of Arunachal Pradesh living in extensive valley lands are least dependent upon the forest (Kumar and Ramakrishnan, 1990). Thus, while the former are essentially jhum farmers, the latter are sedentary agriculturalists engaged in wet rice cultivation. Although both the communities have animal husbandry, the Apatanis are better organized in this respect than the Sulungs, the latter largely depending upon hunting and gathering (Plate 3.2). As a consequence, the Sulungs and the Apatanis represent two extremes in the spectrum of village organization and function. The following discussion, therefore, considers these two contrasting societies, with respect to village organization and function.

Table 3.17 Apatani and Sulung tribal village ecosystem structure in Arunachal Pradesh (from Gangwar and Ramakrishnan, 1987b; Kumar and Ramakrishnan, 1990)

	Tribal/village	
	Apatani/Hari	*Sulung/Tabumah*
Number of households	418	40
Total population	2,021	277
Adult male	626	86
Adult female	646	74
Adolescent 12–21 years	304	60
Children below 12 years	445	57
Total area under cultivation (ha)		
Jhum: Mixed cropping	–	22
Jhum: Monocropping	–	11
Terrace	–	12
Rice cultivation (valley)	356	2
Home garden	24	–
Millet cultivation	5	–
Bamboo garden	47	–
Total animal population		
Mithun	656	4
Cattle	218	–
Goats	–	5
Swine	452	20
Poultry	9,328	200

The Sulungs of Arunachal Pradesh are dispersed over an area of 300 km^2 in the lower Subansiri District of Arunachal Pradesh. These hunter–gatherers emphasize upon jhum, with limited valley cultivation (Table 3.17) and some terrace agriculture introduced into the region. Apart from poultry, swine husbandry and a few goats introduced into the area, Mithuns are owned by the richer families only in the village of Tabumah.

The Apatanis, as at the village of Hari, extensively practise wet cultivation of rice in valley lands (Table 3.17). Millet (*Eleusine coracana*) cultivation is carried out along the raised partition bunds of rice plots. The home gardens maintained by some families have grain and seed crops such as *Eleusine coracana* and *Zea mays*, a few leaf and fruit vegetables, and tuber and rhizome crop species (Kumar and Ramakrishnan, 1990). Bamboo gardens are common. Apart from swine husbandry and poultry, the Apatanis maintain Mithun and cattle.

Table 3.18 Energy and monetary efficiencies (output/input) of the animal husbandry of the Apatanis at Hari in Arunachal Pradesh (from Kumar and Ramakrishnan, 1990)

	Money			*Energy*		
Animal	*Input*	*Output*	*Output/ input ratio*	*Input*	*Output*	*Output/ input ratio*
Mithun	300	1,290	4.3	28,789	1,897	0.07
Cattle	200	750	3.8	21,226	1,017	0.05
Swine	190	1,070	5.6	8,112	1,083	0.13
Poultry	6	35	5.8	137	16	0.12

Agriculture sub-system

With jhum under mono-and mixed cropping and some valley cultivation of rice, the Sulungs at Tabumah have a reasonably wide agricultural base (Gangwar, 1987). Because of good forest cover and a sparsely distributed population, the area under a longer 30-year jhum cycle is common. The mixed cropping system under jhum emphasize upon rice (*Oryza sativa*) and millet (*Eleusine coracana*). Tubers and rhizome species are next in importance. A number of species such as *Amaranthus viridis*, *Chenopodium ambrosioides*, *Coix lacrymajobi*, *Dioscorea* spp., *Panicum miliaceum* and *Perilla ocimoides* are lesser known crops of food value. The economic efficiency (output/input ratio) is in the range of 3–3.6 and the energy efficiency is about 35. Monocropping of rice, *Eleusine coracana* or *Fagopyrum tataricum* is all done under a 10-year jhum cycle, the efficiency of which is lower than that under mixed cropping. The jhum procedures, otherwise are similar to that discussed in Chapter 1. Terrace cultivation of rice recently introduced, and valley cultivation of rice, both had an energy efficiency of 24, and an economic efficiency of about 1.5; this is comparable to monocropping of rice under jhum.

The Apatanis have a highly developed wet rice cultivation method involving late maturing rice varieties closer to the village and an early variety further away (Kumar and Ramakrishnan, 1990). In nutrient-rich areas closer to the village, where waste recycling is more efficient, pisciculture is integrated with rice cultivation. Dryland cultivation of millet and mixed cropping in home gardens contribute to meet the diverse needs of the community. The details are discussed in Chapter 2.

Animal husbandry sub-system

The Mithun, swine and poultry systems essentially are similar for the two tribes and therefore the input/output analysis of the Apatanis alone is discussed here (Table 3.18). Mithuns are cheap to maintain as long as forest resources are abundant for grazing. Since these animals have strong

Plate 3.3 The Sulungs of Arunachal Pradesh use starch obtained from the pith of *Metroxylon sagu* found extensively in the swampy lowlands, as a staple diet

territoriality they need no special care either. With yearly slaughtering of pigs by both the tribes, the efficiencies are better than where slaughtering is done once in 10 years as for the Tsembaga tribe of the New Guinea highlands (Rappaport, 1971).

Domestic sub-system

The Sulungs substantially augment their plant-based food resources through gathering, thus obtaining 45% of the total food energy consumed; hunting contributes 3% of the total. Of the total energy expended by the Sulungs for the food production system, 74% is expended in food gathering; females contribute 45% of the total labour which involves substantial input for agriculture too. Thus the Sulungs are 'partly hunter–gatherers' in contrast to societies of 'gatherers' studied from Africa and Latin America (Lee and De Vore, 1968).

Amongst the plant products gathered by the Sulungs, sago (a starchy food obtained from the pith of *Metroxylon sagu*) is the most important (Plate 3.3). Apart from the animals hunted, a variety of fish, frogs and insects are collected from the wild. Hunting is energy intensive whereas food gathering has a high energy efficiency (Gangwar, 1987; Gangwar and Ramakrishnan, 1987b).

Table 3.19 Seasonal food consumption (MJ × 10^3) by the Sulungs in Tabumah in Arunachal Pradesh (from Gangwar, 1987)

Food item	*September–November*	*December–March*	*April–August*
Cereals and grains	48.7	212.0	178.3
Pulses	8.8	23.3	7.1
Sago	182.9	48.6	173.8
Tachee	–	–	46.9
Bamboo shoots	19.5	15.9	–
Leafy vegetables	7.3	15.3	7.6
Fruits	6.8	15.3	19.9
Roots and tubers	–	25.0	25.3
Fungi	2.8	–	–
Meat	4.6	3.2	12.9
Others	5.1	10.3	10.5
Total	286.5	368.9	482.3

The Sulungs consume two major food items: cereals from agriculture and sago through gathering (Table 3.19). Hunting contributes 80% of the total per capita annual meat consumption in the village. About half of the total food consumed is obtained through gathering. A seasonally differential food consumption pattern was noted. During December–March, cereals and grains from agriculture formed important components of the diet. Sago was more important during the rest of the year along with food collection from the wild and meat obtained through hunting. 'Tachee' (a pulp made after fermenting the pith of *Cyathea gigantea*, a tree fern) is available only during April–August, whereas bamboo shoots and fungi are collected during the rainy season. Thus, the food gathered or hunted is important for the lean season of September–November when traditional food items from agriculture are in short supply.

The village ecosystem of the Sulungs represent one of the cheapest production systems with human labour as the only input, and is based on jhum and hunting/gathering as the chief activities. Their food intake is far above the standard nutritional needs of the Indian societies (Gopalan *et al.*, 1985). Having a highly diversified forest resource base, this community is highly insulated from shortages.

With a strong agriculture and animal husbandry base, the food intake of the Apatanis is met from these two sectors (Table 3.20). The food energy intake by the Apatanis compared favourably with the standard requirements of an average Indian adult (Gopalan *et al.*, 1985). The pathway to meet the food needs of the Apatanis thus differed from that of the Sulungs, the former depending upon their own production systems and the latter largely depending upon hunting and gathering.

Table 3.20 Annual food and protein consumption in an Apatani village at Hari, in Arunachal Pradesh. Values are also given for per capita consumption (from Kumar and Ramakrishnan, 1990)

	Energy (MJ)		*Protein equivalent* (kg)	
Food item	*Consumption*	*Per capita*	*Consumption*	*Per capita*
Consumption				
Plant origin	6,888.1	3.9	36.6	0.02
Animal origin	513.9	0.3	21.7	0.01
Standard requirement	7,529.3	4.2	35.9	0.02

The Apatanis obtain their fuelwood partly from the bamboo garden (about 26% of the total) maintained by them, and the rest from the forest. This contrasts with the Sulungs who exclusively depend upon the forest. Like the Sulungs, fuelwood consumption by the Apatanis is very high being 4 to 5 times the standard requirement (Mitchell, 1979). The Apatanis export rice, bamboo, some fish and eggs. Although fuelwood comes from outside the village boundary, it is a free commodity that is collected from the forest. If this is kept out of the calculation, the efficiency of the village ecosystem would be higher (Fig 3.7).

Conclusions

Even though rice cultivation has, for the Apatanis, an exceptionally high energy efficiency of up to 78 units per unit energy input, this agroecosystem could be intensified through appropriate crop rotation during the winter season when the land is left fallow as at present. Maintaining Mithuns is dependent upon a large forest resource base for grazing, and this husbandry is an important pathway to gain status within the society. Consequently, slaughtering is done on special festive occasions. This low frequency of slaughter reduces the overall energy efficiency of this animal husbandry. This could be adequately taken care of in any redevelopment programme. Similarly, fuelwood – use efficiency needs to be improved through appropriate energy efficient stove technology input. In spite of these possibilities, the Apatani village ecosystem is a good example of a highly developed traditional society that practices ecologically sound sedentary agriculture in north-eastern India.

The more traditional Sulungs, on the other hand, largely depend upon hunting/gathering for meeting their food requirements, from a large forest resource base. In an ethnobiological study of the species used by the four tribal communities of Arunachal Pradesh, that included the Sulungs, a total of 170 plant species and 12 animal species were identified (Gangwar and Ramakrishnan, 1990). Of the plant species, 30% are leafy vegetables and

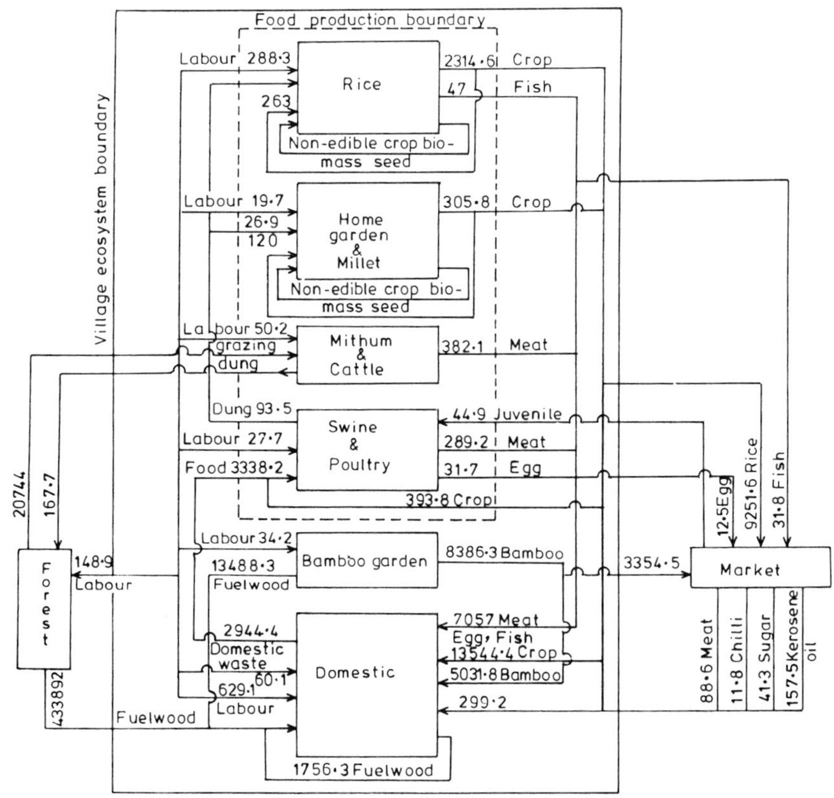

Figure 3.7 Energy flow (MJ × 10^3 family^{-1} year^{-1}) through the Apatani village, Hari, in Arunachal Pradesh (from Kumar and Ramakrishnan, 1990)

28% are edible fruits. Others are used for salt extraction, food, liquor, medicine, household goods, traditional dresses, dyes, tattoos, fish and animal poisons. Such a heavy dependence on wild collections would become more and more difficult in view of the rapidly declining forest resources. It is in this context, cultivation and improvement of some of the more important lesser known plants of food value becomes critical (Gangwar and Ramakrishnan, 1989b).

It is concluded that the Sulungs with a greater dependence on forest-based food resources represent a society that is self-sufficient, this being possible because of the sparsely distributed population structure of this tribe in the remote areas of Arunachal Pradesh. The same degree of self-sufficiency by the Apatanis, a densely populated community of extensive valley lands is due to a highly evolved and intensely operated sedentary valley land agriculture and animal husbandry.

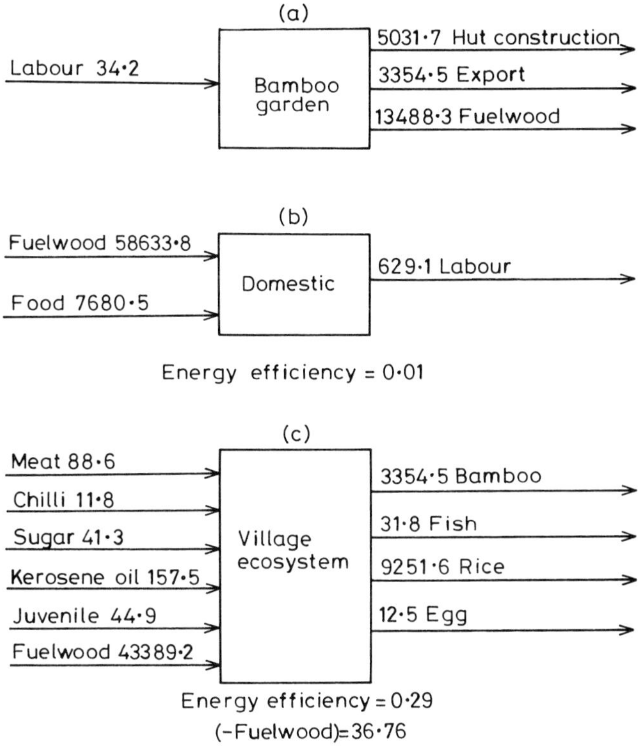

Figure 3.8 Annual energy flow through various sub-systems of the village ecosystem and the market in a desertified Khasi village at Cherrapunji in Meghalaya. Unit MJ × 10^3 (from Gangwar and Ramakrishnan, 1989a)

GENERAL CONSIDERATIONS

We have considered here tribal communities who, through a variety of traditional agricultural and animal husbandry activities, have tried to attain self-sufficiency within their village ecosystem boundary. However, all these societies still have a large forest resource base on which their activities are centred. Where the land is desertified as at Cherrapunji in Meghalaya, the Khasis emphasize upon a plantation based economy (cf Chapter 2). Although the village itself is economically stable under plantations, it is dependent upon imports from outside the village boundary (Fig. 3.8). From a developmental angle, traditional jhum could be replaced by a mixed plantation system, the tree species being arecanut, orange, bay leaf and jackfruit, along with vines of betel leaf and black pepper (Gangwar and Ramakrishnan, 1989a). Where natural resource base is strong, one does find wide variations in the ecological and economic efficiencies in the operation of agriculutre, animal husbandry and domestic sectors. Village functions also differ based upon social and cultural differences between them. This rich variety is an important basis for tribal redevelopment. With increase in population pressure under rapidly depleting forest resource bases, redevelopment should not only be based on the existing technological base, but should also work towards the introduction of new technology. Thus, for example, agriculture should consider jhum or related agroforestry concepts as the basis for redevelopment. Similarly, animal husbandry should be based on traditional practices in this regard, such as poultry and swine husbandry to start with. Intertribal transfers and subsequent adaptation of technologies for the different communities in the region would be an important basis for redevelopment. Any introduced technology from outside the region, such as goat and cattle husbandry should be after careful evaluation of ecological, social and economic aspects. Introduction of new technologies in rapidly changing demographic social, ecological and economic situations is unavoidable. The value system on which they are based and the pace with which such introductions are made would determine their general acceptance by the society.

CHAPTER 4

WEED POTENTIAL IN AGROECOSYSTEMS AND THEIR MANAGEMENT

INTRODUCTION

Weeds in agroecosystems are generally considered to be plants that are out of place, that adversely affect crop growth, and that for a variety of reasons are difficult to control. King (1966) has discussed the problem of definition of weeds and has given an extensive account of their characteristics. Since weeds are often considered to be competitive with crops for limited resources that are available in the system, much of the research efforts have been directed towards looking at the decline in crop yield due to weed–crop competition (Blackman and Templeton, 1938; Zimdahl, 1980; Berkowitz, 1988) and some studies that consider the effect of the crop on the weed growth (Ramakrishnan and Kumar, 1971). Extensive studies on crop yield decline, in any case, seem to suggest a world average loss of about 10% due to weed growth (Zimdahl, 1980).

Although there has been some isolated studies on the useful role of weeds in agroecosystems (Kapoor and Ramakrishnan, 1975), it is only recently that an ecological approach to weed management in agroecosystems is being explored in a serious way (Altierie and Liebman, 1988). Many of the ideas in this direction, looking at weed control have come from a deeper understanding of traditional agroecosystems of the tropics. The Tabasco farmer of the lowland tropics of Mexico manage their weeds in a manner that, those which have a useful purpose are conserved or even encouraged (Chacon and Gliessman, 1982). In the north-eastern Indian jhum system, the traditional farmer conserves about 20% of the total weeds in their plots. He, on the basis of his empirical knowledge, is able to discern when a 'weed' becomes a 'non-weed' in his system. Therefore, he does only a partial weeding rather than total weeding, since they have important nutrient conservation roles in the system (Ramakrishnan, 1984a, 1988a).

In this chapter, therefore, we shall look at the weed potential of the jhum agroecosystem of the tribals of north-eastern India and evaluate the weed management possibilities. In order to evaluate the weed potential during the cropping phase of jhum, we need first to discuss the weed problem during the fallow phase of jhum. While doing so, we shall also compare the weed potential under jhum with some of the other land use systems in the region.

WEED POTENTIAL UNDER JHUM DURING SUCCESSION

The forest farmer in the humid tropics has managed his traditional jhum system for centuries, with the goal of optimizing yield on a long-term basis (Ruthenberg, 1971; Ramakrishnan, 1984a). In doing so, he has taken into consideration the weed potential of the site (Saxena and Ramakrishnan, 1984b) and effectively managed this component for sustainable production (Ramakrishnan, 1988a).

The jhum system has two important phases : (1) the cropping phase and (2) the fallow phase; the slash-and-burn operation is placed between the two. When a forest is converted to cultivable land not only is its original vegetation destroyed due to slash-and-burn, but the perturbation is continued due to the introduction of crop species, weeding, hoeing and other disturbances to the soil when the crops are harvested. During the slash-and-burn operation, the nutrients conserved in the vegetation and those gradually built up in the soil through litterfall and mineralization are released in one flush through the ash deposited on the soil surface, which then is quickly capitalized upon by the farmer through his mixed cropping system (Ramakrishnan, 1980, 1984a). After 1 or 2 years of cropping, often involving over 35 crop species in a small plot of 2.5–3 ha, the plot is fallowed for natural regeneration of vegetation.

Weed communities during succession

During both the cropping and secondary successional phases, the ecological system destabilized by jhum activity tends to regain stability. During the cropping phase, the farmer contributes to this process through crop and weed management procedures (Saxena and Ramakrishnan, 1984b, Swamy and Ramakrishnan, 1988c). During secondary succession after jhum, the weedy species play a major role in the early successional communities of the first 5–10 years (Toky and Ramakrishnan, 1983a; Mishra and Ramakrishnan, 1983a).

A series of perturbations during the jhum operations result in a progressive reduction in species diversity. So the early successional stages contain fewer species, but the number increases gradually as the secondary community develops as shown by us in north-eastern India (Toky and Ramakrishnan, 1983a; Mishra and Ramakrishnan, 1983a) and by others elsewhere (Ross, 1954; Odum, 1969; Whitmore, 1975).

While land use practice history prior to secondary succession was often obtained only from the farmers and the village headmen, the age of the fallow could also be established by observation of the vegetation. Therefore, it was often difficult to know precisely the earlier history. Although all care was taken to ensure similarity in sites in terms of topography and aspect, it is possible that earlier land use history may have been different in many of the areas considered for comparative analysis. This may have affected the secondary succession, particularly the early herbaceous phase. This may partly explain the four distinct weedy communities recognised at Burnihat in Meghalaya (Table 4.1); jhum cycle, weeding intensity and available seed source are all important. In Type I, *Eupatorium odoratum* or *Imperata cylindrica* or both dominate under a 5-year jhum cycle. *Mikania micrantha* may dominate the fallow, as in Type II, under a 10-year cycle. In type III, *Saccharum spontaneum* may dominate under a 5-year cycle, where cropping was done for a year and the rhizomes of this species are not destroyed. *Borreria hispida* dominated fallows (Type IV) under a 30-year jhum cycle (Toky and Ramakrishnan, 1983a).

At higher elevations of Meghalaya, around Shillong *Imperata cylindrica* and *Erigeron linifolius* are common with lower elevation systems. However, many species such as *Eupatorium adenophorum*, *Eupatorium riparium*, *Hypochaeris radiata*, *Oxalis latifolia*, *Dichranopteris linearis* and *Pteridium aquilinum* are specific to higher elevations.

These early successional annual or perennial weeds often produce a large number of seeds which are light in weight and wind dispersed. The importance of this feature for success in an early successional environment was stressed by Hayashi and Numata (1968) and Raynal and Bazazz (1973). Salisbury (1942) emphasized that species with lightweight and highly mobile seeds invade highly disturbed ecosystems, whereas species with heavier seeds, are often animal disseminated and usually enter the succession at a later stage. On the other hand, species such as *Imperata cylindrica* have extensive underground rhizomes protected from fire and they are difficult to eradicate during cultivation. Many of the early successional sprouting species also resort to seed reproduction (Saxena and Ramakrishnan, 1982a).

Weed demography, population dynamics and extinction during early succession

Exotic species are prominent in disturbed sites throughout the world, on the continents as well as islands and in the tropics as well as the temperate zone. The association between disturbed sites and invaders may be due to greater invasiveness on the part of successional species, or it may simply reflect the association between humans and disturbance, which gives early

Table 4.1 Abundance of predominant species in jhum fallows up to 5 years old at Burnihat in Meghalaya (from Toky and Ramakrishnan, 1983a)

	I			*II*		
Fallow type	*Frequency* (%)	*Density* (m^{-2})	*Basal area* (cm^2m^{-2})	*Frequency* (%)	*Density* (m^{-2})	*Basal area* (cm^2m^{-2})
Mikania micrantha	–	–	–	100	165.0	78
Panicum khasianum	48	5.4	6	25	2.2	3
Solanum torvum	–	–	–	15	0.6	7
Eupatorium odoratum	100	18.7	190	5	0.8	4
Cassia tora	–	–	–	10	1.5	4
Sachharum spontaneum	–	–	–	–	–	–
Imperata cylindrica	100	22.76	76	–	–	–
Grewia elastica	12	2.9	9	–	–	–
Brachiaria distachya	–	–	–	–	–	–
Borreria hispida	–	–	–	–	–	–
Panicum maximum	24	15.4	11	–	–	–
Digitaria adscendens	–	–	–	–	–	–
Setaria palmifolia	–	–	–	–	–	–
Paspalidium unctatum	–	–	–	–	–	–
Sataria glauca	40	2.5	3	–	–	–
Cyperus globulosus	40	6.5	11	–	–	–
Ficus hispida	20	2.2	10	–	–	–

	III			*IV*		
Fallow type	*Frequency* (%)	*Density* (m^{-2})	*Basal area* (cm^2m^{-2})	*Frequency* (%)	*Density* (m^{-2})	*Basal area* (cm^2m^{-2})
Mikania micrantha	–	–	–	–	–	–
Panicum khasianum	27	2.4	2	–	–	–
Solanum torvum	–	–	–	–	–	–
Eupatorium odoratum	14	1.4	8	34	1.5	9
Cassia tora	–	–	–	–	–	–
Sachharum spontaneum	40	6.2	108	–	–	–
Imperata cylindrica	13	1.5	5	–	–	–
Grewia elastica	7	0.3	8	–	–	–
Brachiaria distachya	7	0.6	1	–	–	–
Borreria hispida	–	–	–	100	140	112
Panicum maximum	–		–	17	1.0	1
Digitaria adscendens	–	–	–	24	1.5	2
Setaria palmifolia	–	–	–	15	0.8	2
Paspalidium unctatum	–	–	–	10	0.6	1
Sataria glauca	–	–	–	–	–	–
Cyperus globulosus	–	–	–	–	–	–
Ficus hispida	–	–	–	–	–	–

successional species greater opportunities to invade (Ramakrishnan and Vitousek, 1989).

Amongst the exotic weeds that have come to north-eastern India and are noxious under the jhum system are two species of *Eupatorium* – *E.*

Plate 4.1 *Eupatorium odoratum*, an early successional exotic forming almost pure formation of arrested succession under short jhum cycles of 4–5 years

odoratum and *E. adenophorum*. Both are natives of Latin America. *Eupatorium odoratum*, a lowland species, is a relatively recent introduction into India, during the second world war (Kushwaha, *et al.*, 1981a), where as *Eupatorium adenophorum* found at higher elevations was first reported in India at the beginning of the century (Ramakrishnan and Mishra, 1981). As early colonizers, they have contributed to rapid degradation of land, often forcing the jhum farmer to move away from the area where they dominate (Plate 4.1). This has contributed to rapid shortening of the jhum cycle because of reduced land area available for cultivation (Ramakrishnan, 1985b, 1988c; Ramakrishnan *et al.*, 1981b).

During slash-and-burn and cropping phases of the jhum, a number of changes occur in the physical and biological environment. Apart from improving light conditions as a result of slashing, the temporarily nutrient-enriched soil is fast depleted of its nutrients due to run-off and infiltration losses and also due to rapid utilization by crop and weed species during the first few years (Ramakrishnan, 1988a, 1990). It is under these conditions that early successional weeds get themselves established. However, during succession, they soon have to survive under conditions of a thick vegetational cover, increased surface litter, increased evapotranspiration and altered pH

Table 4.2 Population flux of *Eupatorium odoratum* in different jhum fallows at Burnihat in Meghalaya (from Kushwaha *et al.*, 1981a)

	Age of fallow (yr)				
	1	*3*	*5*	*10*	*20*
(a) No. of plants m^{-2}, May 1978	0.0	5	35	3.3	2
(b) No. of plants m^{-2}, May 1979	5.3	153.6	35	2.6	2
(c) Net change (b − a)	5.3	148.6	0.0	−0.7	0
(d) Rate of increase (b/a)	–	30.7	1.0	0.8	1
(e) No. of plants m^{-2} arrived between May 1978 and May 1979	15.7	1,469.2	2,351.8	0.0	0
(f) Total no. plants m^{-2} lost between May 1978 and May 1979	10.4	1,320.6	2,531.8	0.7	0
(g) Plants present May 1978, alive by May 1979	–	5	35	2.6	2
(h) Percentage survival of plants in (a) (g/a × 100)	–	100	100	78	100
(i) Total no. plants m^{-2} recorded during study	15.7	1,474.2	2,566.2	3.3	2
(j) % annual mortality of all individuals (f/i × 100)	66.2	89.5	98.6	21.2	0
(k) Mortality (%) of arrivals between May 1978 and May 1979	66.2	89.8	100	0.0	0

and nutrient status of the soil. With light becoming more and more a limiting factor, they have to survive under more competitive conditions.

Population dynamics of Eupatorium *spp. and* Imperata cylindrica

In a comparative study of the population dynamics of *Eupatorium odoratum* over a successional gradient of up to 20 years of fallow regrowth, sharp differences were observed at lower elevations (296 m altitude at Burnihat) in Meghalaya (Kushwaha *et al.*, 1981a). Since similar results were observed for *Eupatorium adenophorum* growing at higher elevations (1,066–2,130 m) in Meghalaya (Ramakrishnan and Mishra, 1981) and for *Imperata cylindrica* growing over the entire altitudinal range of these two species (Sarma, 1985), what is discussed for *Eupatorium odoratum* generally holds true for the others too.

Seedling survival of *Eupatorium odoratum* was markedly reduced with increase in the age of the fallow (Table 4.2) Recruitment of new individuals to the population reached its peak in the 3-year-old fallow, and in older fallows hardly any recruitment occurred. The size of the existing population declined sharply after 5 years.

The reproductive performance of individuals was inversely proportional to the fallow age (Table 4.3). All plants in 1- and 3-year-old fallows were fertile but only about 50% of the plants in the 5-year-old fallow produced

Table 4.3 Reproductive output of *Eupatorium odoratum* in relation of the age of the fallow at Burnihat in Meghalaya (from Kushwaha *et al.*, 1981a)

	Age of fallow (yr)				
	1	*3*	*5*	*10*	*20*
Plants m^{-2}[a]	2	5	35	3.3	2
% fertile plants m^{-2}	100	100	51	–	–
% sterile shoots m^{-2}	–	4	53	100	100
Capitula fertile $plant^{-1}$	214	2,927	26	–	–
Seeds fertile $plant^{-1}$	6,409	87,237	780	–	–
Seeds m^{-2}	12,818	436,185	13,975	–	–
Reproductive capacity $plant^{-1}$	943	1,107	155	–	–

[a] excluding seedlings

seeds and none did so in the 10- and 20-year-old fallows. The higher seed production in the 3-year-old fallow compared with the 1-year-old-fallow may have been due partly to the improved fertility status of the soil in the former situation (Ramakrishnan and Toky, 1981) and partly to the increased age of the plants. The marked decrease in seed output in the 5-year-old fallow compared to that in the 3-year-old fallow was indicative of the unfavourable conditions created by increased intra- and interspecific competition for the available resources (Harper and White, 1974). Seed numbers per plant reflected the differences in fertility and since the ability of the seeds to germinate from all the populations was the same (39%), the reproductive capacity (Salisbury, 1942) was greatest for plants in the 3-year-old fallow.

The survivorship curves for the shoots and the seedlings of *Eupatorium odoratum* presented in Fig. 4.1 show differences between the fallows. Shoots in the 1-year-old fallow showed no mortality throughout the study period whereas in 3- and 5-year-old fallows, younger shoots survived less well than older shoots. Highest seedling mortality occurred in 5-year-old fallow where all seedlings were dead by January. The heavy seedling mortality in 3- and 5-year-old fallows may be attributed to intraspecific competition (Ramakrishnan and Jeet, 1972) and to the high density of other species that emerged during the monsoon. The highest mortality during June – July is not unexpected, for seedling mortality in its severest form expresses itself during the first few weeks of the growth of a plant (Ramakrishnan and Kumar, 1971). Similar results (Sarma, 1985) whereby survival of introduced seeds of *Imperata cylindrica* was drastically reduced in a 3-year-old fallow compared to a fresh one, and survival was not possible in 5- and 10-year old fallows (Fig. 4.2) agree with that noted for *Eupatorium* species.

Mortality of seedlings and shoots during the mild winter indicates the importance of soil moisture in the survivorship of *Eupatorium odoratum*, because winter is a relatively dry period. The dryness of the soil is

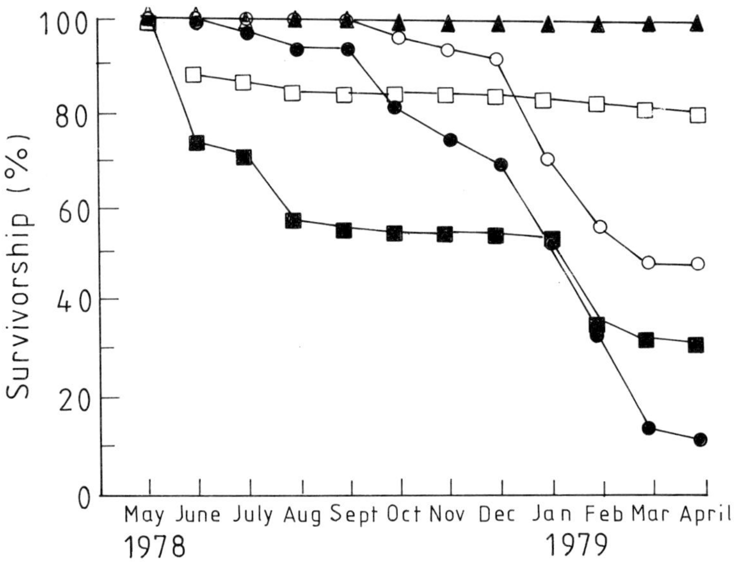

Figure 4.1 Survivorship curves of shoots of *Eupatorium odoratum* in different jhum fallows at Burnihat in Meghalaya. ▲, 1-year-old fallow; □ ■, 3-year-old fallow; ○ ●, 5-year-old fallow. Unfilled symbols, shoots > 1-year-old; filled symbols, shoots < 1-year-old (from Kushwaha *et al.*, 1981a)

accentuated by steep slopes, high infiltration losses and fairly thin soil cover. At higher elevations, where *Eupatorium adenophorum* is the important exotic weed, winter mortality may also be related to lower temperatures with frequent frosts (Ramakrishnan and Mishra, 1981). Winter mortality is higher amongst younger shoots recruited during the preceding monsoon than for the older shoots. This again is in conformity with the general belief that mortality is most severe in the early stages of the growth of a plant, although it is a continuing risk throughout the life cycle.

Population dynamics of Mikania micrantha

In a more recent study (Swamy and Ramakrishnan, 1987b) done at Lailad, 70 km north of Shillong (296 m, altitude) in Meghalaya, the contribution of another exotic weed, namely, *Mikania micrantha*, in early successional communities dominated by it (Toky and Ramakrishnan, 1983a) was considered. *Mikania micrantha* is an exotic weed introduced from Latin America. It grows very rapidly and has well defined ramets. The perennating caudex from the previous year produces an upright stem with six to seven leaves (rosette) in April, and stolons arising out of the axil of leaves. Each ramet has two expanded leaves on a node with a discrete root system. As the stolons extend, they often intermingle with the surrounding vegetation.

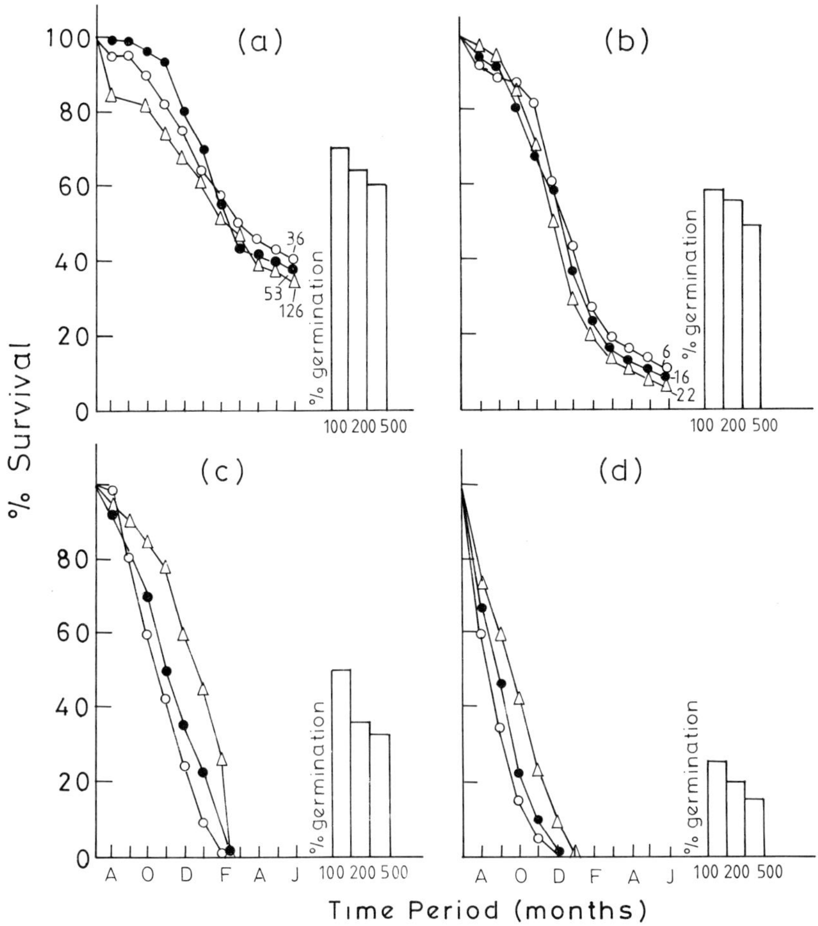

Figure 4.2 Fate of introduced seeds at densities of 100 (○), 200 (●) and 500 (△) at the end of 1-year study period under jhum at Shillong in Meghalaya (a) 0-year-old fallow, (b) 3-year-old fallow, (c) 5-year-old fallow, (d) 10-year-old fallow (from Sarma, 1985)

Flowering occurs from October to late November. The plants die back to ground level during December–January.

While the standing aboveground biomass of *Mikania micrantha* in successional fallows increased significantly with fallow age reaching a maximum of 8.37 kg m^{-2} in a 12-year-old fallow, the contribution by *Mikania micrantha* and by other herbs increased only up to 4 years of fallow age, beyond which it declined drastically (Fig. 4.3). Standing biomass, biomass increment and net primary productivity of *Mikania micrantha* all increased with fallow age up to 4 years, beyond which they declined sharply (Table 4.4). However, biomass accumulation quotient, a ratio of standing

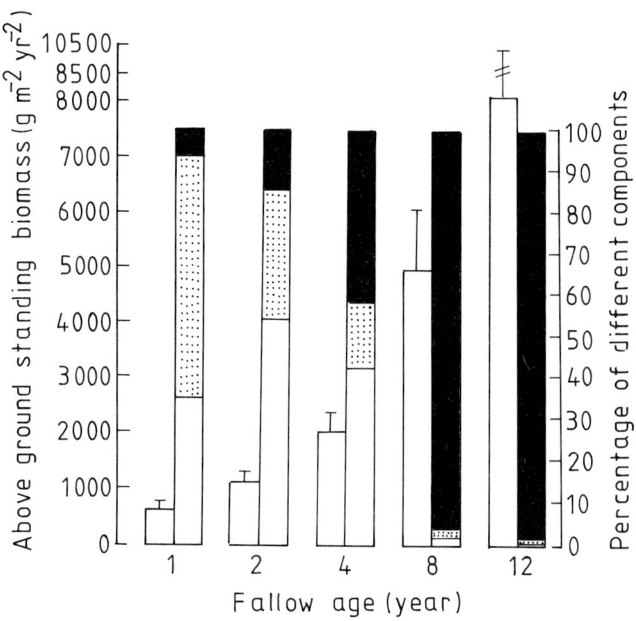

Figure 4.3 Aboveground standing biomass (g m^{-2} years^{-1}) $\pm$ 95% confidence limits (vertical bars) of the vegetation in successional fallows of jhum at Lailad in Meghalaya. Percentage contribution by different vegetation components: *Mikania micrantha* (open column), other herbs (stippled column), and shrubs and trees (dark column) (from Swamy and Ramakrishnan, 1987b)

Table 4.4 Changes in rates of biomass accumulation, litterfall and net production of *Mikania micrantha* in jhum fallows at Lailad in Meghalaya (from Swamy and Ramakrishnan, 1987b)

	Age of fallow (yr)				
	1	*2*	*4*	*8*	*9*
Standing living biomass at the beginning of the growth period (g m^{-2})	9	98	296	56	3.8
Standing living biomass at the end of the growth period (g m^{-2})	210	584	848	90	6.0
Biomass accumulation (g m^{-2})	201	486	552	34	2.2
Losses during the growth period (g m^{-2})	23	48	76	2	–
Net primary production (g m^{-2})	224	534	628	36	2.2
Biomass accumulation quotient	0.9	1.09	1.35	2.5	2.72

biomass divided by net primary productivity, increased with increase in fallow age up to 12 years. This was reflected in the population size too (Swamy and Ramakrishnan, 1987a).

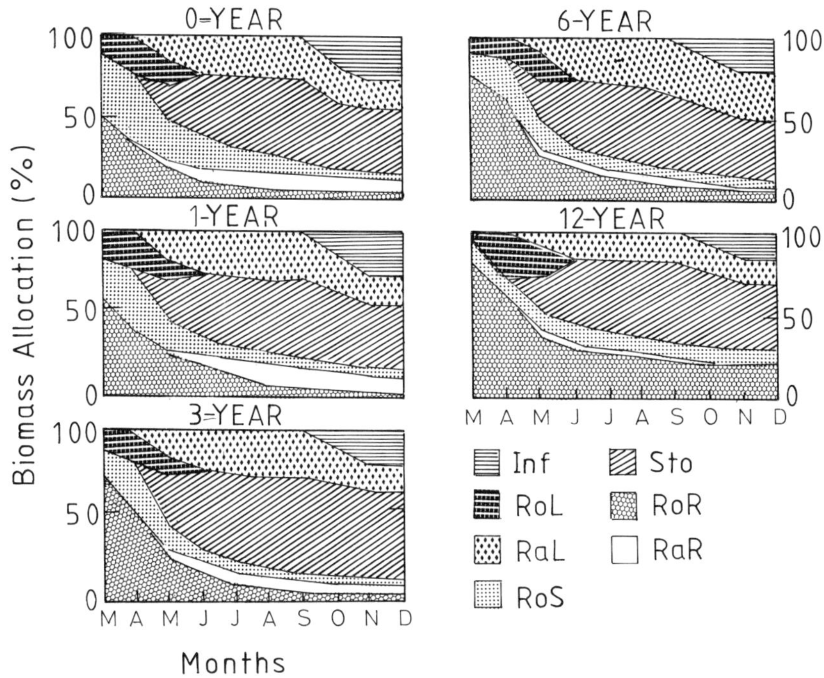

Figure 4.4 Allocation of biomass (expressed as percentage of the total) to different component organs: rosette leaves (RoL), infloresence (Inf), ramet leaves (RaL), stolons (Sto), rosette stem (RoS), rosette root (RoR), and ramet root (RaR), in *Mikania micrantha* in jhum fallows of different ages at Lailad in Meghalaya (from Swamy and Ramakrishnan, 1987a)

Mikania micrantha population size declined in a 6-year-old fallow but the decline was even more pronounced in older fallows. A highly developed vegetative propagation by the ramets and rosettes is advantageous because of the interdependency among ramets, free flow of resources within and the ability of the plant system to regulate position and density of shoots in a clone (Hartnett and Bazzaz, 1985).

Rosettes emerging in March are the chief source for the developing ramets, nutrients being stored in the root system. Since rapid spread and establishment of the species through ramets is important in younger fallows, it is reasonable to find a higher allocation of biomass to the ramets in 0- to 3-year-old fallows when compared with older ones (Fig. 4.4). Greater allocation to the rosette root system in older fallows ensures survival of the species population through late successional communities so that renewed population growth is possible during subsequent slash-and-burn operations.

The larger allocation of biomass to the stolons when compared to other organs of *Mikania micrantha* is to be expected as the species is able to

Table 4.5 Population flux of *Mikania micrantha* in jhum fallows of various ages at Lailad in Meghalaya (from Swamy and Ramakrishnan, 1987a)

	Age of fallow (yr)				
	0[a]	*1*	*3*	*6*	*12*
(a) Rosettes in March 1982	7	17	48	14	5
(b) Rosettes in March 1983	15	22	45	11	2
(c) Net change (b – a)	8	5	–3	–3	–3
(d) Net increase (b/a)	2.2	1.3	0.9	0.8	0.4
(e) Ramets born between March 1982 and March 1983	690	1,775	2,914	1,176	350
(f) Ramets lost between March 1982 and March 1983	675	1,753	2,850	1,155	348
(g) Rosettes surviving from March 1982 to March 1983	4	9	27	6	2
(h) Percentage of rosettes surviving (g/a × 100)	57.2	53	56.3	42.9	40
(i) Expected turnover time in years (1/100-h × 100)	2.3	2.1	2.3	1.7	1.6
(j) Total ramets recorded over the year	697	1,792	2,692	1,190	355
(k) Percentage annual mortality (f/j × 100)	96.8	97.8	96.2	97	98

[a] Zero year fallow indicates first growing season of fallow

spread rapidly in young fallows and place them at high levels in the canopy through rapid vertical growth. Thus, vegetative growth is adapted to horizontal spread under competitive conditions of the younger fallows and under stress (chiefly reduced light) of the older fallows (Abrahamson, 1980). Greater allocation of biomass to the rosette root in the older fallows is advantageous for exploiting deeper soil layers of these fallows where the soil surface has been depleted of nutrients by the developing community (Ramakrishnan and Toky, 1981; Mishra and Ramakrishnan, 1983d).

Reproduction of *Mikania micrantha* through ramets arising from rosettes emerging in March far exceeds seedling regeneration (Table 4.5). The emergence peaked in a 3-year-old fallow. An increase in population size occurred only in 0- and 1-year-old fallows. The population size declined in older fallows, with the decline positively related to the age of the fallow. Both the percentage survival and the expected turnover time were lower in 6- and 12-year-old fallows, compared to others.

Seedling recruitment is particularly important in a species with emphasis on vegetative reproduction, in order to maintain population diversity. Seed numbers were high in younger fallows and declined drastically in 12-year-old fallows. Interestingly, seedling regeneration occurred only in 6- and 12-year-old fallows (Table 4.6). This may be because, in the younger fallows where *Mikania micrantha* forms a mat on the open soil surface through

Table 4.6 Fate of cohorts of seedlings of *Mikania micrantha* in 6- and 12-year-old jhum fallows at Lailad in Meghalaya (from Swamy and Ramakrishnan, 1987a)

	Cohort of the month		
	May	*June*	*July*
6-year-old fallow			
Seedlings recorded	129	83	–
Seedlings established after perennation	38	12	–
Rosettes produced after perennation	2	–	–
12-year-old fallow			
Seedlings recorded	96	73	34
Seedlings established after perennation	21	9	14
Rosettes produced after perennation	1	–	–

horizontal spread, its own seedling regeneration is suppressed; in older 6- and 12-year-old fallows, this weed survives by resorting to vertical growth around other species, thus providing some open space for its seedling establishment in the few open spaces that may still be available.

A rapid decline in biomass, noted for *Mikania micrantha* and for *Eupatorium* spp. discussed earlier, is related to biological suppression due to reduced light availability (Kushwaha and Ramakrishnan, 1982) in communities dominated by bamboo, shrubs and trees (Toky and Ramakrishnan, 1983a; Mishra and Ramakrishnan, 1983a). Indeed, under controlled experimental conditions where potted plants were subjected to full sunlight and 30% of full sunlight (for 11 h light period), drastic reductions in growth of two weeds like *Eupatorium odoratum* and *Imperata cylindrica* were noted in the latter treatment (Table 4.7) (Kushwaha and Ramakrishnan, 1982). This applies to other light-demanding species like *Mikania micrantha* also, of the early successional environment resulting in reduced vigour in fallows older than 8 years.

Conclusions

The foregoing discussion of weed potential in early secondary successional fallows suggests that the weeds here have four distinct recruitment strategies:

(1) Through seedlings as in all species;

(2) Through aboveground ramet formation as in *Mikania micrantha*;

Table 4.7 Growth characteristics of *Eupatorium odoratum* and *Imperata cylindrica* in sun and shade at the final harvest, under controlled conditions (from Kushwaha and Ramakrishnan, 1982)

	E. odoratum			I. cylindrica		
Parameters	*Sun*	*Shade*	*LSD at 5% level*	*Sun*	*Shade*	*LSD at 5% level*
Plant height (cm)	196	167	22	143	61	20
Aboveground dry weight (g)	78.9	13.8	10.2	23.4	4.3	9.8
Belowground dry weight (g)	25.0	5.5	4.8	21.8	2.6	6.2
Aboveground/below ground dry weight ratio	3.1	2.5	0.32	1.0	1.6	0.36
Relative growth rate ($g\,cm^2\,g^{-1}$)	0.008	0.005	0.001	0.003	0.002	0.001
Leaf area ratio ($cm^2\,g^{-1}$)	59.9	59.2	12.2	111.7	111.6	16.2

(3) Through belowground stoloniferous rhizomes as in *Imperata cylindrica*, and;

(4) Through perennating rosettes/root stock as in *Mikania micrantha* and *Eupatorium* spp.

All the species, discussed here and many others of the early successional phase under jhum often have, therefore, both vegetative and sexual reproductive potentials. The realization of this potential, however, is dependent upon the successional age of the fallow. These weeds become less vigorous in growth with declining reproductive potential, in fallows beyond a maximum of about 5–6 years; they often are biologically suppressed beyond 10-year fallow regrowth. Increased competition for available resources, particularly for light, is responsible for their extinction. However, allelopathic factors as in *Mikania micrantha* (Wong, 1964) and *Eupatorium odoratum* (Yadav and Tripathi, 1981) would also contribute. Many of these species closely follow fire events and their presence and indeed dominance in the community are often related to fire events (Toky and Ramakrishnan, 1983a; Swamy and Ramakrishnan, 1987d; Ramakrishnan and Vitousek, 1989), as discussed below. Since the population size of these weeds declines drastically with successional age of the fallow, it is reasonable to assume that the weed potential would decline under longer jhum cycles, an aspect that is discussed in detail subsequently.

FIRE ADAPTATION AND WEED POPULATION STRUCTURE

Emergent properties of ecosystems are linked closely with population level phenomena. In an ecosystem that is subject to frequent perturbations, the

Plate 4.2 *Imperata cylindrica* (the thatching grass) in an important early successional weed reaching peak growth during the first 5–6 years in the fallows after jhum. In the foreground, the exotic climber *Mikania micrantha* could be seen covering over the ground vegetation

effects of these would first be reflected at the population level. Fire is one such perturbation factor that has received much attention (Kozlowski and Ahlgren, 1974; Mooney, *et al.*, 1981). There are many fire adapted plant communities – forests, shrub vegetation and grasslands that are maintained due to fire (Keeley, 1981). However there are many herbaceous species not easily definable as a single community, but which proliferate after fire. One such fire-following species is *Pteridium aquilinum*, which is also common under the jhum system at higher elevations of Meghalaya, which invades recently burnt sites and then spreads vegetatively (Keeley, 1981). This species and others are able to withstand fire because of deeply placed underground rhizomes that can survive scorching surface burn (Flinn and Wien, 1977).

Fire adaptation of *Imperata cylindrica*

Imperata cylindrica var. major is a ruderal exotic weed abundantly found in highly disturbed areas throughout the country. In north-eastern India, it is an early colonizer of jhum fallows (Plate 4.2) and is found elsewhere, in Africa (Richards, 1952; Nye and Greenland, 1960), and in south-east Asia (Eussen and Wirjahardja, (1973).

Imperata cylindrica has an extensive underground rhizome system which helps it greatly in local dispersal. However, regeneration through seeds also occurs. The seeds produced in April–May in north-eastern India are light

Table 4.8 Population flux of *Imperata cylindrica* at different sites at Burnihat in Meghalaya (from Kushwaha *et al.*, 1983b)

	Fallow age (yr)/*treatment*						
	0	*1*	*3*$^{(a+)}$	*3*$^{(a-)}$	*3*$^{(c)}$	*3*$^{(b)}$	*5*
(a) No. of tillers/seedlings per m^2, May 1978	–	149	92	73	0	0	8
(b) No. of tillers/seedlings per m^2, May 1980	(430)	98	14	321	390	413	0
(c) Net change (b – a)	(+430)	–51	–78	+248	+390	+413	–7.6
(d) Rate of increase (b/a)		0	0	4			0
(e) No. of tillers/seedlings per m^2 arrived between May 1978 and May 1980	(2,164)	494	100	582	729 (3,000)	974 (3,000)	35
(f) Total number of tillers/ seedlings per m^2 lost between May 1978 and May 1980	(1734)	544	179	334	339 (3,000)	561 (6,00)	42
(g) Tillers/seedlings present May 1978 alive by May 1980	–	40	2	20	–	–	0
(h) Percentage survival of tillers/seedlings in (a) (g/a × 100)	–	27	2	27	–	–	0
(i) Expected time for complete turnover (yr) (2/100 – h × 100)	–	2.8	2.4	3.1	–	–	2
(j) Total tillers/seedlings per m^2 recorded during study	(2,164)	643	193	655	729 (3,000)	974 (6,001)	43
(k) Annual mortality (%) of tillers/seedlings (f/j × 100)	(80)	85	93	51	46.5 (100)	57.6 (100)	100

In parentheses are the values for seedlings; a+, with associates; a–, without associates; c, clipped; b, burnt.

and easily dispersed by wind. Seedlings come up during the monsoon period and active growth continues up to October, with a steady increase in the tiller population. Flowering occurs during March–April.

Seedlings of *Imperata cylindrica* come up in large numbers only in freshly abandoned fallows under jhum (Table 4.8) (Kushwaha *et al.*, 1983b). However, they may come up in older fallows too, if they are subjected to drastic disturbance such as clipping or burning of the vegetation. However, seedling survival is very poor. Recruitment of individuals is largely through underground vegetative rhizomes. The observation on clipping/burning studies on population dynamics (Kushwaha *et al.*, 1983b) is corroborated by our observation that disturbance also promotes flowering in this species, particularly so when it occurs during the dry season (Kushwaha *at al.*, 1983b). In an experiment performed on clipping/burning effects, flowering

Table 4.9 Reproductive potential of *Imperata cylindrica* at clipped and burnt plots under 3-year jhum fallow at Burnihat in Meghalaya (from Kushwaha *et al*., 1983b)

	Clipped	*Burnt*
Tillers (m^{-2})	390	413
Spikes (m^{-2})	130	160
Seeds per spike	291	306
Seeds (m^{-2})	37,830	48,960
Seeds per tiller	97	119
Reproductive capacity (m^{-2})	34,047	44,064

did not occur in undisturbed plots, but had greater reproductive potential in burnt plots than in the clipped ones (Table 4.9).

In a comparative study of *Imperata cylindrica* growing in jhum fallows at Burnihat and Shillong (Sarma 1985), the survivorship of the population through tillers coming up through four cohorts in a year was found to be markedly enhanced in burnt plots when compared with unburnt plots (Fig. 4.5). These results are also suggestive of adaptation of *Imperata* population to fire.

Fire adaptation of *Mikania micrantha*

Equally pronounced fire effects were also noted in the exotic weed, *Mikania micrantha*. Through a comparative study, on a set of 12-, 4- and 8-year-old fallow plots under jhum that were slashed in February and then dried slash-burnt in March and another set left undisturbed, at Lailad (296 m altitude) in Meghalaya, the manner in which fire alters population density (ramet and seed reproduction) of *Mikania micrantha* was documented (Swamy and Ramakrishnan, 1987d).

The higher density of *Mikania micrantha* in burnt plots (Table 4.10), when compared to unburnt plots of the same fallow could partly be related to increased insolation, enriched soil through ash deposit (Ramakrishnan and Toky, 1981) and possible release from the allelopathic effects of its own individuals (Wong, 1964) and from other species such as *Eupatorium odoratum* (Yadav and Tripathi, 1981). Competition may not be a factor in the increase of *Mikania micrantha* density in burnt plots, as the densities of most associated species are either unaffected by fire or increased after the burn as in an 8-year-old fallow. This is in spite of the general observation discussed earlier that the population size during secondary succession peaks in a 4-year-old fallow (Swamy and Ramakrishnan, 1987d).

Recruitment of individuals into a given site is often density dependent (Ramakrishnan and Kumar, 1971). High birth rates are often accompanied by large-scale mortality. Such a density dependent mortality, preceded by heavy recruitment is a common phenomenon (Sarukhan and Harper, 1973;

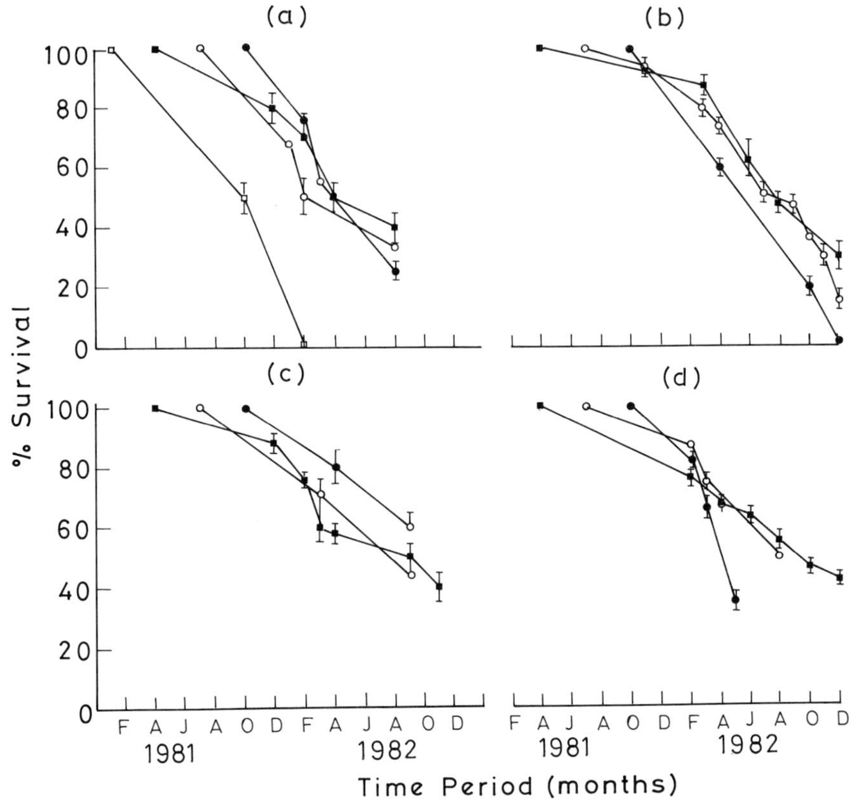

Figure 4.5 Survivorship curve of four tiller cohorts of January, (□); April, (■); July, (○); and October, (●) in a 3-year-old jhum fallow. (a), unburnt plots at Burnihat; (b), unburnt plots at Shillong; (c), burnt plots at Burnihat; (d) burnt plots at Shillong, in Meghalaya (from Sarma, 1985)

Lovett-Doust, 1981). Heavy mortality and the consequent reduced density may result in another wave of heavy recruitment. The double peaking in net population size in burnt sites during May and August as seen in Fig. 4.6 may be explained on this basis. A relatively slower recruitment in unburnt plots may result in a slow increment in net population size and one peak attained during July–August. Indeed, this species responded to burning with increased birth and death rates of the total (seedling plus ramet) population more markedly so in 2- and 4-year-old than in older fallows (Fig. 4.7).

In a species such as *Mikania micrantha* where vegetative reproduction is predominant, sexual reproduction through seedling regeneration is important in maintaining genetic diversity (Harper, 1978). The ruderal nature of this exotic weed is evident from the fact that seedling recruitment

Table 4.10 Importance value indices of dominant plant species associated with *Mikania micrantha* in different burnt and unburnt jhum fallows at Lailad in Meghalaya (from Swamy and Ramakrishnan, 1988a)

	Age of fallow (yr)					
	Burnt			*Unburnt*		
Species	*2*	*4*	*8*	*2*	*4*	*8*
Herbs						
Ageratum conyzoides	41.8	20.6	31.3	44.1	20.4	19.7
Borreria hispida	23.1	7.9	13.3	27.8	13.4	8.7
Erigeron linifolium	13.3	4.8	4.0	15.1	–	11.7
Eupatorium odoratum	56.2	42.9	46.5	51.9	43.5	27.8
Imperata cylindrica	–	14.9	28.3	–	9.2	–
Mikania micrantha	66.1	51.3	28.2	37.4	44.7	4.7
Mucuna bracteata	14.8	13.2	5.1	9.8	10.3	–
Panicum khasianum	12.1	11.4	8.3	6.1	5.4	10.0
Panicum maximum	11.5	13.6	8.1	6.8	8.3	–
Others	22.1	34.2	20.2	18.6	37.0	22.4
Shrubs and trees						
Litsaea assamica	–	1.8	5.0	–	10.2	14.0
Leea spp.	4.5	–	2.7	18.2	–	10.0
Melastoma malabathricum	–	7.3	11.0	3.6	11.9	10.4
Mussaenda roxburghii	0.0	14.5	4.1	0.0	15.1	13.9
Dendrocalamus hamiltonii	13.8	19.7	26.9	17.6	20.7	44.2
Macaranga denticulata	–	8.5	6.1	–	18.2	20.7
Schima wallichii	4.5	–	14.6	–	–	32.1
Others	8.1	28.7	35.9	32.0	34.0	49.2

occurs exclusively in burnt sites in 2- and 4-year-old fallows (Table 4.11). The total absence of seedling recruitment in unburnt plots of 2- and 4-year-old fallows may be due, in part, to rapid vegetative regeneration that would reduce available space and light penetration at ground level. In 8-year-old fallows, seedling recruitment occurred even in unburnt plots; it was largely because of the vertical growth made by the already existing fewer *Mikania micrantha* individuals that would not interfere with light penetration to the ground level. In any case rosette formation occurred only in burnt plots. Such a fire-related seedling recruitment strategy is in spite of the profuse seed production by *Mikania micrantha* (Table 4.12) However, the proportion of sexually reproductive ramets was higher in the burnt plots than in unburnt ones, more markedly so in 8-year-old fallows. Seed output per unit area was, however, maximum in 4-year-old fallows. Unburnt 8-year-old fallow plots had the lowest seed output.

Recruitment through ramets was also influenced by burning (Table 4.13). A net increase in rosette population during a 1-year period was consistently observed in the burnt fallow plots, whereas such an increase only occurred in the 2-year-old fallow plot if unburnt. The percentage of surviving rosettes

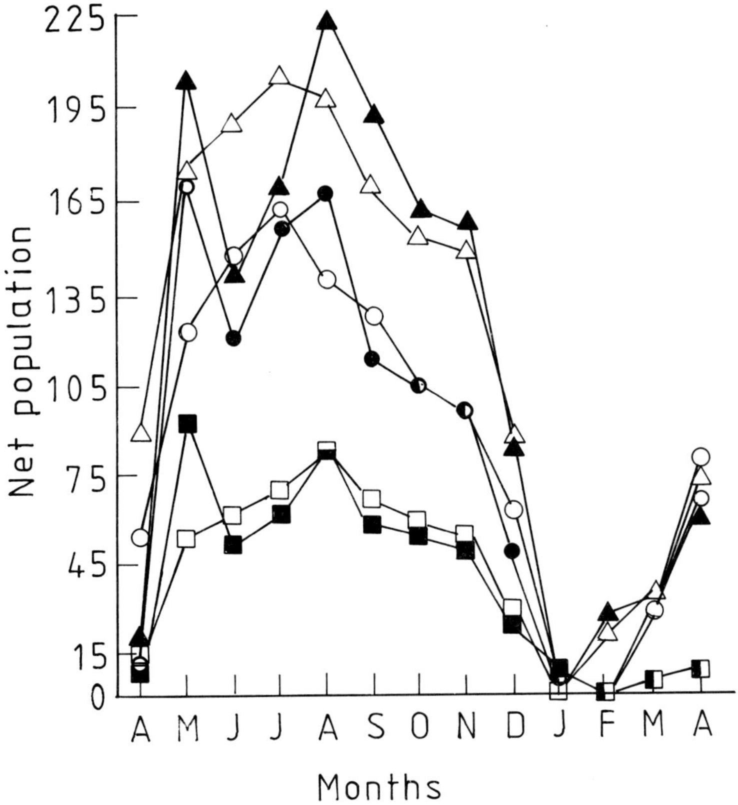

Figure 4.6 Net population size of *Mikania micrantha* in 2-year- (○●), 4-year- (△▲), and 8-year- (□■) old jhum fallows at Lailad in Meghalaya. Closed symbols are burnt sites; open symbols are unburnt sites (from Swamy and Ramakrishnan, 1987d)

and the expected turnover rates increased with the fallow age in burnt plots, whilst it declined with the fallow age in unburnt plots. The ramet mortality was somewhat higher in burnt sites than in unburnt sites. Although the ramet population size in the burnt 2- and 4-year-old fallow plots is not different from the unburnt plots, ramets are largely derived from current seedling recruitment in the former situation, rather than from an already well established rosette population as in the latter case. Thus, there is a qualitative age difference in the population of burnt and unburnt plots.

Conclusions

In the jhum system of north-eastern India, as in all shifting agricultural systems throughout the world, fire is used as a tool for the management of

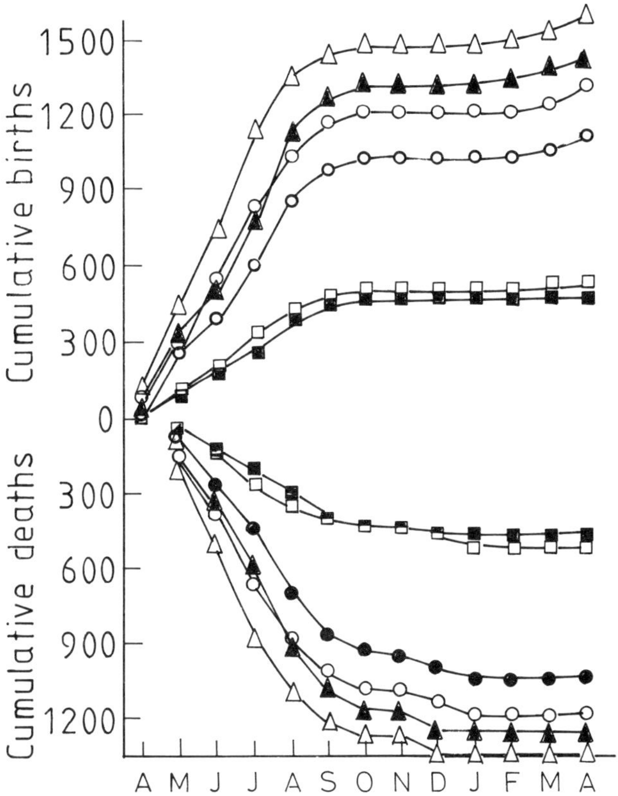

Figure 4.7 Cumulative births and deaths of *Mikania micrantha* population in 2-year- (○●), 4-year- (△▲) and 8-year- (□■) old jhum fallows at Lailad in Meghalaya. Closed symbols are burnt sites; open symbols are unburnt sites (from Swamy and Ramakrishnan, 1987d)

Table 4.11 Fate of seedlings of *Mikania micrantha* in burnt jhum fallows of different ages at Lailad in Meghalaya (from Swamy and Ramakrishnan, 1988a)

Fallow age (yr)	*Seedlings recording*	*Seedlings established before perennation*	*Rosettes produced after perennation*
2	988	59	59
	(0)[a]	(0)	(0)
4	1,752	41	41
	(0)	(0)	(0)
8	24	3	3
	(307)	(15)	(0)

[a] Values in parentheses are for unburnt fallows

Table 4.12 Reproductive potential of *Mikania micrantha* in burnt and unburnt jhum fallows of different ages at Lailad in Meghalaya (from Swamy and Ramakrishnan, 1988a)

	Age of fallow (yr)					
	Burnt			*Unburnt*		
	2	*4*	*8*	*2*	*4*	*8*
No. of ramets per m^2	392	624	200	388	6,000	216
Non-reproductive ramets (%)	12	18	15	17	29	68
Reproductive ramets (%)	88	82	85	83	71	32
Capitula per reproductive ramet	113	109	118	120	108	72
Seeds per capitula	3.3	3.0	3.5	3.3	3.1	2.8
Seeds per reproductive ramet	373	327	413	396	335	202
Capitula per m^2	38,985	48,941	20,060	38,640	46,008	4,968
Seeds per m^2	128,650	146,823	70,210	127,512	142,624	13,910

Table 4.13 Population flux of *Mikania micrantha* in burnt and unburnt jhum fallows of different ages at Lailad in Meghalaya (from Swamy and Ramakrishnan, 1988a)

	Age of fallows (yr)					
	Burnt			*Unburnt*		
	2	*4*	*8*	*2*	*4*	*8*
(a) No. rosettes in April 1982	11	19	5	54	87	13
(b) No. rosettes in April 1983	67	63	8	81	72	8
(c) Net change (b − a)	56	44	3	27	−15	−7
(d) Net increase (b/a)	6.1	3.3	1.6	1.5	0.8	0.6
(e) Ramets born between April 1982 and April 1983	1,117	1,437	483	1,255	1,417	488
(f) Ramets lost between April 1982 and April 1983	1,033	1,245	469	1,196	1,347	493
(g) Rosettes surviving from April 1982 to April 1983	3	6	3	31	43	5
(h) Percentage of rosettes surviving (g/a × 100)	27.3	31.5	60	57.4	49.4	38.4
(i) Expected turnover rate 1/100 − h × 100	1.4	1.5	2.5	2.3	2.0	1.6
(j) Total plants (ramets and rosettes) recorded over the year	1,128	1,456	488	1,330	1,572	508
(k) Percentage annual mortality of ramets (f/j × 100)	99.0	98.7	99.0	94.4	90.1	96.1

this agroforestry ecosystem complex. In other words, fire influences ecosystem function both during the agroecosystem and the forest fallow phases. It is therefore reasonable to find that fire effects during secondary successional fallow phases are most felt in the very early stages of weed communities.

A conclusion that emerges very clearly is that native species such as *Imperata cylindrica* and exotics like *Mikania micrantha* discussed here have fire-related population control mechanisms. Thus, a species such as *Imperata cylindrica* not only has accelerated recruitment of individuals into the population soon after a burn but is able to maintain genetic diversity in the population through seed reproduction after a series of fire events. In this species, a series of perturbations through frequent fire events during the dry season promotes flowering and seed production (Kushwaha *et al.*, 1983b; Sarma, 1985). In the case of *Mikania micrantha*, whilst ramet recruitment is essentially a function of the successional development of the fallow phase after slash-and-burn agriculture (Swamy and Ramakrishnan, 1987a), rosette and seedling recruitment are chiefly determined by fire, as discussed here (Swamy and Ramakrishnan, 1987d). Rosette regeneration is ensured by the underground perennating caudex from the previous year which is protected from fire. Indeed, fire also promotes seed production and the reproductive potential through seeds in this species.

Both fire frequency and intensity could often determine population structure in the early successional phase. A light burn as under a short jhum cycle where fire frequency may occur once in 4–5 years tends to conserve underground rhizomes of species such as *Imperata cylindrica* and *Thysanolaena maxima*, unlike a high intensity burn under a long jhum cycle which may damage at least some of these reproductive structures, particularly those that are superficially placed. The root stocks of *Eupatorium* spp. and the rosette caudex of *Mikania micrantha* could also be damaged by high intensity burn. However, many of these weeds have light wind dispersed seeds that could be wafted in from adjoining sites after a burn, irrespective of the fire intensity.

Because weed potential of a site peaks in about 4–5 years after fallow regrowth during secondary succession, frequent disturbance under a short cycle of 4–5 years would result in an arrested succession (Toky and Ramakrishnan, 1983a; Mishra and Ramakrishnan, 1983a; Saxena and Ramakrishnan, 1984b) at the weed stage. Long secondary successional fallows would effectively suppress the weedy species. Under jhum, the dominance by a given weed during the previous fallow phase would determine, to a certain extent, what dominates during the following fallow phase after a cropping phase. When fire events occur more frequently under shorter jhum cycle, early successional communities are monopolized by the exotic C_3 species (Saxena and Ramakrishnan, 1984c; Ramakrishnan and Vitousek, 1989). In contrast, long jhum cycles impoverish the soil seed bank owing to a more intense fire effect (Saxena and Ramakrishnan, 1984b) thereby affecting the seed propagated C_3 species more than the vegetatively propagated C_4 grasses such as *Imperata cylindrica, Saccharum* spp. and *Thysanolaena maxima.* Contrarily, C_4 grasses are adapted to occupy nutrient-poor sites under short jhum cycles of 4–5 years, unlike the C_3 exotic weeds (Saxena and Ramakrishnan, 1983a, 1984c), because the latter have poor nutrient-use efficiencies, discussed elsewhere (Chapter 8). The ultimate weed community structure under jhum would, therefore, be the product of a complex interaction of demographic and ecophysiological attributes of the species, as a consequence of fire frequency and intensity, altered by the weeding regime of the cropping phase and seed dispersal possibilities from adjoining sites not affected by fire.

WEED POTENTIAL UNDER JHUM DURING THE CROPPING PHASE

The early successional herbaceous species which are established following the slash-and-burn, act as weeds, because of their direct interference with the planted crops during the cropping phase of jhum, competing for the available environmental resources. Whereas the early successional weed

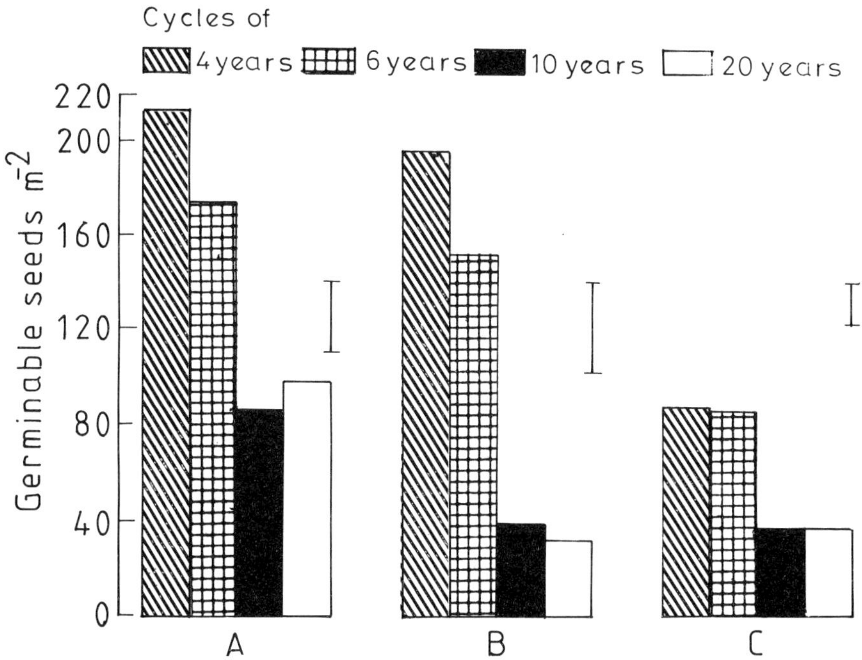

Figure 4.8 Germinable soil seed population of herbaceous species under different jhum cycles at Burnihat in Meghalaya. Vertical lines represent LSD ($p = 0.01$). (a), just before slash-and-burn; (b), uncropped sites; (c), cropped sites (from Saxena and Ramakrishnan, 1984b)

community structure during the fallow phase of jhum is a function of the frequency and intensity of perturbation events during the slash-and-burn agricultural phase as discussed earlier, the weed potential of the site during the cropping phase under jhum is essentially a function of the weed potential of the fallow phase, just prior to the slash-and-burn event. Of this reciprocal relationship, we shall now consider the latter component and look at the weed potential during the cropping phase as influenced particularly by the length of the fallow phase (jhum cycle).

We have shown through many studies during the cropping phase under different jhum cycles that the weed population is generally higher under shorter cycles than under longer ones (Mishra and Ramakrishnan, 1981; Swamny and Ramakrishnan, 1988c,d). In a study conducted at Burnihat (296 m altitude) in Meghalaya, under four different jhum cycles of 4–20 years, the germinable soil seed populations of weeds obtained at sites: (a) before slash-and-burn, (b) after the burn but uncropped sites and (c) cropped sites following the burn, were measured (Saxena and Ramakrishnan, 1984b). The germinable weed seed bank was higher under 4- and 6-year jhum cycles than under 10- and 20-year cycles (Fig. 4.8). There was a relatively greater effect of burning on weed potential under longer cycles. By contrast, the weeding effect was much larger on sites under shorter cycles. It may however, be noted that the germinable seed bank recorded here was low in comparison to the seed production capacity of the weeds (Saxena and Ramakrishnan, 1982a,b, 1983a,b).

The species which form early successional communities are commonly found as weeds in agricultural fields (Woodwell and Whittaker, 1968; Perozzi and Bazzaz, 1978; Toky and Ramakrishnan, 1983a; Saxena and Ramakrishnan, 1984b). Continuous imposition of short jhum cycles of 4–6 years leads to 'arrested succession' at an early stage where the community is exclusively composed of herbaceous weedy species characterized by high reproductive potentials (*r*-strategists). In contrast, under longer cycles of 10 and 20 years, these weedy species are eventually replaced at lower elevations of Meghalaya, for example, by bamboos (*Dendrocalamus hamiltonii*), with a few dicotyledonous trees, and shade-tolerant herbaceous species such as *Oplismenus compositus*, *Centotheca lappacea* and *Oryza granulata* with low reproductive potential (*s*-strategists) which may dominate the herbaceous community (Saxena and Ramakrishnan, 1982a,b; 1984b). This change in the vegetation structure results in the burn being more intense under the long cycles due to heavier fuel load as compared to that under the shorter cycles. A high reproductive potential of the early successional species and consequently high propagule production during the intervening fallow period coupled with a low intensity burn may account for a high weed population under short cycles of 4 and 6 years compared to longer cycles of 10 and 20 years. This may also be associated with comparatively less loss of viability in buried soil seed banks during short fallow periods than the longer ones where the rate of loss may be more severe (Livingston and Allessio, 1968).

Following burning, the total density and biomass values of all herbaceous plants was markedly higher under 4- and 6-year jhum cycles than under 10- and 20-year cycles (Fig. 4.9). Weed population sharply declined in the cropped sites in comparison with the uncropped sites under all cycles. This conspicuous decline after a year of cropping following the burn under all the cycles may be due to weeding during the cropping phase, and may also be due to some loss of seeds through water and sediment losses from steep slopes (Toky and Ramakrishnan, 1981b; Mishra and Ramakrishnan, 1983c). These results are somewhat contrary to the belief that the need to go into fallow is partly because weeds increase to serious proportions because weeding is not adequate during cropping. However, it may be emphasized that the fallow phase becomes effective for weed control only if it is longer than 10 years, as discussed earlier.

Coefficient of community indices (CC) which expresses the floristic similarity of two communities based solely on species presence or absence are given in Table 4.14. A CC value of 100 indicates identical composition

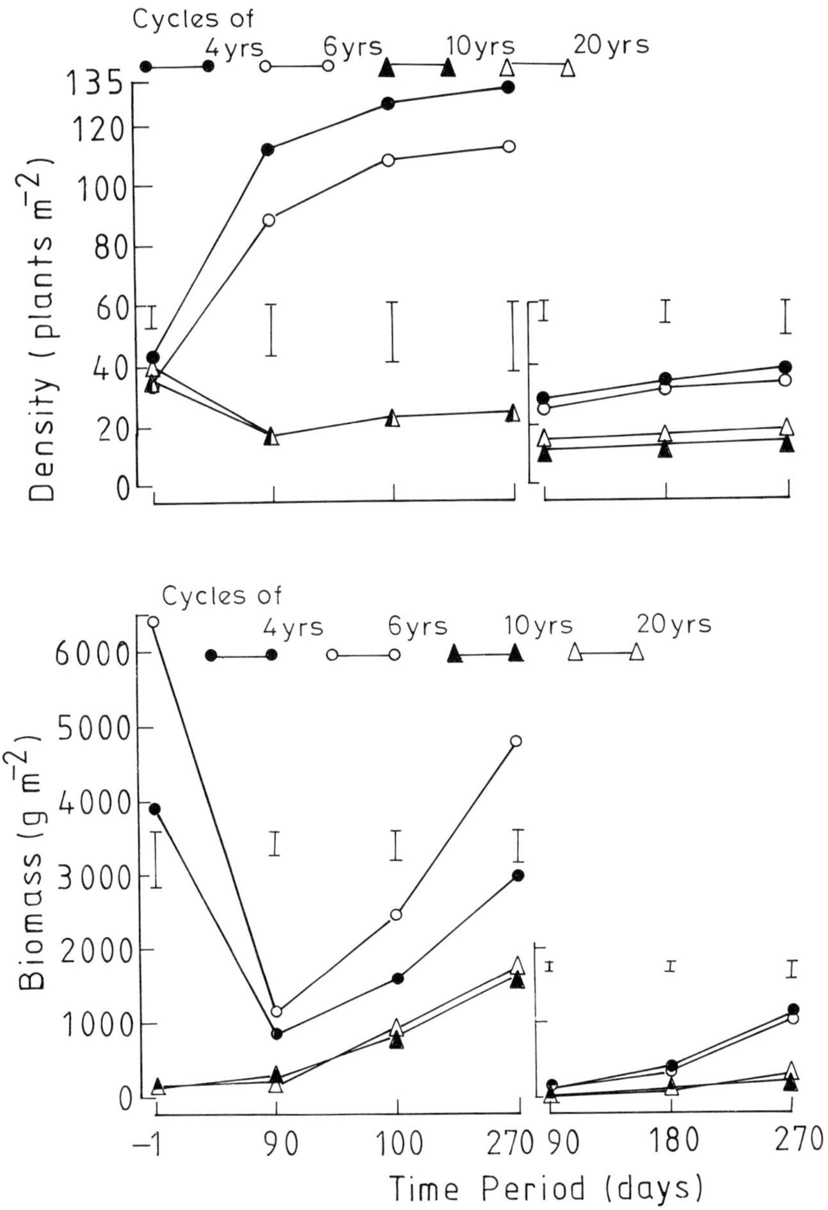

Figure 4.9 Density and biomass of herbaceous communities under different jhum cycles at Lailad in Meghalaya. (a), uncropped sites; (b), cropped sites; –1, a day before slash-and-burn; 90, 180, 270 are the days of fallow regrowth after the burn. Vertical lines represent LSD ($p = 0.01$) (from Saxena and Ramakrishnan, 1984b)

Table 4.14 Floristic similarity based on the coefficient of community index of preburn herbaceous vegetation with that of uncropped and cropped sites under different jhum cycles at Burnihat in Meghalaya (from Saxena and Ramakrishnan, 1984b)

	Treatments	
Jhum cycle (yr)	*Preburn and uncropped sites*	*Preburn and cropped sites*
4	98.0	74.1
6	66.7	70.6
10	14.8	20.0
20	14.3	20.0

of two communities, the degree of departure from 100 is a measure of change immediately after the burn (uncropped sites) or after cropping following the burn (cropped sites). The CC value between the preburn herbaceous vegetation and that developed in the uncropped sites after burning was highest under 4-year cycles (98.0) followed by 6-year cycles (66.7), but low values were obtained under 10- and 20-year cycles. A similar difference between the short cycles of 4 and 6 years and long cycles of 10 and 20 years was also found when the preburn herbaceous community was compared with that developed after burning and 1-year cropping in the cropped sites.

Most of the studies on postfire succession support the 'initial floristic composition' model of Egler (1954) according to which it is the species present before the burn which re-establish after the burn (Christensen and Kimber, 1975; Gill, 1975; Purdie and Slatyer, 1976). On the other hand, the classical 'relay floristics model' (Clements, 1916; Odum, 1969) assumes that successive suites of species progressively occupy the site under consideration during succession. The data presented above (Table 4.14) show that under short cycles of 4 and 6 years, where the disturbance of fire is more frequent, early succession following fire conforms more closely to the 'initial floristic composition' model. In contrast, the 'relay floristics' model was only found to operate in long cycles of 10 and 20 years, with a relatively low frequency of fire disturbance. Such differences in patterns of vegetation development under long and short jhum cycles may be expected because of differences in the degree and frequency of perturbations (Cornell and Slatyer, 1977).

Conclusions

The question of weed potential under shifting agriculture has been discussed by many workers (Nye and Greenland, 1960; Ruthenberg, 1971; Watters, 1971; Zinke *et al.*, 1978), though quantification of weed potential under different cycles (Saxena and Ramakrishnan, 1984b), and their role in nutrient budget (Swamy and Ramakrishnan, 1988c) discussed below, is not available. The increased weed potential under short jhum cycles of 4–6 years is obviously due to the presence of the same weed species in the 4- to 6-year-old fallow plots that are slashed, which helps to build up the soil seed bank. Longer jhum cycles of 10 years or more, on the other hand, may be viable both from the ecological and economic viewpoint as under these situations weeds decline naturally during the long fallow periods as shown here, and at the same time soil fertility is built up (Ramakrishnan and Toky, 1981; Mishra and Ramakrishnan, 1983d), as discussed in Chapter 8.

The presence of weeds may cause rapid depletion of resources often resulting in reduced crop yield. The shading of crops by weeds is also an important factor. The reduction in monetary yield (see Chapter 1) by 1.9 times under a 5-year jhum cycle compared to a 10-year cycle (yield per 5-year cycle versus yield per 10-year cycle) may partly be related to the weed problem along with poorer soil fertility recovery under shorter cycles (Ramakrishnan and Toky, 1981). The depletion of resources by weeds suggested by some workers is only a short-term diversion from the crop. It is surely the harvesting of crops along with nutrient losses associated with drastic hydrological changes (Toky and Ramakrishnan, 1981b; Mishra and Ramakrishnan, 1983c) that actually deplete the nutrient resources in the soil. Indeed, as discussed below, the weed population under jhum has often a resource conservation role under the given climatic and topographic conditions.

WEED MANAGEMENT

Until recently, agricultural scientists have always talked of weed control in croplands. It is only recently that weed management rather than total control is being considered as an approach to overall agroecosystem management (Aldrich, 1984; Altirie and Liebman, 1988). Many management guidelines have been evolved based on sound ecological principles. In all these, the empirical knowledge of the traditional farmer of the tropics has played a key role (Ramakrishnan, 1984a, 1988a; Gliessman, 1988). The jhum in north-eastern India is one such system which has much to offer in that the traditional jhum farmer has manipulated the weeds in a manner that not only does not interfere with the crop but also has some positive role in agroecosystem function.

The 'non-weed' concept

In many traditional systems, the incomplete removal of weeds from croplands may not be a deliberate act and may merely be due to lack of

Table 4.15 Use pattern of biomass by the Garos at Lailad in Meghalaya (after Ramakrishnan, 1984a)

Jhum cycle (yr)	*No. of weedings*	*Total weed biomass produced* ($kg\ ha^{-1}$)	*Weed biomass ploughed back* ($kg\ ha^{-1}$)	*Weed biomass retained* ($kg\ ha^{-1}$)
20	1	9,357	7,110(76.0)	2,247(24.0)
	2	10,180	8,257(81.1)	1,923(18.9)
10	1	8,085	6,189(76.5)	1,896(23.5)
	2	8,608	6,930(80.5)	1,678(19.5)
5	1	4,378	3,160(72.2)	1,218(27.8)
	2	5,780	4,815(83.3)	965(16.7)
	3	4,242	3,592(84.7)	650(15.3)

Percentage values given in parentheses

labour and poor returns. All the same, the partial weeding done by the jhum farmer in north-eastern India may often be a consequence of his intuitive expertise based on long tradition. Any other explanation would seem untenable since such a partial weeding is practised by many diverse traditional societies that are linguistically and culturally separated (Maikhuri and Ramakrishnan, 1990a) such as the Garos and the Khasis, but yet not done by non-traditional jhum farmers such as the Assam plain tribals, Mikirs, and the immigrant Nepalis. Such a weed management is also characteristic of spatially separated societies in the region. Indeed, in all the cases, the jhum farmer retains about 20% of the weed biomass in his plot. The remaining 80% is weeded and put directly back into the plot (Ramakrishnan, 1984a; Maikhuri and Ramakrishnan, 1990a) or via the compost pit (Mishra and Ramakrishnan, 1982).

Such an approach to weed management is followed by the Garos at Lailad (296 m) in Meghalaya (Table 4.15). The entire weed biomass is utilized either as organic waste put back into the agroecosystem or with about 20% biomass retained *in situ*, which, as shown in the following pages, is of value in many ways.

Weeding regime and crop yield

In the study done at Lailad in Meghalaya to evaluate the effect of residual weed populations on agroecosystem function, jhum plots under 20- and 5-year cycles had : (1) traditional weeding (TrW) which is a partial weeding as practised by the Garo farmer, (2) total weeding (ToW) and (3) no weeding (UnW). Both partial and total weeding regimes involved weeding by hand-hoeing, four times during the cropping period, except that in the former case, part of the current weed population (about 20% of weed biomass) was left undisturbed (Swamy and Ramakrishnan, 1988d).

Table 4.16 Mean economic yield (kg ha^{-1} ± SE) of crops under different weeding regimes under 20- and 5-year jhum cycles (from Swamy and Ramakrishnan, 1988c)

	Category		
	Grain and seed	*Leaf and fruit vegetables*	*Tuber crops*
20-year jhum cycle			
TrW	2,464 ± 139	308 ± 16	244 ± 12
ToW	2,812 ± 156	354 ± 18	251 ± 15
UnW	355 ± 21	172 ± 11	156 ± 11
5-year jhum cycle			
TrW	1,291 ± 83	172 ± 8	121 ± 8
ToW	1,444 ± 91	174 ± 14	125 ± 6
UnW	224 ± 17	135 ± 6	103 ± 4

Table 4.17 Biomass (× 10^3 kg ha^{-1}) (mean ± SE) of the crop and the weed under different weeding regimes under 20- and 5-year jhum cycles at Lailad in Meghalaya (from Swamy and Ramakrishnan, 1988c)

	Biomass		*Biomass retained*
	Crop	*Weed*	in situ
20-year jhum cycle			
TrW	16.8 ± 1.1	2.0 ± 0.1	0.4
ToW	19.3 ± 1.2	2.0 ± 0.2	
UnW	8.7 ± 0.7	10.2 ± 1.0	
5-year jhum cycle			
TrW	10.0 ± 0.9	1.2 ± 0.1	0.3
ToW	10.2 ± 0.9	2.0 ± 0.2	
UnW	3.6 ± 0.2	5.8 ± 0.4	

Values in parentheses are biomass retained *in situ*

Retention of weeds during traditional weeding seems to have little adverse effect on crop yield, but as is to be expected, the yield on unweeded plots was drastically reduced (Table 4.16). Grain and seed crops were more susceptible for weed competition with the consequent sharper decline in yield than others.

The weed biomass under traditional and total weeding regimes remained at a similar low level, unlike in the unweeded plots (Table 4.17). This is suggestive of the effective weed control in spite of traditional weeding, with special advantage occurring because of total weeding effort. The exceptionally high weed biomass under an unweeded 20-year cycle regime is largely because of bamboo, shrub and tree sprouts here, unlike under the 5-year cycle where the biomass is exclusively composed of herbaceous weeds. Otherwise the weed potential under the 5-year jhum cycle was greater.

Table 4.18 Energy (MJ $ha^{-1} yr^{-1}$) and economic (Rupees $ha^{-1} yr^{-1}$) analysis under 20- and 5-year jhum cycles with different weeding regimes. Values in parentheses are for money (from Swamy and Ramakrishnan, 1988c)

	Jhum cycle (yr)					
	20			*5*		
	TrW	*ToW*	*UnW*	*TrW*	*ToW*	*UnW*
Input total	971 (1,384)	1,028 (1,504)	883 (1,200)	487 (817)	499 (841)	363 (553)
Labour for weeding	87 (184)	145 (304)	– –	123 (264)	136 (288)	–
Output total	49,948 (6,318)	56,186 (7,093)	10,842 (1,487)	25,063 (2,551)	27,637 (2,860)	6,850 (554)
Output/input ratio	51 (4.6)	55 (4.7)	12 (1.2)	52 (3.1)	15 (3.4)	19 (1)

A high labour input for weeding under a short 5-year cycle is also reflected in the cost–benefit analysis (Table 4.18). Labour input for total weeding was more or less the same under the two jhum cycles, except that traditional weeding under a 5-year cycle was more costly than under a 20-year cycle. Under a 20-year cycle, the shrub and tree sprouts check herbaceous weed growth through shading; the control of shrub and tree weeds are less costly because they can be checked through slashing, unlike herbs that need hand-hoeing. While there was only a marginal reduction in economic output under traditional weeding when compared with total weeding, the reduction was very pronounced in unweeded plots. While net gain under a 5-year jhum cycle was lower than under a 10-year cycle, this was so low under the unweeded 5-year cycle regime that the input equalled the output from the system. Whereas monetary and energy efficiencies (output/input ratios) are similar under total and traditional weeding, they were drastically reduced only in unweeded plots. These results suggest that residual 20% weed biomass of traditional weeding does not interfere with crop yield. Indeed the assumption here is that weeds below a particular density level have useful functions to perform in the agroecosytem.

Weeding regime and nutrient conservation

In the same study done at Lailad in Meghalaya, the role of the weeding regime to conserve nutrients in the agroecosystem was evaluated (Swamy and Ramakrishnan, 1988d). The data presented here for jhum under a 20-year cycle alone (Table 4.19) suggests that the losses from the system through water were very different under the three weeding regimes namely, total (ToW), traditional (TrW) and no weeding (UnW). With retention of about 20% of the weed biomass, the loss of sediment and a labile elements

Table 4.19 Losses through water (mean ± SE) under a 20-year jhum cycle with different weeding regimes during the monsoon (total rainfall received 183 cm) (from Swamy and Ramakrishnan, 1988d)

	20-yr jhum cycle		
Category of loss	*TrW*	*ToW*	*UnW*
Run-off water (cm)	30.4 ± 1.8	34.3 ± 1.4	22.1 ± 1.0
Percolation water (cm)	11.6 ± 0.8	15.4 ± 1.0	18.8 ± 0.0
Sediment ($t\ ha^{-1}$)	16.4 ± 1.1	21.1 ± 1.4	7.2 ± 0.7
Nutrients through water ($kg\ ha^{-1}$)			
NO_3–N	4.3 ± 0.2	5.3 ± 0.2	2.8 ± 0.08
	(6.9 ± 0.4)	(9.6 ± 0.7)	(4.9 ± 0.2)
PO_4–P	1.2 ± 0.06	1.6 ± 0.05	0.6 ± 0.003
	(0.04 ± 0.002)	(0.05 ± 0.003)	(0.21 ± 0.01)
K	81.4 ± 5.6	106.2 ± 8.1	51.3 ± 3.1
	(12.2 ± 0.9)	(16.2 ± 1.1)	(8.2 ± 0.04)
Ca	23.7 ± 2.0	31.5 ± 2.4	14.6 ± 0.9
	(5.0 ± 0.2)	(6.4 ± 0.3)	(3.0 ± 0.1)
Mg	10.6 ± 0.7	15.3 ± 1.0	9.0 ± 0.6
	(2.3 ± 0.01)	(3.1 ± 0.01)	(1.5 ± 0.01)

Values in parentheses are for nutrient losses through pecolation water

such as potassium through run-off is reduced by about 20%, when compared with total weeding. Similar reductions in losses for other elements was also noted. Leaching losses of nutrients were also curtailed. Obviously, the losses would be least in the unweeded plots, which would, to a large extent, be negated through weed–crop competition and the consequent reduced crop yield.

During the cropping phase, the ecosystem loses the ability to hold nutrients because of a series of perturbations (Toky and Ramakrishnan, 1981b; Mishra and Ramakrishnan, 1983c). These perturbation effects would largely be mitigated when the crop cover is established later in the season. It is therefore, during the early part of the cropping season (May–June) that the traditional weeding regime is expected to play a crucial role in nutrient conservation. This is reflected in the more exaggerated differences during May–June, between the weeding regimes under 20- and 5-year jhum cycles (Fig. 4.10). Indeed, the nutrient budget, as for potassium, under traditional weeding was closer to that under total weeding, whilst the budget under unweeded plots was markedly altered (Table 4.20). Similar results were obtained for other nutrients too (Swamy and Ramakrishnan, 1988c). The consequence of this is that the net gain in the soil pool, at the end of cropping (only the 20-year cycle is given here) was higher under traditional weeding than under total weeding, except for nitrogen (Swamy and Ramakrishnan, 1988c) where the nitrogen pool is altered by fire too

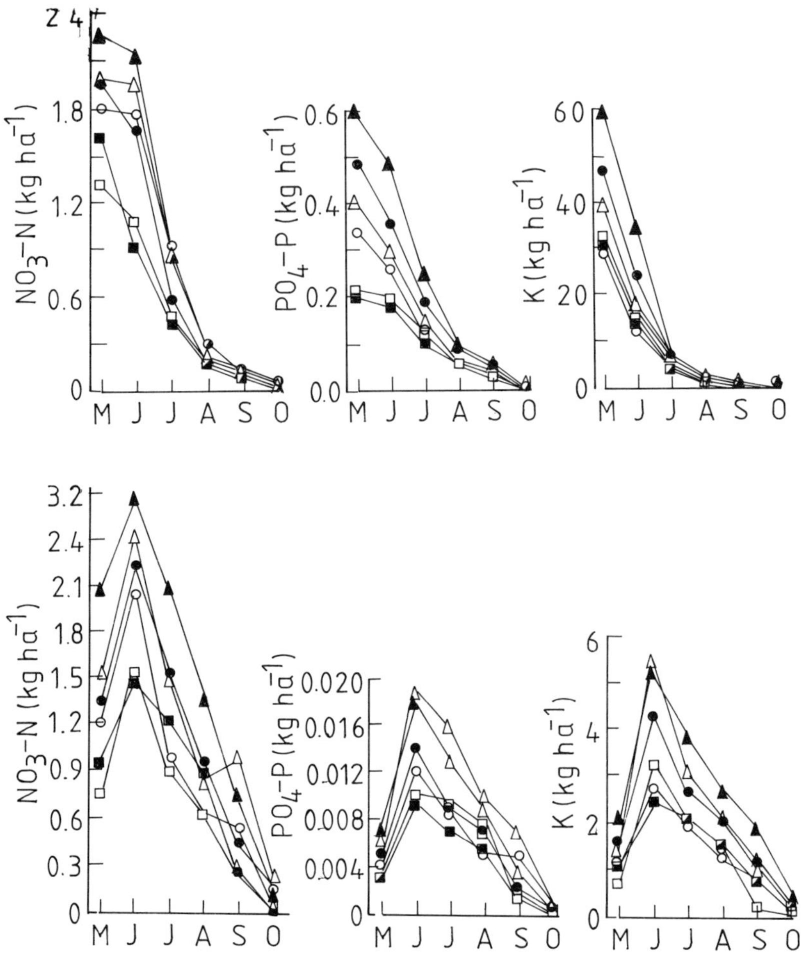

Figure 4.10 Monthly loss of NO_3–N, PO_4–P and potassium in run-off (a), and perculation (b), water during the monsoon under traditional weeding (●, ○), total weeding (▲, △), and no weeding (■, □) plots under jhum at Lailad in Meghalaya. Closed symbols, 20-year jhum cycle; open symbols, 5-year jhum cycle (from Swamy and Ramakrishnan, 1988c)

because of volatilization (Allen, 1964; Knight, 1965; White *et al.*, 1973; Ramakrishnan and Toky, 1981; Mishra and Ramakrishnan, 1983d).

Conclusions

Weeds in the jhum system form an important component of the agroecosystem with management implications. Under longer jhum cycles of 10 years or more, the agroecosystem is closer to natural ecosystems where nutrient cycling is based on efficient internal controls; this contributes to its stability.

Table 4.20 Potassium input and output (kg/ha^{-1}) for different weeding regimes under 20- and 5-year jhum cycles, at Lailad in Meghalaya (from Swamy and Ramakrishnan, 1988c)

	Jhum cycle (yr)					
	20			*5*		
	TrW	*ToW*	*UnW*	*TrW*	*ToW*	*UnW*
Input						
Precipitation	6.9	6.9	6.9	6.9	6.9	6.9
Slash-burn	930.0	930.0	930.0	760.0	760.0	760.0
Weeds put back	31.7	42.7	–	28.3	32.8	–
By-products put back	175.6	182.1	134.8	100.1	107.1	29.5
Output						
Sediment	48.7	62.6	21.4	26.7	32.8	12.0
Run-off	81.4	106.2	51.3	49.3	60.7	55.8
Percolation	12.2	16.3	8.2	7.9	13.6	8.3
Weed removal	31.7	42.7	135.7	28.3	32.9	65.5
Crop removal	181.7	183.3	132.6	107.7	114.9	34.5

With depletion in soil fertility under shortened jhum cycles, weeds left *in situ* not only help in effectively conserving soil nutrients but recycle them more effectively through dead biomass, since the weed potential here is higher than under a longer cycle. It is not unlikely that, if it was not for the integrated weed management practices of the traditional jhum farmer, the highly shortened 4–5-year cycles prevalent now would have resulted in even more distorted ecosystem function than at present, through adverse effects on nutrient cycling processes.

WEED POTENTIAL UNDER TERRACE AGROECOSYSTEM

In a study on terrace agroecosystem done at Nayabunglow (960 m altitude) located about 30 km north of Shillong (Patnaik, 1988; Patnaik and Ramakrishnan, 1990), an aspect that was considered was weed potential on terraces that were maintained for different time periods.

As discussed earlier (cf Chapter 2), 4- and 12-year-old terraces were compared. The former had two croppings done in a year whilst the latter had only one cropping in a year. The economic and energy efficiencies (cf Table 2.24) declined under older terraces. Apart from problems of maintaining soil fertility in older terraces (Patnaik and Ramakrishnan, 1990), the increased weed potential in them (cf Table 2.23) is an important factor for the declining crop yield. In the jhum system, we have seen how weed potential would become aggravated under shorter cycles (Saxena and Ramakrishnan, 1984b). Therefore, it is reasonable to expect further aggravation of the weed problem under continuous cropping on terraces. This is

Table 4.21 Biomass and nutrient (kg ha^{-1}) contribution through weed in terrace agroecosystem during cropping in north-east India. Values in parentheses are for grasses (from Patnaik, 1988)

	Terrace age (yr)					
	4-yr-old terrace			*12-yr-old terrace*		
	Biomass	*Nitrogen*	*Phosphorus*	*Biomass*	*Nitrogen*	*Phosphorus*
1st Cropping						
Total weed biomass	2,050.2 (741.1)	22.6 (6.31)	0.2 (0.07)	2,759.1 (254.5)	43.8 (2.5)	0.4 (0.03)
Weed put back	547.5 (167.4)	7.8 (2.4)	0.05 (0.02)	1009.3 (142.5)	8.5 (0.9)	0.13 (0.01)
2nd Cropping						
Weed slash put in prior to 2nd cropping	1,502.7 (573.7)	14.9 (3.9)	0.2 (0.06)	– –	– –	– –
Weed biomass	1,038 (628.3)	12.5 (5.5)	0.1 (0.05)	– –	– –	– –

obvious when the weed biomass under the first cropping phase is compared between 4-and 12-year-old terraces.

Because of a higher weed potential on the site during the first cropping on a 12-year-old terrace compared to a 4-year-old one, the nitrogen and phosphorous accumulation by the weed in the former situation is almost double (Table 4.21). This is also reflected in the cation budget (Patnaik and Ramakrishnan, 1990). More interestingly, whereas during the first cropping on a 4-year-old terrace the weeds are largely dicots of the family compositae, during the second cropping on the same site the graminaceous component increased. With a drastic decline in the nutrient status of the soil after the first crop removal, the C_4 grasses have an advantage over C_3 dicots because of the higher nitrogen-use efficiency of the former category of species. This is reflected in the proportion contributed by the two categories in terms of biomass and nitrogen. Thus the C_4 grasses growing in less fertile soils have lower nitrogen concentration, unlike others that may have luxury consumption but with reduced growth (Epstein, 1972; Chapin, 1980).

Conclusions

With a higher weed potential on older terraces, the weed biomass recycled through this component of the agroecosystem was twice that under a 12-year-old terrace than under a 4-year-old one, when first cropping alone is considered. While rapid nutrient depletion of the soil with declining physical qualities contributes to reduced crop yield on old terraces (Patnaik, 1988; Patnaik and Ramakrishnan, 1990), the increasing weed potential on older terraces is an important factor. On a poorly developed mountain soil, often all these factors could be critical. However, at Nayabunglow, the Nepali farmer does terracing on a well-developed soil that has a deeper soil profile. Here he is able to sustain terrace agroecosystem over a period of time, through heavy organic matter input such as cow dung, inspite of increased weed potential.

GENERAL CONSIDERATIONS

Although weeds are traditionally considered to be undesirable (Baker, 1965) adversely affecting crop yield in agroecosystems, yield reductions of crops are caused by a variety of complex factors of which weed species and density are important. The integrated weed management concept of the traditional jhum farmer as analysed here indicates that the farmer knows precisely how intense the weeding should be so that the weed stops interefering with crop yield, and yet the beneficial effects are manifest. In other words, the ingenuity of the jhum farmer lies in his ability to distinguish the 'weed' and the 'non-weed' status of the same species or a set of species in his jhum plot. Weeds have been suggested to control insect population and the resultant damage to the crop (Altierie, 1983; Andow, 1988). This and many other suggestions (Altierie and Liebman, 1988) need rigorous experimental verification. What we know, however, from the present case study is that the weeds growing with the crop, if properly managed as done by the farmer and recycled back into the agroecosystem as dead biomass, could contribute to soil fertility maintenance in the jhum system; this is particularly so when the system has started breaking down due to a shortened jhum cycle.

With weed potential becoming aggravated under terrace farming, the jhum farmer has to face this, as well as rapidly declining soil fertility which he cannot sustain. That is why sedentary agriculture on terraces has not found acceptance, except in specific situations where the necessary nutrient inputs are available and the soil is well developed to sustain such inputs. Valley cultivation, in spite of a high weed potential, is possible because of good soil which receives nutrient input from hill slopes. Another traditional system like the home garden, being a highly organized agroforestry system, would have very low weed potential because of the shading effect by trees. Indeed, any agroforestry system would be appropriate for the north-eastern hill region in that the tree component, as discussed here, can effectively suppress weed growth, largely through the effect of shading.

CHAPTER 5

SOIL FERTILITY PATTERNS AND NUTRIENT BUDGET UNDER JHUM AND OTHER LAND USES

INTRODUCTION

A major difference, with respect to soil fertility, between natural ecosystems and a man-managed ecosystem such as agriculture lies in the fact that substantial quantities of soil nutrients removed through crop harvest never get recycled back into the system, apart from other losses that may occur. Whilst the depletion of soil fertility through plant removal is perhaps the largest pathway, the removal through run-off water and leaching processes could be substantial under situations of uneven topography and poor soil physical qualities. In the humid tropics, the soils are highly variable. Oxisols, ultisols, inceptisols and entisols are some of the more important soil types. The most widespread soil constraint for plant growth in the humid tropics has often been suggested to be chemical, with low nutrient reserves (Sanchez, 1989). The chemical constraints could be overcome by application of organic/inorganic fertilizers or through crop rotation and agroforestry, but the physical constraints are more difficult to manage.

In specific terms, what then are the chief limiting factors for agricultural production in the humid tropics and, how these constraints could be handled for sustainable yield? Nitrogen is one of the key elements. Being very labile, this is often in short supply for tropical agriculture (Mishra and Ramakrishnan, 1984; Swift and Sanchez, 1984). There are three pathways that could be the basis for improving nitrogen staus: (1) inorganic fertilizer, (2) biological nitrogen fixation and (3) nitrogen mineralization from the organic pool. Phosphorus deficency is reported to be one of the most widespread soil chemical constraints, particularly well studied in tropical America (Sanchez and Cochrane, 1980); this deficiency is further compounded by high phosphorus fixation related to soil acidity. Sanchez and Salinas (1981) have discussed

(1) Possible methods of application of phosphorus sources;

(2) Use of moderate quantities of lime to increase soluble phosphorus sources;

(3) Mycorrhizal associations to improve phosphorus uptake by plants; and

(4) Selection of plant species and varieties that grow well at low phosphorus levels.

Potassium is another labile element like nitrogen. Potassium losses through surface run-off and leaching in the humid tropics, under land of uneven topography could be very high (Toky and Ramakrishnan, 1981b; Mishra and Ramakrishnan, 1983c; Ram and Ramakrishnan, 1988b). Effective recycling of this element could be done, particularly by using species that would concentrate and conserve this element, such as *Mikania micrantha* (Swamy and Ramakrishnan, 1987c) and *Dendrocalamus hamiltonii* (Toky and Ramakrishnan, 1982b, 1983b), as shown under the jhum system in north-eastern India, through grass fallows in Africa (Lal *et al.*, 1979) or by using tree species such as *Musanga cecropioides* in the African forest fallows developed after shifting agriculture (Bartholomew, *et al.*, 1953). Besides these, there are problems related to soil acidity, that would adversely affect availability of calcium and magnesium and may often cause aluminium and manganese toxicities (Sanchez and Salinas, 1981).

When a forest is cleared and burnt for shifting agriculture, the soil properties are drastically altered. Because of fire effects, surface soil temperature increases and due to high insulation, moisture levels fluctuate. Much of the soil nitrogen is lost through volatilization along with sulphur. With ash deposition, the availability of nutrients would increase. Increased soil pH would also decrease micro-element toxicity, particularly aluminium and manganese. Under cultivation not only the original vegetation is destroyed but the site is subject to continuing perturbations due to fire, the introduction of crop species, weeding, hoeing and crop harvest. During these events, the remaining components of the ecosystem lose their ability to hold the nutrients in the soil. A period of rapid depletion of nutrients occur during the slash-and-burn and the cropping phase and this depletion process continues through the early secondary successional phases. The soil fertility recovery starts only subsequently. We shall consider here the patterns and processes involved in soil fertility changes and nutrient budgets during the jhum operations in north-eastern India. We shall also consider the merits and demerits of jhum *vis a vis* other land uses of north-eastern India, as related to soil fertility. We shall further discuss possible land use management strategies in the context of maintaining soil fertility through low-input technology. Maintenance of soil fertility through biological means is attractive in this context.

SOIL FERTILITY PATTERNS UNDER JHUM

In the jhum system of north-eastern India, as elsewhere under similar situations of shifting agriculture, the farmers utilize the nutrient accumulation of several years in the forest fallow biomass to fertilize the agricultural plots through slash-and-burn (Nye and Greenland, 1960; Watters, 1971; Ramakrishnan *et al.*, 1981b; Ramakrishnan, 1984a). The farmer capitalizes upon the nutrients released in a single flush after the fire, by growing a mixture of crop species for a year or two after which the land is reverted back to its natural vegetation so as to restore soil chemical fertility and to improve its physical properties. The long-term success of this practice of agriculture depends upon the recovery and maintenance of soil fertility. If the nutrients lost or displaced during the short period of cultivation are approximately balanced by those replaced during the fallow period, the system could continue indefinitely. The maintenance of soil fertility in hot humid and high rainfall areas is a serious problem and becomes more severe in situations when the cycle becomes short, due to more frequent perturbations and elemental losses related to it, poorer soil fertility recovery under the shorter fallow phase, and increased intensity of weed competition (cf Chapter 4). These in turn result in reduced crop yield under short jhum cycles (cf Chapter 1; Toky and Ramakrishnan, 1981a; Mishra and Ramakrishnan, 1981).

Hydrology and nutrient losses

Low elevation jhum system

Under the typical jhum as practised at Burnihat (100 m altitude) in Meghalaya, after the slash-and-burn operation, the ash is deposited on the soil surface. The burning of dried slash is usually done a few weeks before the onset of the monsoon in order to ensure that the burning is complete. This period also happens to have strong wind-blow in north-eastern India. Therefore, a considerable proportion of the ash released after the burn is lost from the jhum plot due to wind-blow, this being deposited in the adjoining fallow plots or in the valley lands. In view of the large quantities of ash liberated under a 30-year jhum cycle, the losses also tend to be high, this being 4.2 times more than under a 5-year cycle; the percentage of the total production that was lost from the site declined with shortening of the cycle (Table 5.1). The ash released under a 30-year cycle contained a relatively high concentration and quantity of calcium owing mainly to broad-leaved trees having been slashed, whereas under a 10-year cycle the ash had a preponderance of potassium. This was due to the large bamboo, *Dendrocalamus hamiltonii*, which was the main component of the slash

Table 5.1 Total quantity of nutrients liberated through ash, and the amount lost through blow-off ($kg\,ha^{-1}$), in agroecosystems under various jhum cycles at Burnihat in Meghalaya (from Toky and Ramakrishnan, 1981b)

	Jhum cycle (yr)		
	30	*10*	*5*
Ash			
released	17.4	13.8	6.9
blown off	8.2	8.2	1.9
P			
released	313.0	262.2	150.7
blown off	147.1	155.6	42.7
K			
released	1,739.0	2,070.0	685.0
blown off	817.0	1,228.5	194.0
Ca			
released	956.5	193.2	116.5
blown off	449.4	114.7	33.0
Mg			
released	208.7	151.8	113.7
blown off	98.0	90.1	32.2

Table 5.2 Loss of water from agroecosystems under various jhum cycles and jhum fallows (total rainfall during the monsoon period in 1978 was 142.1 cm) (from Toky and Ramakrishnan, 1981b)

Site	*Run-off water* (cm)	*Percolation water* (cm)	*Sediment loss* ($t\,ha^{-1}\,yr^{-1}$)
Agroecosystem under 5-yr jhum cycle	36.64	22.92	30.05
Agroecosystem under 10-yr jhum cycle	33.94	18.98	23.08
Agroecosystem under 30-yr jhum cycle	29.37	14.35	22.50
5-year fallow	26.90	21.30	1.13
10-year fallow	18.50	14.20	0.76

under a 10-year-old fallow, which was shown to be an accumulator of potassium (Toky and Ramakrishnan, 1982b, 1983b).

Sizeable increases in the volume of run-off and percolation water after the burn are chiefly due to lack of plant cover, which binds soil particles and contributes to utilization of water through uptake and canopy interception (Table 5.2). The loss of water through run-off and percolation, and the consequent loss of sediment increased with the shortening of the

Table 5.3 Total loss of nutrients ($kg\,ha^{-1}\,yr^{-1}$) in run-off and percolation (within parentheses) water under different jhum cycles at Burnihat in Meghalaya (from Toky and Ramakrishnan, 1981b)

Element	*Jhum cycle* (yr)		
($kg\,ha^{-1}\,yr^{-1}$)	*30*	*10*	*5*
NO_3-N	3.7(8.8)	4.2(10.7)	5.3(9.2)
PO_4-P	1.1(0.1)	1.3(0.1)	0.9(0.1)
K	64.7(15.1)	91.2(21.2)	51.0(13.7)
Ca	15.1(5.3)	15.9(4.9)	13.8(4.6)
Mg	6.3(2.5)	5.4(2.1)	9.5(2.3)

jhum cycle. This may partly be related to poor physical characteristics of the soil and partly to poorer crop cover under shorter cycles (Toky and Ramakrishnan, 1981a). The poorer quality of the soil under short jhum cycles may be due to frequent cropping after every 4–5 years. It is particularly interesting to note here that percolation losses of water are often in excess of 50% of that lost through run-off. From a conservational viewpoint, this implies that percolation losses may still be heavy even if erosion is checked through terracing (as has been advocated by some agricultural scientists, as an alternative to jhum).

During the cropping phase, the loss of potassium through run-off and leaching was much greater compared to other cations (Table 5.3). This is because potassium is a very labile element; this confirms the findings of others (Allen, 1964). The exaggerated losses under a 10-year jhum cycle is because of the bamboo (*Dendrocalamus hamiltonii*), the predominant fallow species that is slashed and burnt and which is an accumulator of potassium, as indicated earlier. As discussed for water loss, the loss of nutrients through percolation water is substantial in spite of the steep slope of 30–40° angle on which jhum is done. This has management implications for terracing because the leaching losses here would only be exaggerated.

When fire releases the cations in a single pulse, their availability may exceed the retention capacity of the jhum agroecosystem. Hence many of the cations are lost through sediment and water before they could be incorporated into the plant biomass. The large amount of cations released after the burn increases the soil pH which in turn may promote nitrification (Ahlgren and Ahlgren, 1965; Wells, 1971). This, together with an increased concentration of bicarbonate ions, results in an increase of anions, which are balanced by free cations so that the rate of nutrient loss through water may increase (Bormann *et al.*, 1968; Lewis, Jr., 1974). During the early months of the monsoon, heavy losses of cations through run-off are due to heavy nitrate losses, and the high volume of water loss is because of poor plant cover (Fig. 5.1). Later in the season, the nutrient losses are halted to some extent by an effective vegetation cover of crops and associated weeds.

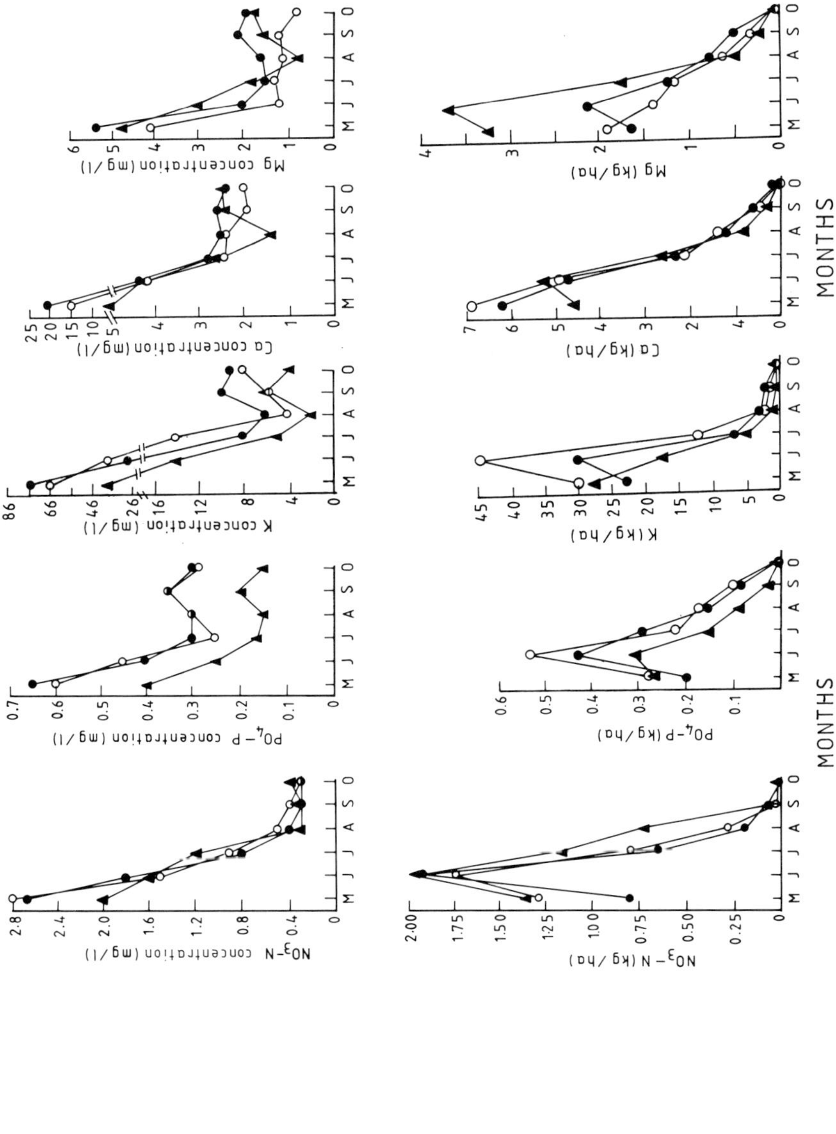

Figure 5.1 Changes in the concentration and quantities of nutrients lost through run-off water from jhum plots at Burnihat in Meghalaya, during the monsoon. ●, 30-yr; ○, 10-yr; and ▲, 5-yr jhum cycles (from Toky and Ramakrishnan, 1981b)

Table 5.4 Concentration and total loss of nutrients during the monsoon period in run-off and percolation water under jhum fallows at Burnihat in Meghalaya (from Toky and Ramakrishnan, 1981b)

	Element				
	NO_3-*N*	PO_4-*P*	*K*	*Ca*	*Mg*
Concentration (mg/1)					
5-yr fallow					
run-off water	0.30	0.05	0.35	0.75	0.50
percolation water	0.56	0.01	0.25	1.25	0.40
10-yr fallow					
run-off water	0.25	0.03	0.90	0.60	0.45
percolation water	0.35	0.01	0.15	1.10	0.35
Total loss (kg/ha/yr)					
5-yr fallow					
run-off water	0.807	0.134	0.941	2.071	1.345
percolation water	1.065	0.021	0.532	2.662	0.852
10-yr fallow					
run-off water	0.462	0.055	1.665	1.110	0.832
percolation water	0.497	0.014	0.213	1.562	0.497

A similar pattern for percolation losses during the monsoon was also noted (Toky and Ramakrishnan, 1981b).

The biogeochemical recovery of a forested ecosystem depends upon the re-establishment of biotic regulation of ecosystem functions such as uptake of nutrients, mineralization, release of nutrients and withstanding erosion. Under jhum fallows at Burnihat, rapid recovery of these characteristics occurs, owing to favourable temperature, moisture and nutrient conditions in the soil. The concentration of different nutrients in run-off and percolation water was markedly lower under 5- and 10-year-old fallows (Table 5.4), compared with jhum agroecosystems (cf Fig. 5.1), as a result of which the total losses from the fallows were also low. Thus, the losses through water from jhum under a 10-year cycle was about 15-fold for NO_3-N, 20-fold for PO_4-P, 60-fold for potassium, 8-fold for calcium and 6-fold for magnesium, as compared to 10-year-old fallow. This rapid reduction in losses in the fallows may be accounted for by interception of rain by the vegetation, a lowered rate of nitrification and the low concentration of nutrients in the soil pool resulting from their rapid uptake by the fast developing plant cover. Further, ecosystems with nutrient-saturated high exchange-capacity soils would lose relatively more of their nutrients than those with nutrient-unsaturated and/or low exchange-capacity soils (Jordan *et al.*, 1972). The latter is the case for most of the soils of north-eastern India, such as that at Burnihat.

Modified version of jhum of higher elevations

In the modified version of jhum at Shillong (1,500 m altitude), in spite of the land being prepared into alternating ridges and compacted furrows running down the slope, and the slash on ridges being subjected to a controlled burn, the loss of sediment and nutrient load through run-off water alone was substantial (Table 5.5). This is particularly obvious for a labile element such as potassium when this is compared to the losses under the typical version of jhum of lower elevations. If to this is added the nutrient removal through the sediment itself and through percolation water (Mishra and Ramakrishnan, 1983c), then the losses become very high compared to the low elevation jhum system (typical version) discussed earlier. Here again, the higher quantities of sediment and nutrient losses under a 5-year jhum cyle could be related to poorer physical quality of the soil under more frequent cropping and the resultant lesser crop cover, compared to a 10-year cycle (Mishra and Ramakrishnan, 1981). During the second year of cropping under a 5-year jhum cycle, the losses increased further which only strengthens the argument given above. It is also interesting to see a drastic reduction in sediment and nutrient losses during the first year of the fallow phase, which is indicative of the conservation role of weedy communities developing soon after the cropping phase.

This conservation role of weeds during the cropping phase is significant and this is discussed elsewhere under the 'non-weed' concept in Chapter 4. Suffice to say here that retention of about 20% weed biomass, as is done under traditional weeding by the jhum farmer, would reduce losses through run-off during the cropping phase by about 20%, when compared with total weeding (Swamy and Ramakrishnan, 1988d). Conservation of soil and nutrients during the cropping phase, even to this extent is significant because heavy losses from frequent jhum operations often lead to site desertification (Ramakrishnan 1984a, 1985b).

Soil fertility and nutrient budget under jhum agroecosystem and fallows

The jhum procedures involving low or high intensity burning, which depends on the quantity of slash under different jhum cycles, brings about a number of changes in the physico-chemical properties of the soil (Ramakrishnan and Toky, 1981; Mishra and Ramakrishnan, 1983d). This is known to alter the availability of some of the nutrients important for plant growth (Ahlgren and Ahlgren, 1960; Nye and Greenland, 1960). It also affects the amount of water passing through the soil sub-system, dissolved substances, and the particulate matter lost from the top soil as discussed earlier (Toky and Ramakrishnan, 1981b; Mishra and Ramakrishnan, 1983c). During the cropping phase under jhum, nutrients are taken

Table 5.5 Sediment and nutrient losses through run-off water under the modified version of jhum at Shillong in Meghalaya (from Mishra and Ramakrishnan, 1983c)

	Jhum cycle		*Age of the jhum fallow* (yr)		
	10-yr	*5-yr*	*1*	*5*	*10*
Sediment ($t\,ha^{-1}\,yr^{-1}$)	49.7 ± 1	54.9 ± 1.1 (56.3 ± 1.3)	7.4 ± 0.2	3.5 ± 0.1	1.9 ± 0.1
Nutrients ($kg\,ha^{-1}\,yr^{-1}$)					
NO_3–N	1.7 ± 0.02	2.3 ± 0.02 (2.6 ± 0.02)	1.3 ± 0.002	1.0 ± 0.01	0.5 ± 0.01
PO_4–P	0.9 ± 0.01	0.8 ± 0.01 (1.1 ± 0.01)	ND	ND	0.04 ± 0.003
K	88.1 ± 0.2	104.6 ± 0.4 (110.0 ± 0.4)	26.0 ± 0.1	19.6 ± 0.1	2.3 ± 0.1
Ca	10.3 ± 0.1	10.6 ± 0.1 (12.7 ± 0.1)	3.6 ± 0.04	2.2 ± 0.03	0.3 ± 0.01
Mg	3.1 ± 0.03	3.6 ± 0.03 (2.0 ± 0.02)	ND	ND	ND

Values in parentheses indicates losses during the second year of croping
ND = Not detectable

Table 5.6 Changes in soil pH under jhum fallows of various ages at Burnihat in Meghalaya (from Ramakrishnan and Toky, 1981)

Soil depth (cm)	*Age of the fallow* (yr)					
	0	*1*	*5*	*10*	*15*	*50*
0–7	6.0	6.5	6.1	5.9	6.1	5.7
7–14	5.8	6.0	5.6	5.6	5.5	5.8
14–28	5.7	5.8	5.4	5.3	5.6	5.7
28–40	5.3	5.4	5.3	5.2	5.4	5.8

up by crops and weeds, some of which are recycled back into the system, and substantial quantities are removed through crop harvest and weed removal from the plot (Mishra and Ramakrishnan, 1984; Swamy and Ramakrishnan, 1988c). Some nutrient input may also come through precipitation and activities related to soil organisms (Saxena and Ramakrishnan, 1986), apart from the major input coming through slash burning. The net consequence of these input/output events is often a net loss from the system and a decline in soil fertility at the end of the cropping period. This decline would, to a large extent, depend upon the length of the jhum cycle. Since soil fertility during the cropping phase is closely linked up with soil fertility patterns during the following fallow phase, we shall consider the two together in this discussion.

An important change that would occur as a consequence to burning the slash is a rapid increase in soil pH (Table 5.6) with implications on soil biological activities and nutrient availability. This increase in pH, being related to the ash deposition, is more marked under longer jhum cycles. Further, the effect is more pronounced in the surface soil. Such an increase in soil pH is significant for these otherwise acid soils (Ramakrishnan and Toky, 1981), particularly when the soil pH is normally in the range of 5–5.5 as at higher elevations, under pine forests (Mishra and Ramakrishnan, 1983c).

Soil carbon

The significant reduction in soil carbon content immediately after fire under 30- and 10-year jhum cycles at Burnihat could be attributed to high temperatures (Fig. 5.2), an observation also recorded for longer jhum cycles at higher elevations (Mishra and Ramakrishnan, 1983d). Under a 5-year jhum cycle, however, the burn is also of low intensity since the quantity of slash is much less, and therefore the surface soil humus is less affected. The site under longer jhum cycles always starts with higher carbon and the high intensity fire depletes it rapidly in the top soil.

At the time of cultivation, loss of organic matter from the soil due to decomposition is likely to be faster due to high insolation and also due to

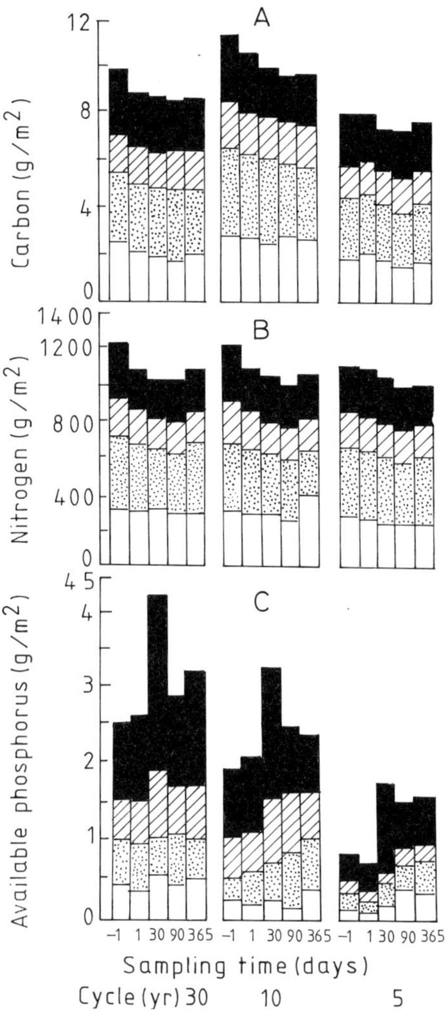

Figure 5.2 Changes in total quantity of carbon (A), nitrogen (B), and available phosphorus (C), within a soil column of 40 cm depth a day before burn (−1) and after burn during cropping under 30-, 10- and 5-yr jhum cycles at Burnihat in Meghalaya. Dark column, 0–7 cm; hatched column, 7–14 cm; stippled column, 14–28 cm; open column, 28–40 cm soil depths (from Ramakrishnan and Toky, 1981)

rapid surface run-off. After an initial period of depletion of organic matter during cultivation, there is a period of recovery of humus in the soil, derived from slash of the harvested crops and weeds. Such an input of crop biomass is spread over a period of time between July and December, because of sequential harvest, which would ensure a steady supply of organic matter and its effective decomposition. The crop biomass thus cycled during one cropping season could be as high as about 8,200 $kg\,ha^{-1}\,yr^{-1}$ of crop biomass (Toky and Ramakrishnan, 1981a) and about 6,900 $kg\,ha^{-1}\,yr^{-1}$ of weed biomass (Ramakrishnan, 1984a) for the lower elevation jhum system of Meghalaya, under a 10-year cycle. Addition of organic and inorganic manure could be a further source of nutrients as at higher elevations of Meghalaya (Mishra and Ramakrishnan, 1981, 1984).

Because of higher levels of organic content maintained after a year of cropping, sites under long jhum cycles may be able to sustain another year of cropping. However, at the end of the cropping period, there was a net loss of carbon compared to that before the burn, an observation also made by other workers (Nye and Greenland, 1964; Zinke, *et al.*, 1978). Juo and Lal (1977) showed that under shifting agriculture in west Nigeria, due to a fast rate of decomposition of organic matter under continuous cropping, 16 $t\,ha^{-1}\,yr^{-1}$ of dry plant material is required to maintain soil organic matter in the surface soil at a level comparable to soil under secondary forest. The depletion of organic matter depends upon the intensity of cropping, type of fallow vegetation and the ratio of cropping to the fallow period. With an optimum cropping and fallow period, the humus in the soil could be maintained at a relatively high level even after many years of shifting cultivation (Coulter, 1950; Birch and Friend, 1956).

The depletion of organic matter in the surface soil continued through the early successional fallows up to about 5 years (Fig. 5.3), as shown at Burnihat. This was mainly due to the low return of litter during the initial colonization and growth of the vegetation in the fallows dominated by species such as *Imperata cylindrica* and *Eupatorium odoratum*, and also due to faster decomposition in these open sites. This trend in reduction of soil humus in up to 5-year fallow regrowth could be one of the reasons that could be held against a short jhum cycle which has become so common in the present times. Maximum level of soil humus is reached under a 10-year-old fallow, with a slightly lower level maintained under a 50-year-old fallow. Laudelot and Meyer (1954) and Ewel (1976) have made similar observations and concluded that organic matter production in a mature stand is less compared to a secondary seral forest.

Similar patterns to those observed during the cropping and fallow phases at lower elevations at Burnihat was also observed for the high elevation modified version of the jhum system at Shillong except that the carbon levels in the soil were much lower (Mishra and Ramakrishnan, 1983d). This is to be expected under the lower temperature conditions of higher elevations where the pine litter decomposition is a slower process, the pine trees themselves are sparsely distributed, and the soils are more acidic (Mishra and Ramakrishnan, 1983a).

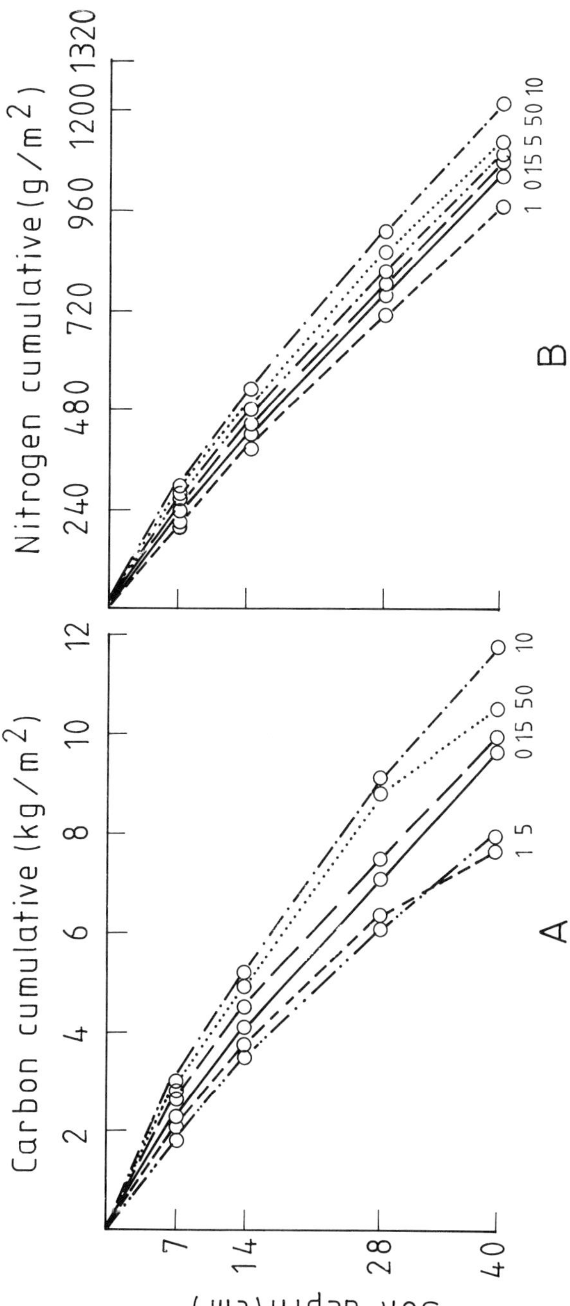

Figure 5.3 Changes in cumulative quantity of carbon (A) and nitrogen (B) within a soil column of 40 cm depth under 0-, 1-, 5-, 10- and 50-yr jhum fallows (from Ramakrishnan and Toky, 1981)

Soil nitrogen

Similar to the pattern for humus in the surface soil, total soil nitrogen declined sharply after the burn under longer jhum cycles compared to a short 5-year cycle at Burnihat (Fig. 5.2). This is partly due to low nitrogen build-up in the fallows before the burn under the 5-year cycle and partly due to the low intensity burn. Under longer cycles, the build-up of nitrogen was higher in the soils at the end of the fallow period and the soil experiences a high intensity burn resulting in greater loss of nitrogen through volatilization from the surface soil. A decline in soil nitrogen after the burn has been observed by others as well (Wells, 1971; White *et al.*, 1973) and could be attributed to the conversion of organic nitrogen to volatile forms during the pyrolysis (Allen, 1964; Knight, 1965; Debell and Ralston, 1970). Obviously, the degree of volatilization would depend upon the intensity of the burn.

Although there was a net loss of nitrogen from all sites at Burnihat after a year of cropping compared to that before the burn, longer jhum cycle sites had higher nitrogen status compared to that under a 5-year cycle and therefore, could probably sustain cropping for a longer duration. This agrees with the observation of Nye and Greenland (1960) that response to nitrogen is small where fallows are long (10 years or more), but in land more intensively cultivated with shorter fallows, the responses become larger. Such a conclusion is further supported by our nitrogen budget studies (Mishra and Ramakrishnan, 1984; Swamy and Ramakrishnan, 1988c).

The decline in nitrogen level up to 5 years during fallow regrowth could be partly due to low litter production and rapid utilization of nitrogen by the fast growing plant cover (Fig. 5.3). As for carbon, nitrogen level in the soil also reached its peak in a 10-year-old fallow. The humus level in the soil generally remained low due to its faster rate of mineralization. Only a fraction of the organic matter is incorporated into the soil (Nye and Greenland, 1960) and this accounts for the low C/N ratio during all phases of the jhum cycle (Table 5.7).

Nitrogen budget under jhum

The nitrogen budget of the jhum system (15-, 10- and 5-year cycles of the Khasi tribe at Shillong (1,500 m) in Meghalaya is indicative of the issues involved in nitrogen fertility maintenance during the cropping phase, under different jhum cycles (Mishra and Ramakrishnan, 1984). The soil nitrogen budget in the preburn and the postburn stages indicate (Table 5.8) that nitrogen addition through slash was higher, nitrogen in the soil pool was higher and the losses through volatilization was also higher, under longer jhum cycles.

Table 5.7 Changes in per cent carbon and nitrogen and C:N ratio in the soil profile under jhum fallows of various ages at Burnihat in Meghalaya (from Ramakrishnan and Toky, 1981)

	Fallow age (yr)					
Soil depth up to 40 cm	*0*	*1*	*5*	*10*	*15*	*50*
0–7 cm						
C	2.08	2.01	1.82	2.75	2.50	2.60
N	0.221	0.217	0.220	0.268	0.246	0.236
C:N	9.1	9.2	8.6	10.0	10.1	11.0
7–14 cm						
C	1.80	1.53	1.48	2.14	1.73	2.05
N	0.172	0.159	0.200	0.240	0.178	0.215
C:N	10.4	9.4	7.0	8.7	9.5	9.5
14–28 cm						
C	1.36	1.12	1.21	1.78	1.35	1.76
N	0.154	0.142	0.170	0.170	0.148	0.180
C:N	8.3	7.7	7.6	10.0	8.7	9.8
28–30 cm						
C	1.37	0.81	0.94	1.42	1.18	0.86
N	0.158	0.138	0.150	0.162	0.145	0.116
C:N	8.6	5.8	6.0	8.6	7.5	7.4

Table 5.8 Loss of nitrogen (kg ha^{-1}) through fire with SE values, under jhum at Shillong in Meghalaya (from Mishra and Ramakrishnan, 1984)

Jhum cycle (yr)	*Preburn soil pool* (a) ($\times 10^3$)	*Addition through slash* (b)	*Total in soil pool before burning* (a + b) $\times (10^3)$	*Post pool soil pool* (c) ($\times 10^3$)	*Loss* (a + b) – (c)
15	7.68 ± 0.11	43.6 ± 0.6	7.73 ± 0.11	7.22 ± 0.11	510.2 ± 14
10	7.74 ± 0.09	38.6 ± 0.9	7.78 ± 0.09	7.32 ± 0.13	462.1 ± 22
5	6.40 ± 0.12	29.3 ± 0.7	6.43 ± 0.13	6.17 ± 0.10	262.8 ± 16

Table 5.9 indicates major aspects of the input/output nitrogen budget for three jhum cycles. Although slash is shown as an input, the nitrogen volatilized through fire is shown as an output. Organic matter forms a major input under 10- and 5-year cycles. Inorganic fertilizer is a small fraction of the total input under a 5-year cycle only. The crop by-products ploughed back decreased with shortening of the jhum cycle, since this biomass declined under shorter cycles (Mishra and Ramakrishnan, 1981). The organic manure addition largely comes from the agriculture and animal husbandry wastes via the compost pit of the village ecosystem (Mishra and Ramakrishnan, 1982). The component ploughed back through weed increased with decrease in the jhum cycle. This is related to the increased weed potential of the site under shorter cycles (Saxena and Ramakrishnan, 1984b; cf Chapter 4). Thus the transfer of nitrogen from the crop to the

Table 5.9 Nitrogen input:output budget (kg ha^{-1} yr^{-1}) under jhum at Shillong in Meghalaya (from Mishra and Ramakrishnan, 1984)

	15-yr jhum cycle	*10-yr jhum cycle*	*5-yr jhum cycle*	
			1st yr crop	*2nd yr crop*
Inputs				
Precipitation	3.6	3.6	3.6	3.6
Slash	43.6	38.6	29.3	–
Organic manure	–	8.4	14.0	25.9
Inorganic fertilizer	–	–	0.7	1.4
Weed ploughed back	0.8	2.6	3.5	3.4
By-products ploughed back	0.8	0.6	0.6	0.3
Total	48.8	53.8	51.7	34.6
Outputs				
Fire	510.2	462.1	262.8	–
Sediment	119.1	128.5	172.9	176.3
Run-off	13.0	8.2	8.9	10.0
Percolation	0.9	0.5	0.8	0.7
Weed removal	4.0	13.0	17.3	16.8
Crop removal	43.3	31.3	19.7	9.7
Total	690.5	643.6	482.4	213.5
Net differences	642.7	589.8	430.7	178.9

– represents absence of that input/output from the system

Table 5.10 Net change of nitrogen (10^3 kg ha^{-1} yr^{-1}) in the soil under jhum at Shillong in Meghalaya (from Mishra and Ramakrishnan, 1984)

	15-yr fallow cycle	*10-yr fallow cycle*	*5-yr fallow cycle*	
			1st yr crop	*2nd yr crop*
Soil pool before burning	7.68	7.74	6.40	5.98
Soil pool at the end of cropping	7.04	7.15	5.98	5.60
Net difference	0.64	0.59	0.42	0.38

non-crop (weed) compartment, then back again, becomes an important mechanism for optimum utilization of this labile element. This is apart from the role of standing weed biomass of about 20% under traditional weeding regimes, in conserving soil nitrogen and other elements through good plant cover (Swamy and Ramakrishnan, 1988c). This value of weeds in tropical agriculture by providing a high degree of integrated management capability has already been discussed in Chapter 4. It may be noted here that the input/output analysis has not considered at least three aspects, namely, nitrogen fixation, mineralization and denitification. While the net differences may change if these aspects are considered, the conclusions may not be very different.

The net change in the nitrogen pool (Table 5.10) suggests that the shortening of the jhum cycle results in a lower nutrient capital at the preburn stage as well as at the end of the cropping period. The loss of nitrogen at the end of 1 year of cropping under the 5-year jhum cycle is higher only if the values are put on a time scale of 15 years: these then would lose three times more than that from a plot under a 15-year cycle and twice that from a plot under a 10-year cycle.

If the plots under a 5-year jhum cycle now, have longer fallow cycles preceeding the 20-year period, for which land use history at the Shillong site was available (Mishra and Ramakrishnan, 1984), then the system seems to have lost, in terms of initial capital, about 1.28×10^3 kg of nitrogen (7.68–6.40×10^3 kg) during the last 20 years. Thus, while 15- or 10-year jhum cycles are long enough to restore the original nitrogen status in the soil before the next cropping, it seems unlikely that about 600 kg of nitrogen lost (the difference between the nitrogen capital before and after two croppings) could be restored under a 5-year cycle. One of the important disadvantages of the 5-year jhum cycle lies in the reduced nitrogen capital with which the agroecosystem has to operate due to increased frequency of fire and cropping with too short a fallow phase. Similar conclusions also arise for the jhum systems at lower elevations, at Lailad in Meghalaya (Swamy and Ramakrishnan, 1988c).

Nitrification under jhum

Prediction of nitrification rates on the basis of nitrate or ammonium pool size and nitrifier populations which have been a general practice in ecological studies, may sometimes be misleading. Consistently low levels of nitrate in soil may not essentially imply low rates of nitrification, as nitrate ions are readily absorbed by the existing plants or are leached out. Thus both nutrient pool size and nitrifying activity are important for a precise evaluation of nitrogen during ecosystem development (Robertson and Vitousek, 1981). Further, although nitrifier population size may be strongly correlated with nitrification rates, the reliability of MPN (most probable number) counts has been suspect (Belser and Schmidt, 1978). In this analysis, we shall consider the temporal variations in this activity during the cropping and fallow phases of jhum under varied jhum cycles.

Drastic changes in the physico-chemical properties of the soil after slash-and-burn would obviously influence the structure and function of the existing soil microbial communities. High rates of nitrification after slashing the vegetation is recognized as due to more favourable microenvironmental conditions (Smith, *et al.*, 1968). During the burn, although ash release occurs and soil pH increases, the thin-walled nitrifying bacteria are killed between 35 to 58°C. The period of inhibition of nitrification following the burn (Fig. 5.4) depends upon fire intensity and subsequent environmental conditions affecting the recovery of the nitrifier population (Raison, 1979).

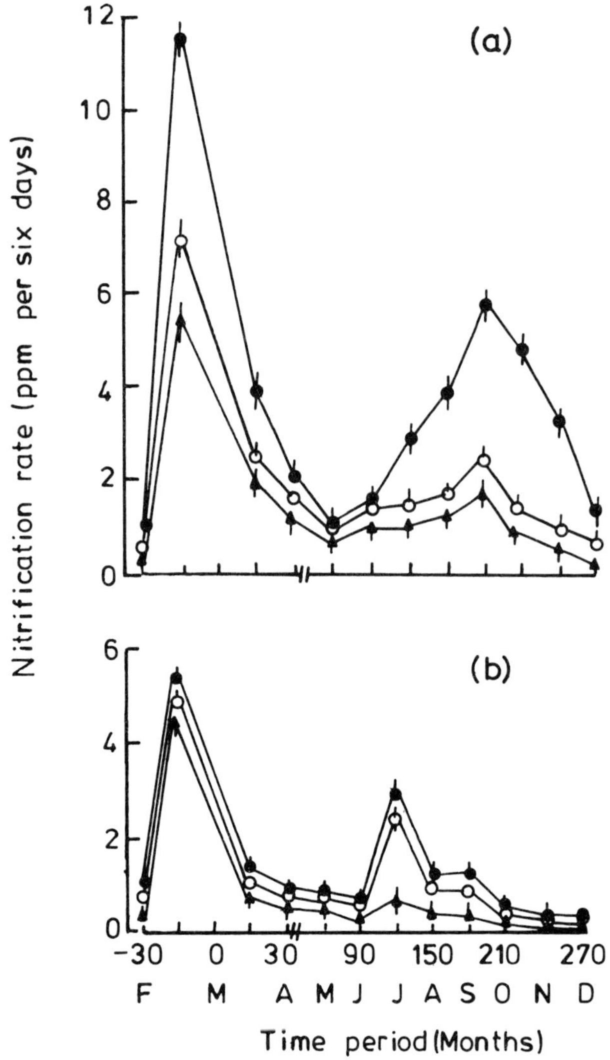

Figure 5.4 Changes in nitrification rate during cropping under different jhum cycles (a) and under jhum fallows (b) at Burnihat in Meghalaya. Time period in days after the burn, −1 being a day before the burn and O being the day of the burn. ●, 1-yr; ○, 6-yr and ▲, 10-yr (from Saxena and Ramakrishnan, 1986)

On the other hand, heterotrophic activity becomes stimulated after heating treatments (Simmon-Sylvestre, 1967). Nitrifiers, therefore, would be more adversely affected by the heating effects of fire when compared to heterotrophs. Thus, although soil physio-chemical properties become more favourable for nitrifiers following the burn (Woodmansee and Wallach,

1981), the low residual viable nitrifier population could, perhaps, be responsible for a steep decline in nitrification soon after the burn.

A decrease in nitrification with increasing jhum cycle is possibly related to the structural variation in the microbial communities before the burn and the differential fire regimes. The increase in heterotrophic potential during seral development would reduce ammonium availability to nitrifiers so that nitrification under a longer cycle would automatically start at a lower level compared to that under a shorter cycle. This hypothesis which needs further testing, along with the fire intensity affecting microbial population size, would explain the differences between different jhum cycles.

A gradual decrease in nitrification rate with increasing fallow age during secondary succession may account for a similar trend in NO_3-N concentration in the soil but with a reverse trend for NH_4-N levels (Fig. 5.5). Similar findings also reported elsewhere (Todd *et al.*, 1975) may be associated with a progressive shift from a less to a more intense heterotroph–nitrifier interaction during ecosystem development rather than any change in allelopathic potential as envisaged by Rice and Pancholy (1974). Deka and Mishra (1982) reported that heterotrophic microbial populations increase during seral development following jhum, which is also apparent from the higher organic carbon content in the older fallows. Low competitive vigour of nitrifiers compared to heterotrophs (Jones and Richardson, 1977) would lead to their low population levels in older fallows resulting in low nitrification potential.

On the other hand, if the hypothesis of Rice and Pancholy (1974) that allelopathic inhibition of nitrification increases during succession holds true, then the differences in nitrogen mineralization potential amongst different fallow fields should be consistent even in the long-term incubation studies of soil supplied with additional ammonium sulphate (30 ml of 0.01 mol/l ammonium sulphate solution per 100 g of soil). Contrarily, a rapid conversion of added ammonium (Table 5.11) irrespective of successional maturity of the fallow suggests that nitrifiers are adequately present and it is the availability of ammonium ions which is more crucial for the observed trend instead of any allelochemical inhibition. This observation is further strengthened by the results of nitrogen mineralization in soils mixed with leaf material of different species (*Eupatorium odoratum* alone is discussed here). In these treatments, nitrification was arrested in the initial stages of incubation (8–16 days) probably because of nitrogen immobilization by the more vigorous heterotrophs on the added organic matter. Gradually, where the heterotrophic activity decreases owing to carbon mineralization and the release of ammonium ions as a by-product of decomposition, nitrifiers are activated which is reflected in a sharp increase in NO_3-N levels after 24 hours of incubation. The possibility of allelopathic inhibition of nitrification could also be ruled out in view of maximum nitrifying activity

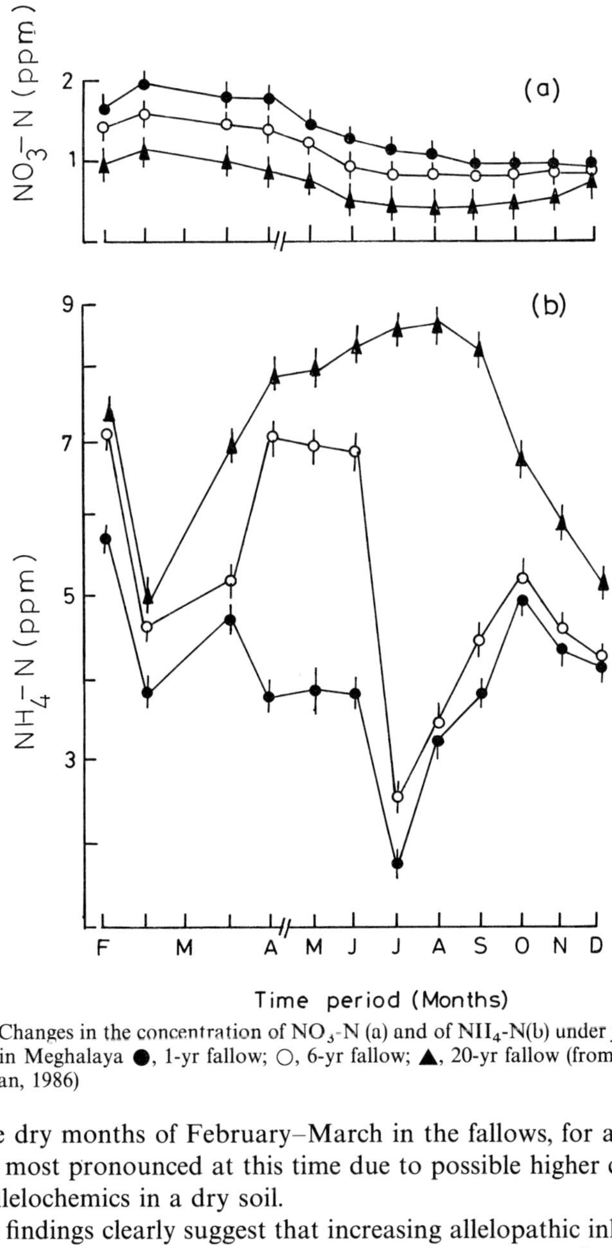

Figure 5.5 Changes in the concentration of NO_3-N (a) and of NH_4-N(b) under jhum fallows at Burnihat in Meghalaya ●, 1-yr fallow; ○, 6-yr fallow; ▲, 20-yr fallow (from Saxena and Ramakrishnan, 1986)

during the dry months of February–March in the fallows, for allelopathy should be most pronounced at this time due to possible higher concentrations of allelochemics in a dry soil.

Present findings clearly suggest that increasing allelopathic inhibition of nitrification as a mechanism for improving nutrient conservation efficiency during ecosystem development might have not evolved in all the environments. Competitive interference by heterotrophs, as discussed here, may be more important in the hot and humid climates, as also suggested by other

Table 5.11 Increase in the concentration of nitrate–nitrogen during incubation of soils from different sites mixed with ammonium sulphate or leaf material, under jhum at Burnihat in Meghalaya (from Saxena and Ramakrishnan, 1986)

Sites and treatments	*Concentration of nitrate–nitrogen* (ppm) (days of incubation)			
	8	*16*	*24*	*32*
Control				
4-year (J)	2.00	4.51	11.69	20.88
6-year (J)	1.90	4.20	10.02	19.42
10-year (J)	1.80	3.20	8.35	17.54
1-year (F)	1.98	3.74	6.21	11.42
3-year (F)	1.86	3.64	5.92	9.46
20-year (F)	1.25	3.55	5.02	5.45
Ammonium sulphate				
4-year (J)	13.36	26.72	36.90	51.77
6-year (J)	8.98	19.60	29.42	47.08
10-year (J)	7.56	18.43	26.72	45.42
1-year (F)	8.02	19.72	30.24	48.80
3-year (F)	5.34	17.20	24.52	45.92
20-year (F)	4.62	15.03	18.87	42.43
Leaf material of *Eupatorium*				
4-year (J)	5.12	8.60	29.25	38.80
6-year (J)	5.98	9.80	31.10	40.44
10-year (J)	6.51	9.92	39.30	42.45
1-year (F)	1.67	6.01	26.15	40.46
3-year (F)	2.24	12.53	18.35	46.42
20-year (F)	1.80	16.37	40.08	48.42

(J), jhum cycle : cropping sites; (F), jhum fallows

investigations (Lamb, 1980; Vitousek, *et al.*, 1982; Ramakrishnan and Saxena, 1984). The present results also suggest that soil inorganic nitrogen pools may not always be indicative of the nitrogen mineralization rates in all situations as has often been implicated in ecological studies.

Nitrogen is a critical element for plant growth and is often limiting in terrestrial ecosystems. Under the jhum system of north-eastern India, under high rainfall conditions of a monsoonic climate, with cropping done on steep slopes and under drastically reduced jhum cycles of 4–5 years, this is even more critical. The importance of intact vegetation in nitrogen conservation is evident from what is discussed here. Indeed, a minimum of 10 years of fallow period is required for adequate recovery of nitrogen fertility.

Soil phosphorus

At lower elevations, at Burnihat in Meghalaya, the increase in available phosphorus after the burn was slight (Ramakrishnan and Toky, 1981) (Fig.

5.2). The rapid build-up of available phosphorus level, a month after the burn could be due to : (1) release from the ash after the rainfall, and (2) increase in soil pH and the consequent increase in microbial activity (Ahlgren and Ahlgren, 1965), which may result in rapid mineralization of the residual humus. This build-up of available phosphorus was maximum under longer jhum cycles and minimum under a 5-year cycle. This may be related to the amount of slash, intensity of the burn and the consequent lower pH level and the generally low humus content under a shorter cycle. The sharp decline in available phosphorus 3 months after the burn may partly be due to run-off losses before the plant cover is established and removal by the developing crops and weeds. After a year of cropping, the available phosphorus level was as that, or sometimes even slightly higher than, before the burn (Nye and Greenland, 1960; Zinke *et al.*, 1978).

While the general pattern for the available phosphorus budget in the soil under the modified version of jhum at Shillong in Meghalaya remained the same (Mishra and Ramakrishnan, 1983d) as that discussed above, a few major differences were noticed. The decline in phosphorus during the cropping phase was such that the level was lower at the end of 1 year cropping than that at the preburn stage, which is at variance from the low elevation jhum system. This may partly be related to phosphorus fixation into unavailable forms reported for soils rich in sesquioxides of iron and aluminium in their clay fraction (Gebhardt and Coleman, 1974; Tinker, 1977). Another major difference was related to the significant decline in soil phosphorus levels soon after the the burn.

Some workers have reported no significant effect of the burn on phosphorus levels (Allen 1964; Viro, 1974) or even an increase after the burn (Nye and Greenland, 1960; Stark, 1971). Although there are no obvious mechanisms for volatilization of phosphorus, substantial losses of this element were reported through many studies, due to burning (Lloyd, 1971; Ashton, 1976) which agrees with the present results obtained for a modified version of jhum at high elevations (Mishra and Ramakrishnan, 1983d). Although our studies at lower elevations, at Burnihat in Meghalaya, (Ramakrishnan and Toky, 1981) did not show any obvious phosphorus losses from the system due to fire, the study in the same general area at Lailad did show some loss occuring through fire (Swamy and Ramakrishnan, 1988c) (Table 5.12). This loss was more under an intense burn of a longer 20-year cycle, although the available phosphorus level itself was lower to start with under the shorter cycle.

During early phases of development of vegetation up to 5-year fallow regrowth, there is a gradual depletion in available phosphorus, at Burnihat, throughout the soil profile, reaching a minimum after 5 years (Fig. 5.6). This is followed by a rapid build-up through 10-, 15- and 50-year-old fallows. The accumulation of phosphorus, particularly in the surface soil

Table 5.12 Loss of total nitrogen and available phosphorus (kg ha^{-1}) through fire ($\pm$ SE) under 20- and 5-year jhum cycles at Lailad in Meghalaya (from Swamy and Ramakrishnan, 1988c)

	20-jr jhum cycle		*5-yr jhum cycle*	
	Nitrogen	*Phosphorus*	*Nitrogen*	*Phosphorus*
Preburn soil pool (a) ($\times 10^3$)	10.09 $\pm$ 0.12	0.024 $\pm$ 0.001	10.45 $\pm$ 0.15	0.007 $\pm$ 0.007
Addition through slash (b)	489.0 $\pm$ 13.0	64.0 $\pm$ 3.0	26.0 $\pm$ 0.0	6.0 $\pm$ 0.01
Total in soil pool before burning (a + b) ($\times 10^3$)	11.43 $\pm$ 0.16	0.088 $\pm$ 0.01	10.48 $\pm$ 0.16	0.013 $\pm$ 0.006
Postburn soil pool (c) ($\times 10^3$)	9.77 $\pm$ 0.11	0.023 $\pm$ 0.001	9.93 $\pm$ 0.12	0.006 $\pm$ 0.002
Loss (a + b) – (c)	1,660.0 $\pm$ 2.0	63.0 $\pm$ 3.0	550.0 $\pm$ 27.0	7.2 $\pm$ 0.001

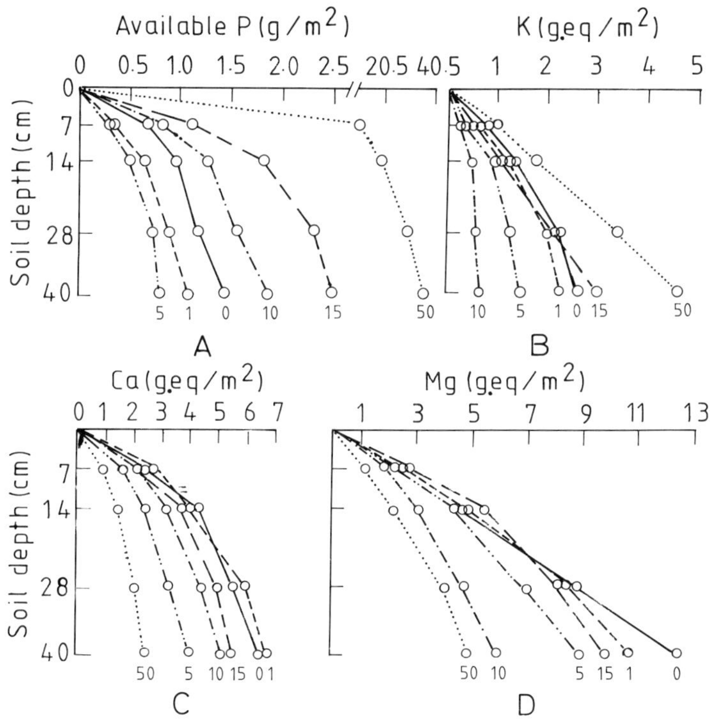

Figure 5.6 Changes in cumulative quantity of available phosphorus (A), potassium (B), calcium (C), and magnesium (D) within a soil column of 40 cm depth under 0-, 1-, 5-, 10-, 15- and 50-yr old jhum fallows (g.eq/m^2 = equivalent g/m^2) (from Ramakrishnan and Toky, 1981)

layers of older forests may be due to transfer of phosphorus from the deeper layers of the soil to the upper stratum through litterfall (Nye and Bertheux, 1957).

Table 5.13 Net change of soil nitrogen and available phosphorus (kg ha^{-1}) under jhum at Lailad in Meghalaya (from Swamy and Ramakrishnan, 1988c)

	20-yr jhum cycle		*5-yr jhum cycle*	
	Nitrogen	*Phosphorus*	*Nitrogen*	*Phosphorus*
Soil pool before burning ($\times 10^3$)	10.9	0.02	10.45	0.0073
Soil pool at the end of cropping ($\times 10^3$)	9.78	0.032	9.91	0.012
Net change	1.12	0.012	0.54	0.0047

Phosphorus budget under jhum

The budget analysis of available phosphorus for the jhum system at Lailad (Table 5.13) indicates that the soil sub-system lost about 17 kg of available phosphorus: ($0.024–0.0073 \times 10^3$ kg ha^{-1}) during a 20-year period under a 5-year cycle (cf also Table 4.19 of Chapter 4 for a budget analysis for phosphorus). Such losses of phosphorus and nitrogen discussed earlier, under continuous imposition of short jhum cycles on the same sites have serious implications for site desertification (Ramakrishnan, 1985b,d,e; Ramakrishnan and Saxena, 1984).

Soil cations

There are four sources of basic cations in the upper layer of the soil: from weathering of parent material, decay of soil humus, decay of surface organic matter and ash from burning the slash. The importance of fire under jhum is seen from the fact that there is a quick release of cations after the burn on the surface layer of the soil (Nye and Greenland, 1960; Zinke, *et al.*, 1978; Ramakrishnan and Toky, 1981; Mishra and Ramakrishnan, 1983d).

Immediately after burning the slash, the concentration of all cations in the surface soil increased (Ramakrishnan and Toky, 1981; Mishra and Ramakrishnan, 1983d; Swamy and Ramakrishnan, 1988c). This is reflected in the changes in the total quantity of potassium, calcium and magnesium, as shown through one of the studies at Burnihat in Meghalaya (Fig. 5.7). The greater quantity of cations released after the burn under longer jhum cycles is understandable due to the larger quantities of slash that were burnt compared to that under a 5-year cycle.

Qualitative differences were observed in cation release after the burn, depending upon the jhum cycle. The amount of potassium and calcium released under a 30-year jhum cycle was fairly high due to high accumulation of calcium in the dicot trees and potassium in the common bamboo at Burnihat, namely, *Dendrocalamus hamiltonii*; this bamboo is an accumulator of potassium (Toky and Ramakrishnan, 1982b, 1983b). With this bamboo species being also a predominant component in a 10-year-old fallow, the

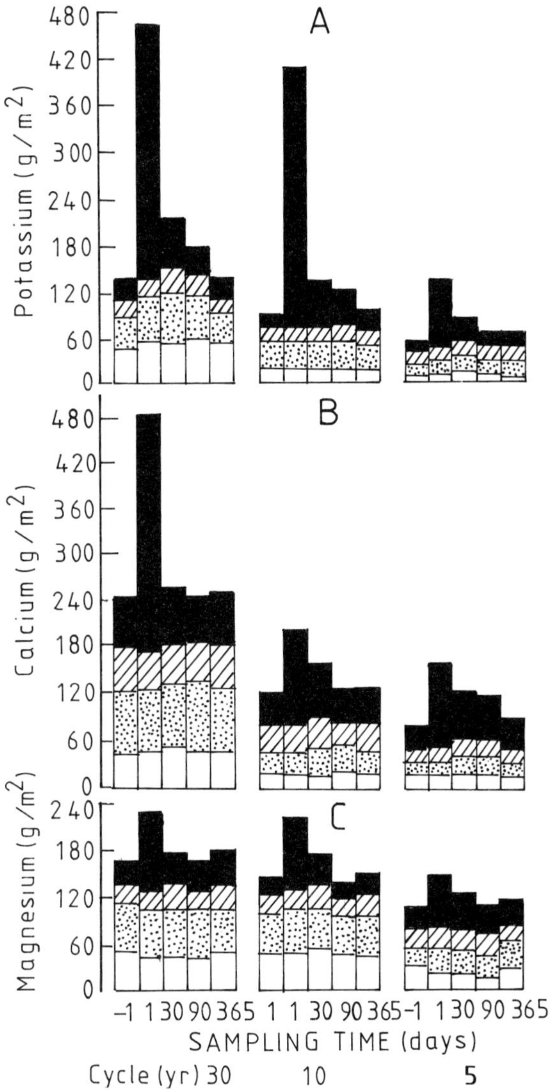

Figure 5.7 Changes in total quantity of potassium (A), calcium (B), and magnesium (C), within a soil column of 40 cm depth, a day before burn (−1) and after burn during cropping under 30-, 10- and 5-yr jhum cycles at Burnihat in Meghalaya. Dark column, 0–7 cm; hatched column, 7–14 cm; stippled column, 14–28 cm; open column, 28–40 cm soil depths (from Ramakrishnan and Toky, 1981)

level of potassium is much higher compared to other cations after the burn of the site under a 10-year jhum cycle. It is worth noting here that potassium conservation could occur even under a 5-year jhum cycle, if the 5-year-old fallow has the exotic weed *Mikania micrantha* as a major component; this weed also accumulates potassium in its biomass (Swamy and Ramakrishnan, 1987c; Ramakrishnan, 1989).

Increase in cationic content under different jhum cycles was for a brief period after the burn. The abrupt decline in levels was due to heavy losses of ash due to blow-off by strong wind prevalent during the dry months of March–April (see earlier discussion in this chapter). Subsequent declines during the cropping period are related to surface run-off and percolation losses through water and by crop and weed uptake. The cation level at the end of 1-year cropping was equal to, or somewhat higher than, the preburn soil level. Further, with the level of cations at the end of 1-year cropping under a 30-year jhum cycle being much higher than the level at any time under 5-year cycle, a second year of cropping could be sustained under longer jhum cycles (Nye and Greenland, 1960).

The exchangable cations are depleted from the top column of the soil, at Burnihat in Meghalaya (Ramakrishnan and Toky, 1981), as the forest fallow develops, due to the rapid absorption by the developing vegetation (Fig. 5.6) Potassium accumulation by the bamboo, *Dendrocalamus hamiltonii* in a 10-year-old fallow is an example of rapid transfer of this element from the soil pool to the living biomass compartment. Indeed, the level of potassium in the soil of a 10-year-old fallow was the lowest during fallow regrowth; in fact the soil potassium level and that in the living biomass could be inversely related in a 10-year-old fallow.

As the age of the jhum fallow increased beyond 10 years, the soil cationic level and particularly that of potassium increased due to faster returns of nutrients into the soil through litterfall from a well-developed stand of vegetation containing species like *Dendrocalamus hamiltonii.* An older forest fallow like a 50-year-old one showed low levels of calcium and magnesium in the soil, with rapid depletion of both with soil depth. This is because of the greater accumulation of these two elements in the boles of dicot trees and shrubs which are dominant at this stage, an observation that agrees with those made by others (Greenland and Kowal, 1960; Grubb and Edwards, 1982). Thus the recharge of calcium and magnesium in the soil pool seems to be more fire-dependent, whereas potassium release due to litterfall recharges the soil pool resulting in a high level in a 50-year-old fallow. It may be noted here that *Dendrocalamus hamiltonii,* which is a dominant feature in the vegetation of lower elevations of only up to about 20–30 years of fallow regrowth of the forest, accumulates potassium in the biomass up to this stage and discharges it into the soil pool when this species dies out and is replaced by dicot trees during secondary succession. Zinke *et al.*, (1978) concludes, on the basis of his study of the 'Lua' forest fallow system, that the sum of basic cations is highest at the beginning of the cycle at the time of slash-and-burn, with gradual discharge of soil calcium and magnesium as the forest fallow regrows, in contrast to

Table 5.14 Cation budget (kg ha^{-1}) under jhum agroecosystem at Lailad in Meghalaya (from Swamy and Ramakrishnan, 1988c)

	20-yr jhum cycle			*5-yr jhum cycle*		
	K	*Ca*	*Mg*	*K*	*Ca*	*Mg*
Input						
Precipitation	6.9	8.5	4.4	6.9	8.5	4.4
Slash-burn	930.0	1,640.0	400.0	760.0	528.0	255.0
Weeds put back	31.7	79.2	13.1	28.3	12.7	10.0
By-products put back	175.6	53.2	59.0	100.1	21.3	25.7
Output						
Sediment	48.7	35.5	13.8	26.7	26.7	12.9
Run-off	81.4	23.7	10.6	49.3	12.9	10.5
Percolation	12.2	5.0	2.3	7.9	3.1	1.5
Weed removal	31.7	19.2	13.2	28.3	12.8	10.6
Crop removal	181.7	60.3	60.9	107.7	25.9	26.5
Soil pool before burn	1,400	2,300	1,400	480	870	930
Soil pool after burn	3,710	4,380	2,030	1,430	1,480	1,250
Soil pool after cropping	1,444	2,490	1,520	580	910	960

potassium which had a high level in the soil under an old forest. This broadly agrees with our own observations in north-eastern India.

The patterns of cation depletion and recovery in the soil under the modified jhum at Shillong in Meghalaya was essentially similar to that discussed for the typical version at Burnihat, except that the levels in the soil were much lower (Mishra and Ramakrishnan, 1983d).

Cation budget under jhum agroecosystem

The budget analysis for the cations indicate that slash burning is a major input into the jhum system at Lailad in Meghalaya (Swamy and Ramakrishnan, 1988c) (Table 5.14). However, this input alone cannot account for the cation gains in the soil pool soon after the burn. Input, through accelerated weathering of parent material, decay of soil humus and surface organic matter, is also important for the gain in the soil pool after burning. Further the exchange between non-exchangeable and exchangeable forms due to burning (Stromgaard, 1984) is an aspect that needs further study. Two major outputs from the system during the cropping phase under jhum are losses related to crop and weed removal, and sediment- and water-related losses. For all the cations, a net gain occurred at the end of the 1-year cropping phase. The gain was maximum for calcium followed by potassium and magnesium.

Earthworm population dynamics and their contribution to soil fertility under jhum agroecoystem

Earthworms have been shown to play an important role in litter decomposition and maintenance of soil fertility in the surface layers of the soil (Syers and Springett, 1984). Most of the studies available on earthworm ecology are from temperate countries, where species largely belong to the family Lumbricidae (Edwards, 1983). In contrast, the earthworm species of tropical soils belong to a wider variety of families, such as the Almidae, Kynolidae, Megascolecidae, Enditridae and Ocnerodrilidae about which little is known except for some preliminary reports on natural ecosystems such as grasslands (Dash and Patra, 1977) and woodlands (Krishnamoorthy, 1985).

In the humid tropics, maintenance of soil fertility is a major concern (Ramakrishnan, 1985b). The importance of earthworms in nutrient cycling within the soil sub-system is well documented through many studies on temperate agroecosystems (Edwards and Lofty, 1982), pasture lands (Syers, *et al.*, 1979) and forests (Satchell, 1983); similar studies on tropical systems are limited (Lee, 1985; Krishnamoorthy, 1985). Wormcasts contain more nutrients than the soil matrix, but less than plant litter (Barley and Jennings, 1959). Their major role is in locally increasing the rate of nutrient cycling in the soil sub-system. In this discussion, we shall confine ourselves to the earthworm population dynamics and their role in nutrient cycling within the soil sub-system, largely as related to the modified version of jhum at Shillong, at higher elevations in Meghalaya (Bhadauria and Ramakrishnan, 1989). However, similar results apply also to the typical version of jhum in the region (Mishra and Ramakrishnan, 1988).

During jhum, the wide fluctuations that occur in the physico-chemical properties of the soil discussed earlier, are all confined to surface soil layers. These fluctuations, under the modified version of jhum at higher elevations of Meghalaya, are more pronounced on the ridges rather than the grooves (Mishra and Ramakrishnan, 1983d). Since soil physical and chemical properties are more favourable under a long jhum cycle, it is reasonable to expect higher earthworm activity here. Further, since all these fluctuations are more drastic in the surface soil layers, the worms located there are most affected.

Of the three species found during the cropping phase, *Tonoscolex horaii* occurred under 15-year jhum cycles, *Nelloscolex strigosus* occurred under 5-year cycles, and *Amynthas diffringens* occurred under both (Fig. 5.8). Of these, *Nelloscolex strigosus* occurred in up to 5-year-old fallow periods after jhum, *Tonoscolex horaii* in up to 15-year-old fallow and *Amynthas diffringens* declined in population vigour with fallow age of up to 15 years of fallow regrowth (Bhadauria and Ramakrishnan, 1989). On the ridges, *Nelloscolex strigosus* was dominant under 5-year jhum cycles whereas *Tonoscolex horaii*

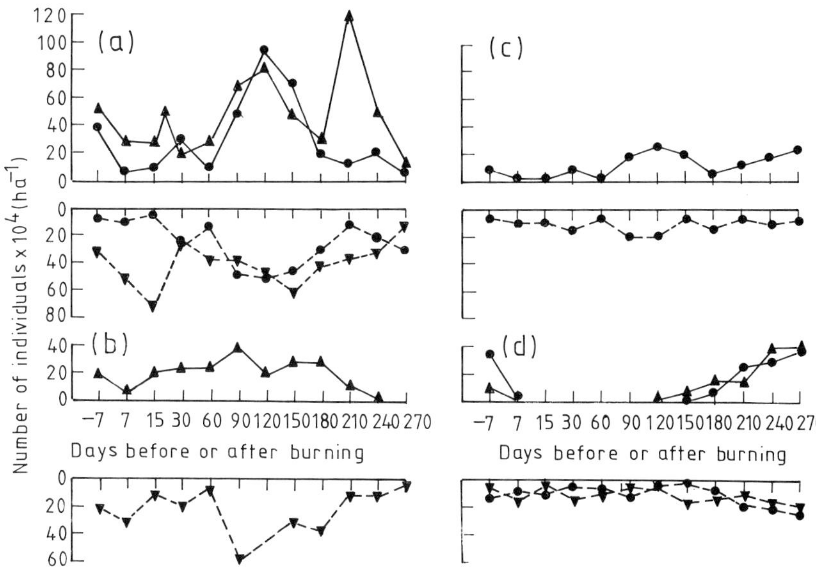

Figure 5.8 Population fluctuations of earthworm species 1 week before burning (−7) and 1 week after burning (+7) and during subsequent cropping under 5-yr (●) and 15-yr (▲) jhum cycles on the ridges (——) and furrows (-----) at Shillong in Meghalaya. (a) juveniles, (b) *Tonoscolex horaii*, (c) *Nelloscolex strigosus*, and (d) *Amynthas diffringens* (from Bhadauria and Ramakrishnan, 1989)

Table 5.15 Earthworm population structure under jhum at Shillong in Meghalaya (from Bhadauria and Ramakrishnan, 1989)

	Amynthas diffringens	Nelloscolex strigosus	Tonoscolex horaii	*Juvenile*
5-yr jhum cycle				
Earthworms				
(m^{-2}) (± SE)	62 ± 1.2	109 ± 2.7	–	414 ± 3.5
Dominance (%)	36.0	64.0	–	–
Frequency (%)	33	75	–	100
15-yr jhum cycle				
Earthworms				
(m^{-2}) (± SE)	102 ± 1.6	–	197 ± 2.6	608 ± 4.6
Dominance (%)	34.0	–	66.0	–
Frequency (%)	42	–	83	100

– indicates absence

was dominant under 15-year cycles (Table 5.15). *Amynthas diffringens* had lower frequency and dominance values under both jhum cycles. More juveniles occurred in under 15-year jhum cycles than under 5-year cycle. Such perturbation-related dominance by some species with narrow or wide ranges of distribution has been seen under sedentary agricultural systems in temperate regions (Hay, *et al.*, 1978; Edwards and Lofty, 1982).

Fire during jhum affected population recovery differently for different species: *Tonoscolex horaii* recovered within a few days whereas *Nelloscolex strigous* and *Amynthas diffringens* recovered only after a few months. Some migration of individuals from the ridges to the unburnt furrows is suggested from these results.

Earthworm population size fluctuated during the cropping season. High moisture in August–September with good crop cover was perhaps responsible for a large juvenile population during this period (Fig. 5.8). Higher temperature and moisture levels favour the growth of *Nelloscolex strigosus* and *Tonoscolex horaii*, as shown by Satchell (1967) for Lumbricidae. In contrast, the ability of *Amynthas diffringens* to adjust to lower temperatures and moisture levels may confer an advantage on this species in avoiding competition, while others dominate during the monsoon season. The adaptation of *Amynthas diffringens* to lower temperatures is also reflected in the absence of winter diapause in this species. However, with a reduction in the total population size in winter, earthworm activity was also reduced. None of the species occurred below a depth of 14 cm and there was no evidence of vertical migration during unfavourable periods, as reported in temperate soil systems (Grant, 1955). The population density of earthworms (mature) ranging from 4 to 47 m^{-2} during cropping was less than that reported for temperate sedentary agroecosystems (Edwards, 1983).

In general, cast production on the ridges was higher under 15-year jhum cycles than under 5-year cycles (Fig. 5.9). The cast production pattern during the cropping phase declined sharply on the ridges soon after the burn. However, it picked up, peaking after 120 days (in May) followed by a subsesquent decline. Annual cast production on ridges was 14.3 and 18.3 $t\ ha^{-1}\ yr^{-1}$ under 5- and 15-year jhum cycles, respectively. In the furrows, the corresponding values were 8.5 and 10.4 $t\ ha^{-1}\ yr^{-1}$, respectively. As far as we know, there are no reports on cast production by earthworms under shifting agriculture. In temperate arable land under wheat, Watanabe and Ruaysoongnern (1984) reported 24.5 $t\ ha^{-1}\ yr^{-1}$ of cast material, greater than that reported here.

The cast soil had a higher pH than the normal soil. Such a dramatic increase in soil pH was also observed by others (Nye, 1955; Sharpley and Syers, 1976). This would increase microbial activity and nitrogen fixation in the soil so that nitrogen gain in the cast may be due, at least in part, to this.

Changes in the nutrient concentration in the soil due to worm casts are shown in Fig. 5.10. A generally greater nutrient concentration in casts than in normal soil was observed throughout. However, this was more

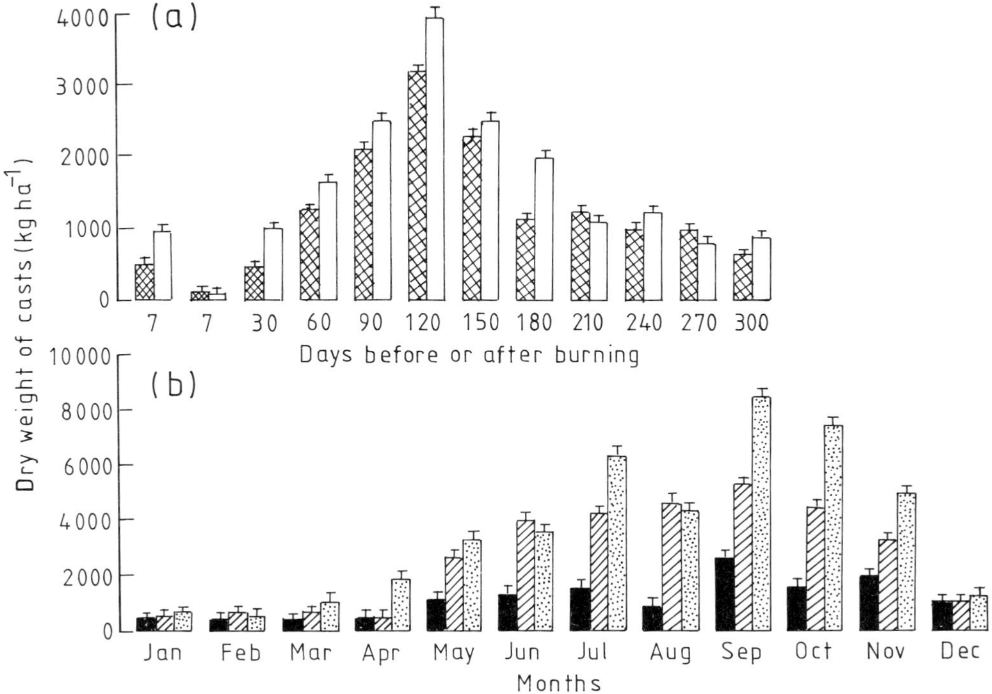

Figure 5.9 Dry weight of wormcasts produced during (a) cropping under 5-yr (▩) and 15-yr (□) jhum cycles starting with a week before the burn as compared with 5-yr (■), 10-yr (▨), and 15-yr (⛆) old jhum fallows (−7) and subsequent to the burn at Shillong in Meghalaya (from Bhadauria and Ramakrishnan, 1989)

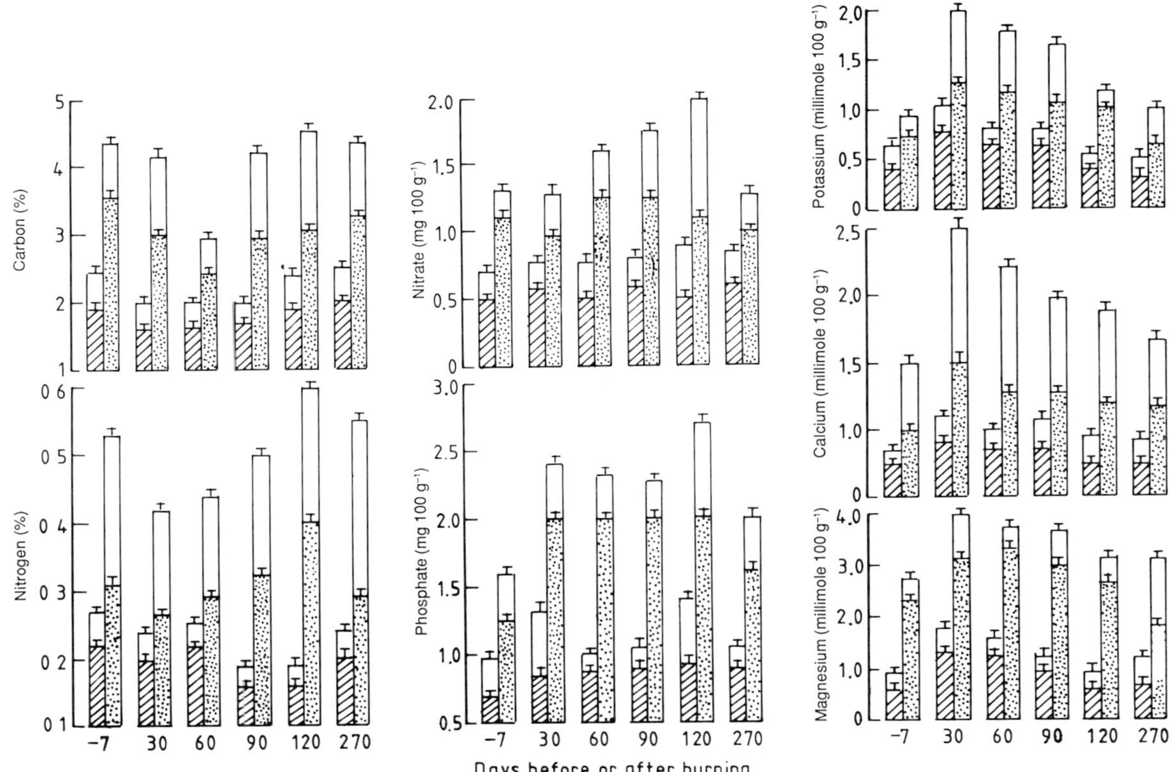

Figure 5.10 Variation in nutrient concentration of earthworms casts and surface soil (0–10 cm) a week before the burn (−7) and subsequent to the burn during cropping under 5-yr (▨) and 15-yr (▩) jhum cycles at Shillong in Meghalaya. Open columns represent additional nutrient concentrations in wormcasts (from Bhadauria and Ramakrishnan, 1989)

Table 5.16 Mean (± SE) values for nutrients (kg ha^{-1} yr^{-1}) returned to the soil in dead worm tissue, mucus and wormcasts during the cropping phase under jhum at Shillong in Meghalaya (from Bhadauria and Ramakrishnan, 1989). Values are also given for nitrogen contributed through mucus production. ND = no data

	5-yr jhum cycle		*15-yr jhum cycle*	
	Dead tissue	*Worm casts*	*Dead tissue*	*Worm casts*
Organic C	ND	413.5 ± 8.6	ND	448.4 ± 6.3
Total N	9.5 ± 0.3	27.0 ± 0.4	12.1 ± 0.1	65.6 ± 1.3
Nitrogen from mucus	59.9 ± 2.4		95.3 ± 2.0	
NO_3-N	ND	0.2 ± 0.03	ND	0.2 ± 0.01
P	0.7 ± 0.01	0.2 ± 0.04	0.9 ± 0.02	0.3 ± 0.03
K	4.8 ± 0.4	4.1 ± 0.5	5.8 ± 1.2	4.4 ± 0.5
Ca	0.8 ± 0.1	2.7 ± 0.2	1.0 ± 0.1	4.0 ± 0.3
Mg	0.8 ± 0.1	2.3 ± 0.7	0.8 ± 0.7	3.6 ± 0.2

pronounced for nitrogen, calcium, potassium, organic carbon and phosphorus, more so under 15-year jhum cycle than under 5-year cycles. Further, in general, nutrient concentration under the longer cycle was more than under the short cycle.

Total nutrient contribution through earthworm activity was higher under 15-year jhum cycles than under 5-year cycles, for all elements except nitrate-N and phosphate-P (Table 5.16). The addition of nutrients through dead worm tissue was substantial for nitrogen and potassium. Total nitrogen in mucus was higher under the longer cycle and this addition was substantial (Krishnamoorthy, 1985) and contributes to more than half of the total.

The chief role of earthworms lies in their ability to concentrate nutrients in the casts (Satchell, 1967) so that their localized availability on the surface soil is increased. Of all the elements concentrated, nitrogen and calcium are the most significant, whereas magnesium was the least influenced. Jakoby and Bhat (1958) also reported an increase in calcium status of the cast through the activity of calcium-oxalate-decomposing bacteria. Since worm activity is generally greater during the wet season, the differential between the casts and normal soil would be greater than in the winter. Earthworms, thus help rapid mineralization of detritus and locally concentrate nutrients in the surface worm casts, so that nutrients are available in the surface soil where a larger biomass of the root system is located during the cropping phase.

Conclusions

One of the chief conclusions arising from of the present discussion relates to the poorer nutrient status of the plots under a 5-year jhum cycle which would partly account for the poorer crop yield under this cycle compared to longer cycle lengths. The soil nutrient depletion process which continues through the cropping phase under jhum also continues through the fallow phase up to about 5–10 years. The rapid transfer of nutrients from the soil to the living biomass, during the early successional phase, is reversed when litterfall occurs either through leaf drop or complete turnover of the early herbaceous vegetation, which happens only after about 10 years of fallow regrowth. This implies that a 10-year jhum cycle is the critical one from the point of view of sustainable cropping under this system. Shorter jhum cycles not only do not permit nutrient recovery in the soil but would also accelerate a variety of losses from the system under frequent perturbations. More frequent losses under low soil fertility levels under short jhum cycles of 4–5 years would eventually lead to site desertification (Ramakrishnan, 1985b,c) Indeed, traditional jhum systems with longer cycles of 10 years or more alone are closer to natural ecosystems where nutrient cycling and maintenance of soil fertility are based on efficient internal controls, this contributing to their stability.

Earthworms play an important role in the jhum agroecosystem. Worm casts are local concentrations of nutrients. In highly leached soils of the humid tropics, earthworm activity is beneficial because of rapid incorporation of detritus into the mineral soil. Vermiculture and introduction of suitable species during the cropping phase into jhum plots offers potential for biological maintenance of soil fertility.

SOIL FERTILITY UNDER OTHER LAND USES

Fallow and sedentary systems

With rapid shortening of the jhum cycle of the modified version to less than 5 years, the jhum agroecosystem is replaced by fallow systems (FAO/SIDA, 1974) as discussed in Chapter 2. Unlike jhum, the fallow system involves no burn but rotation of the crop with a natural fallow regrowth phase of a year or 2, before the next set of rotation of crops. Further evolution in the land use is towards sedentary agriculture. In these systems, maintenance of soil fertility is a major constraint. Of the lesser-known crop species used by the Khasis around Shillong in Meghalaya *Flemingia vestita* plays a major role in building up soil fertility. Nitrogen, a highly labile element in the soils of the humid tropics, is a critical element for crop productivity. The use of *Flemingia vestita* as one of the species in the crop mixture or its use in a pure stand for crop rotation with more traditional species (Gangwar and Ramakrishnan, 1987a, 1989b) is an approach evolved by the Khasis for replenishing the lost soil nitrogen under continuous cropping.

During one cropping season under *Flemingia vestita* in pure stands and mixed with cabbage, a net gain in nitrogen was observed owing largely to

Table 5.17 Change in soil nitrogen status (kg ha^{-1} yr^{-1} under *Flemingia vestita* (pure or mixed with cabbage) cultivated by the Khasis at Shillong in Meghalaya (from Gangwar and Ramakrishnan, 1989b)

	Flemingia vestita			Flemingia vestita + *cabbage*		
Soil depth (cm)	*After sowing*	*Before harvest*	*Gain*	*After sowing*	*Before harvest*	*Gain*
0–10	312 ± 9	320 ± 18	8 ± 1	248 ± 8	256 ± 8	8 ± 1
10–20	299 ± 10	340 ± 14	59 ± 3	174 ± 8	224 ± 15	50 ± 2
20–30	183 ± 12	305 ± 17	122 ± 8	139 ± 9	226 ± 17	87 ± 5
30–40	180 ± 9	198 ± 10	18 ± 2	135 ± 10	144 ± 9	9 ± 1
Total (0–40)	974 ± 67	1,163 ± 92	207 ± 10	696 ± 43	850 ± 38	154 ± 9

In pure stands *Flemingia vestita* had 511 nodules per plant and 4.1 × 10^7 nodules per hectare

Table 5.18 Nitrogen economy (kg ha^{-1} yr^{-1}) under *Flemingia vestita* cultivated in pure or mixed stands by the Khasis at Shillong in Meghalaya (from Gangwar and Ramakrishnan, 1989b)

	Pure	*Mixed with cabbage*
Accumulation in the crop biomass		
Shoot	19 ± 0.7	16 ± 0.6 (3 ± 0.1)
Root	10 ± 0.5	10 ± 0.3 (2 ± 0.1)
Economic yield removed	31 ± 2.5	23 ± 1.0 (10 ± 0.1)
Crop biomass recycled	29 ± 1.7	26 ± 1.5
Wood biomass recycled	9 ± 0.6	13 ± 0.9
Nitrogen fixation	207 ± 7.2	154 ± 5.5
Net gain in soil	245 ± 18.3	193 ± 15.6

Values for cabbage are in parentheses

nitrogen fixation by the nodules of this legume (Table 5.17). This gain was more in pure stands. Maximum gain in the soil nitrogen occurred at a depth of 20–30 cm followed by 10–20 cm. With a range of values between 65 and 224 kg ha^{-1} yr^{-1} for many traditional leguminous species (Nutman, 1976) which could go up to 500 kg ha^{-1} yr^{-1} in some cases (National Research Council, 1979), the nitrogen fixation by *Flemingia vestita* (about 250 kg ha^{-1} yr^{-1}) compares favourably with the reported values. With a matty growth form and dense crop cover, this species would also check losses of nutrients during the monsoon through run-off water (Mishra and Ramakrishnan, 1983c).

About 13% of the total nitrogen added to the soil during one cropping season comes from the recycled root and shoot biomass (Table 5.18). The weed biomass recycled was significantly lower in pure stands; the nitrogen fixation in the soil and that recycled through crop biomass was higher.

While in the developed countries, nitrogen in the soil is maintained through intense use of nitrogen fertilizers, attempts to transfer this technology to the humid tropics have been unsuccessful (Greenland, 1977). Under the jhum system at higher elevations of Meghalaya, we have seen that during one cropping season, the loss of nitrogen may be as high as 600 kg ha^{-1} yr^{-1} (Mishra and Ramakrishnan, 1984). Under shorter jhum cycles of 5 years, not more than half of what is lost is put back into the system. Therefore, the role of a nitrogen fixing legume species such as *Flemingia vestita* in a sedentary system becomes particularly significant. With a gain of over 200 kg of nitrogen per hectare during one cropping season, this species offers possibilities of improving yield under fallow and sedentary systems, at the same time meeting the energy and protein needs of the tribals during the lean season when traditional crop species are in short supply (Gangwar and Ramakrishnan, 1989b).

Cash crop ecosystems

With rapid shortening of the jhum cycle, a shift towards a cash crop economy is a possibility that has often been suggested as an alternative (Ramakrishnan, 1985a,c). With pineapple and ginger as traditional cash crops in the region, coffee and tea plantations have been tried on an experimental basis at Nayabunglow in Meghalaya by the local Soil Conservation Department of the Government (Kumar and Ramakrishnan, 1989b). Apart from social considerations, if approached correctly from the point of maintenance of soil fertility, this land use system has potential for adaptation on a larger scale. It is in this context that we evaluated the nutrient losses from these land use systems at Nayabunglow in Meghalaya (Kumar and Ramakrishnan, 1989b).

Considering that substantial amounts of organic and inorganic manure may go into some of these cash crop plantations (for coffee it is NP, 5:3 at the rate of 155 kg ha^{-1} twice a year; for tea it is NPK, 2:1:2 at the rate of 555 kg ha^{-1} twice a year; for ginger it is NPK, 1:1:1 at the rate of 500 kg ha^{-1} twice a year along with 720 kg ha^{-1} of organic manure once a year) at Nayabunglow, it is reasonable to expect substantial losses of nutrients from the soil during the monsoon season. Further, coffee and tea plantations also receive weedicide/ pesticide applications that would help to keep the herbaceous ground cover under check leading to sediment losses. Apart from tea grown on slopes, others considered here are all grown on terraces. In spite of terracing, the losses of water through run-off and percolation and sediment loss were substantial (Fig. 5.11). This may be related to frequent disturbances of the soil during weeding and harvest operations. Pineapple mixed cropping systems along with *Colacasia antiquorum*, *Curcuma longa* and *Mainhot esculentus* were least affected.

The loss of nutrients was also minimal under the pineapple mixed cropping system (Table 5.19). Since all other systems except the pineapple mixed cropping system received heavy inputs of inorganic fertilizers, the

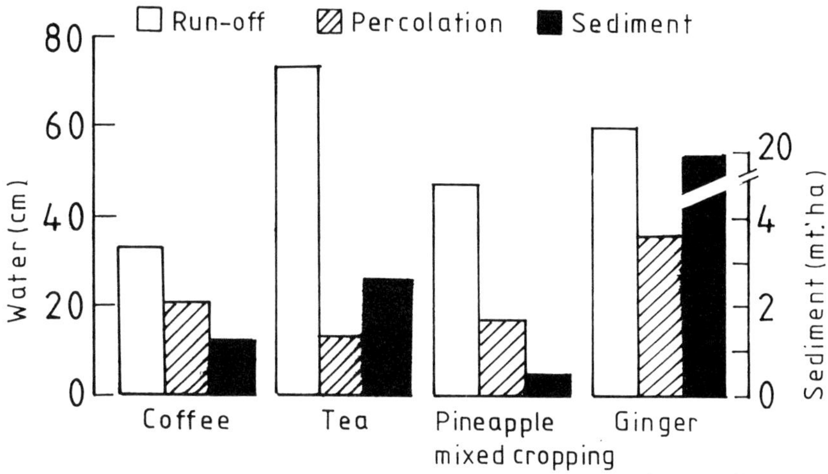

Figure 5.11 Annual loss of water (cm) and sediment (metric tonne ha^{-1}) from different cash crop ecosystems at Nayabunglow in Meghalaya (from Kumar and Ramakrishnan, 1989b)

Table 5.19 Total loss of nutrients ($kg\,ha^{-1}\,yr^{-1}$) in run-off and percolation water from different plantation/cash crop ecosystems at Nayabunglow in Meghalaya (from Kumar and Ramakrishnan, 1989b)

Element	*Coffee*	*Tea*	*Pineapple mixed cropping*	*Ginger*
NO_3-N	3.03 ± 0.18	6.61 ± 0.12	1.05 ± 0.08	6.10 ± 0.35
	(3.80 ± 0.17)	(2.20 ± 0.14)	(2.28 ± 0.13)	(8.38 ± 0.44)
PO_4-P	0.59 ± 0.05	2.56 ± 0.15	0.61 ± 0.04	1.68 ± 0.07
	(0.22 ± 0.02)	(0.30 ± 0.02)	(0.19 ± 0.01)	(1.0 ± 0.05)
K	22.01 ± 0.96	54.67 ± 1.61	15.15 ± 0.68	41.03 ± 1.57
	(9.45 ± 0.35)	(5.90 ± 0.35)	(3.80 ± 0.22)	(15.86 ± 1.20)
Ca	10.83 ± 0.64	26.0 ± 1.36	8.01 ± 0.40	12.94 ± 0.56
	(3.84 ± 0.77)	(3.0 ± 0.21)	(2.33 ± 0.12)	(6.56 ± 0.40)
Mg	8.97 ± 0.41	36.94 ± 1.95	6.32 ± 0.40	10.86 ± 0.49
	(2.33 ± 0.11)	(4.55 ± 0.25)	(1.62 ± 0.17)	(5.58 ± 0.35)

Values in parentheses are for percolation losses

minimal loss under the pineapple system is understandable; the mixed cropping system here also provides good ground cover.

A comparison of the losses of nutrients under cash crops with the jhum system at Burnihat (Toky and Ramakrishnan, 1981b) suggests that the losses under the former are very close to losses under jhum, the pineapple cropping system being an exception. Further, in spite of terracing the land, the losses through percolation water could be a third to as high as two times that of the run-off loss, as compared to 50% of the total under jhum. Two conclusions seem to be obvious. Mixed cropping systems and greater dependence on organic manure could be more appropriate for nutrient conservation in the north-eastern hill region. Indeed, traditional societies of the north-east are based upon these ecological concepts.

Terrace agroecosystem

Terrace agroecosystem, as discussed earlier has not been found to be a viable alternative to jhum. In our earlier discussions in Chapters 2 and 5, we have implicated reduced crop yield to be attributable to increased weed potential of the site under continuous cropping, in spite of high fertilizer inputs.

In a comparative analysis of jhum under 10- and 5-year cycles and terrace agroecosystems at higher elevations in Meghalaya (Mishra and Ramakrishnan, 1983c) the sediment load under terracing was shown to be $34\,t\,ha^{-1}\,yr^{-1}$ during the 1st year of observation, which increased to $42\,t\,ha^{-1}\,yr^{-1}$ during the 2nd year, as compared to $50\,t\,ha^{-1}\,yr^{-1}$ loss under a 10-year jhum cycle. Nitrogen and available phosphorus losses through sediment, run-off water and percolation water were comparable to that under the jhum system. Cation losses, although lower than under the jhum system, increased during the 2nd year of cropping. Indeed, all nutrient losses from the terrace agroecosystem increased during the 2nd year of cropping.

A similar study done on soil fertility changes at Shillong in Meghalaya, suggested that in spite of heavy input of inorganic (1:1:1 ratio of NPK at the rate of $74\,kg\,ha^{-1}$) and organic ($3{,}000\,kg\,ha^{-1}\,yr^{-1}$) fertilizers, the soil under terraces started off with a nutrient capital generally comparable to that under a 5-year jhum cycle (Mishra and Ramakrishnan, 1983d). At the end of the 2-year cropping period, the soil nutrient status was much lower than that under a 5-year jhum cycle. This suggests that terrace agroecosystem, in spite of heavy fertilizer demand, is not sustainable in this region where the soil is very weak; the terraces often are abandoned after 6–8 years of continuous cropping.

This is not to suggest that terrace cultivation on a long-term basis is not sustainable at all in the region. We have been able to study terraces that are up to about 12 years old, under continuous cropping at Nayabunglow in Meghalaya as discussed in Chapter 2. These terrace systems have been sustained by the Nepalis who have easy access to cow dung, because they are cattle farmers.

In a comparative analysis of terrace agroecosystems under 4- and 12-year-old terraces, the nutrient budget analysis was done (Patnaik, 1988; Patnaik and Ramakrishnan, 1990). The input/output budget for nitrogen and phosphorus is given in Table 5.20. Since a light burn of weed biomass of the fallow phase between the two successive croppings in a year is done, nitrogen and phosphorus losses were noted after fire, as discussed under

Table 5.20 Input/output patterns for nitrogen (N) and phosphorus (P) ($kg\,ha^{-1}\,yr^{-1}$) under terrace agroecosystem at Nayabunglow in Meghalaya (from Patnaik, 1988)

	4-yr-old terrace		*12-yr-old terrace*	
	N	*P*	*N*	*P*
Input				
Precipitation	4.3	1.1	4.3	1.09
Addition through plant biomass before fire	31.5	0.57	65.5	0.81
Thinned crop biomass	8.5	0.07	3.9	0.03
Weeds ploughed back during				
1st crop	7.8	0.05	8.5	0.13
2nd crop	14.9	0.13	–	–
Organic manure	13.8	8.3	11.7	7.0
Total	80.8	10.21	93.8	9.06
Output				
Fire	126.1	0.48	239.5	0.94
Sediment	4.1	0.01	5.2	0.02
Run-off	1.8	0.5	2.2	0.5
Percolation	2.1	0.3	0.99	0.2
Weed removal				
1st crop	22.6	0.2	43.8	0.41
2nd crop	12.5	0.09	–	–
Crop removal				
1st crop	105.5	9.98	73.9	10.28
2nd crop	29.6	0.71	–	–
Total	304.4	12.3	365.6	3.29
Net difference	223.76	2.06	271.74	12.35

Table 5.21 Net change in nitrogen (N) and available phosphorus (P) ($kg\,ha^{-1}\,yr^{-1}$) under terrace agroecoystem at Nayabunglow in Meghalaya (from Patnaik, 1988)

	4-yr old terrace		*12-yr-old terrace*	
	N	*P*	*N*	*P*
Soil pool before burning	7,715	21.30	7,674	22.19
Soil pool at the end of cropping	6,977	19.78	6,754	20.28
Net difference	738	1.52	920	1.91

the jhum agroecosystem, earlier in this chapter. Percolation losses are substantial when associated with run-off losses. Although, the Nepali farmer does only one cropping on an old terrace, the loss through uptake is substantial from it because of intensified weed potential. The net loss increased with the age of the terrace. This was also reflected in the nitrogen and phosphorus soil pools at the end of a 1-year cropping (Table 5.21); in

Table 5.22 Net change in nutrients (kg ha^{-1} yr^{-1}) under terrace agroecosystems at Nayabunglow in Meghalaya (from Patnaik and Ramakrishnan, 1989)

	Soil pool before burning (a)	*Soil pool at the end of cropping (b)*	*Net difference (b–a)*
4-yr-old terrace			
K	510 ± 25.1	552 ± 50	+42
Ca	2,122 ± 150	1,473 ± 50.5	−649
Mg	792 ± 43.1	1,307 ± 57.7	+515
12-yr-old terrace			
K	982 ± 34.1	643 ± 45.0	−339
Ca	1,385 ± 57.6	1,670 ± 120.3	+285
Mg	1,086 ± 47.8	1,547 ± 30.1	+461

spite of a 1-year of cropping under a 12-year-old terrace, the loss from the soil was much higher.

The cation budget analysis for the 4- and 12-year-old terrace at Nayabunglow (Patnaik and Ramakrishnan, 1990) shows that in spite of a heavy input of fertilizers, the net potassium loss, a highly labile and critical element, from the system after a year of cropping under a 12-year-old terrace was 339 kg ha^{-1} year (Table 5.22). This is in sharp contrast to a gain of 42 kg ha^{-1} yr^{-1} under a 4-year-old terrace.

From a soil fertility point of view, these terraces developed at Nayabunglow can be sustained only by the Nepalis because of their cattle husbandry system that could provide the much needed cow dung for maintaining both soil physical and chemical qualities. The local Khasi tribe prefers to carry out jhum because they are unable to obtain the organic manure in sufficient quantities, since they are not cattle farmers. This limited sustainability of terraces at Nayabunglow is in spite of a deep and well-developed soil profile. Elsewhere, where the soil is weak, this land use system is not at all viable.

Conclusions

A few major conclusions emerge from this discussion. Land use options, as alternatives to jhum have to be based upon mixed cropping systems where species such as legumes and tree crops are incorporated. The example of the pineapple mixed cropping system at Nayabunglow, for example, illustrates the minimal loss that occurs here compared to other land uses based upon monoculture. This system also illustrates the value of organic matter recycling in maintenance of soil fertility, unlike many others that heavily depend upon inorganic fertilizer that are expensive and difficult to procure. Terrace cultivation has only limited value as an alternative to jhum. Mixed cropping systems involving nitrogen fixing legumes could be further strengthened if a well conceived crop rotation practice is evolved. The fallow and sedentary systems of agriculture where *Flemingia vestita* is rotated as a pure crop, after a few years of mixed cropping indicates the value of this lesser-known crop that the Khasi farmer uses for maintenance of soil fertility through biological means, based upon his empirical knowledge.

GENERAL CONSIDERATIONS

A major outcome from this analysis is that, under the ecological and socio-economic conditions that prevail in the north-eastern hill region, the management options for maintaining, indeed improving, soil fertility has to be by manipulation of biological processes. Ecologically, this high rainfall area with weak mountain soils is less amenable to petro-based energy-subsidized agricultural system management. Fertilizer use-efficency in the region is very low, much of it being wasted through hydrology-related losses. Also, the economically weaker sections of the society here cannot afford the high cost involved, particularly when the efficacy of fertilizer use is questionable. These considerations and the non-availability of an alternate viable technology from agricultural scientists and governmental agencies have forced the tribals to carry out jhum although in a distorted form under very short jhum cycles of 4–5 years. In spite of these constraints, the jhum farmer has optimized production from his system by capitalizing upon the limited soil fertility under the short cycle, which in any case is highly transient. The reflection of this is seen in his ability to adapt to fallow systems or sedentary systems of agriculture by using appropriate crop mixtures and crop rotation involving leguminous crops. Indeed, in many of the traditional jhum practices, the farmer conserves many tree species such as the Nepalis alder (*Alnus nepalensis*), a nitrogen fixer, while doing his slash-and-burn operation, the implications of which are discussed in Chapter 7.

There are a variety of ways in which different communities handle decline in soil fertility. This is evident from the variety of agroecosystem types adapted to varied ecological conditions. Thus, the shift towards tuber crop at higher elevations under pine forests and the elaborate land preparation into ridges and furrows so as to conserve soil fertility and optimize production from the system is such an example. Eventual shift to sedentary fallow system agriculture with crop rotation using the lesser known legume crop *Femingia vestita* is to cope with extreme site desertification. In the same site, a given tribal community may change the cropping pattern to cope with decline in soil fertility. The shift towards tuber and rhizomatous crops under short jhum cycles is indicative of the optimal use of soil fertility by the communities concerned.

Crop placement on a given slope with more nutrient-use efficient species being placed at the top of the slope under the jhum system represents a finer level of appreciation of plant soil interaction by the farmer. Nutrient release from organic matter synchronizing with nutrient uptake by the crop species comes out elegantly through the sequential harvesting of crops and the management of crops and weed biomass, such that biomass from the earlier crop is made available to the next crop at its peak growth in the mixed cropping jhum system. The management of organic matter on the soil surface in all these systems helps in conserving soil moisture and controlling weed growth.

Keystone species such as bamboo (*Dendrocalamus hamiltonii*) for conserving potassium and Nepalese alder (*Alnus nepalensis*) for nitrogen economy in the Jhum system in north-eastern India are also species that are traditionally valued and conserved by the jhum farmer. In this manner, traditional knowledge could be linked with soil biological process studies.

Suffice to say here that the technology input for maintenance of soil fertility has to be based, to some extent at least, on traditionally evolved technologies. Jhum is an agroforestry system (Ramakrishnan, 1986a, 1988a). Traditionally, the home garden is another agroforestry system. Soil fertility maintenance in these ecologically and socially fragile ecosystems should be based upon agroforestry concepts, introducing it where it is absent and strengthening it where it is already weakened. Agriculture, in the forseeable future, in many areas of the tropics has to be based on low-input technology that works. Legume intercropping, minimum tillage agriculture and agroforestry systems are all essential ingredients of the jhum system. Building up on these concepts not only would help to conserve soil fertility at a relatively low economic cost, but would conserve and strengthen natural biogeochemical processes in the soil. In the ultimate analysis, redevelopment based on these concepts if built upon a value system with which the traditional farmer is familiar, would be easy to implement.

SECTION 2

Secondary Successional Patterns and Processes

CHAPTER 6

VEGETATION DYNAMICS IN JHUM FALLOWS

INTRODUCTION

Succession is a process of continuous colonization and extinction of species on a given site and, therefore, it is a time-dependent integration of ecological processes. While earlier studies on succession were all descriptive, considering the species present at various stages, it is only recently that attempts have been made to analyze functional attributes of species over a successional gradient. Many attempts have been made to look for an underlying and unifying scheme for all successional processes. Clements (1916) saw succession as a form of ontogeny, while Gleason (1926) emphasized an individual species approach to succession. Many attempts have been made to look for general community attributes. Properties such as species diversity (Margalef, 1968) were considered important. Odum (1969) has listed many parameters such as species diversity, energetics, biomass, productivity and nutrient cycling as attributes that change during succession.

Secondary sucession has often been thought of as an orderly process of development with a relay of species following one another, eventually leading to a climax vegetation (Dansereau, 1957). On the other hand, Egler (1954) suggested that the initial floristic composition contributes to subsequent vegetational structure; this was supported by the studies of Bormann and Likens (1979) through their studies on the secondary succession after clear-cutting of northern hardwoods of the USA. We shall consider here the secondary successional patterns and processes related to vegetation structure, biomass and productivity after jhum in north-eastern India. We shall relate these patterns and processes to the frequency and intensity of perturbations that are responsible for the initiation of the secondary succession itself.

VEGETATION, BIOMASS AND PRODUCTIVITY CHANGES

Over large areas of north-eastern India, the secondary successions on abandoned farms after jhum rehabilitate the land. The structure and composition of the fallow vegetation indicate the site's potential soil fertility, when the area would be farmed again. As discussed earlier, if the fallow is farmed after only 4–5 years, degradation sets in and crop yield declines (Toky and Ramakrishnan, 1981a; Mishra and Ramakrishnan, 1981). This recent shortening of the jhum cycle makes an understanding of the structure of the fallow vegetation, particularly its early phases, of considerable importance.

Detailed studies on ecosystem structure and function during secondary succession after clear-cutting temperate forests are available (Bormann and Likens, 1979) and some work on these aspects has also been carried out in the humid tropics (Bartholomew *et al.*, 1953; Nye and Greenland, 1960; Ewel, 1971). The jhum systems considered here represents one of the first studies of its kind from the Asian tropics (Toky and Ramakrishnan, 1983a; Ramakrishnan, 1988b).

Early succession in up to 20-year-old fallow at Burnihat in Meghalaya

Vegetation

When a forest is converted to cultivable land, as under jhum, not only is its original vegetation destroyed, but the site is subject to continuing perturbations due to fire, the introduction of crop species, weeding, hoeing and other disturbances to the soil when the crops are harvested. These result in a progressive reduction in species diversity and so the early successional stages contain few species but the number increases gradually as the secondary community develops (Ross, 1954; Odum, 1969; Whitmore, 1975). Thus, species diversity was very low in the first 5 years; it increased considerably between 5 and 10 years and showed further but less increase in the next 10 years at Burnihat (100 m altitude) in Meghalaya (Fig. 6.1). Conversely, dominance was maximal in the early phases of succession and decreased sharply with the age of the fallow. Essentially a similar pattern was seen for the high elevation jhum systems too, at Shillong (1,500 m) in Meghalaya (Mishra and Ramakrishnan, 1983a). This somewhat contrasts the finding of Aweto (1981) who reported stabilization in species diversity and tree density in the 7th year of fallow in south-western Nigeria.

In the study done at the low elevation jhum systems at Burnihat in Meghalaya, we have been able to discern a variety of herbaceous weedy communities developing after jhum during the first 5 years (Toky and Ramakrishnan, 1983a), as also considered in detail in Chapter 4. These differences were related by us to the land use history, and the time since farming ceased was the predominant factor governing the succession in the region as a whole.

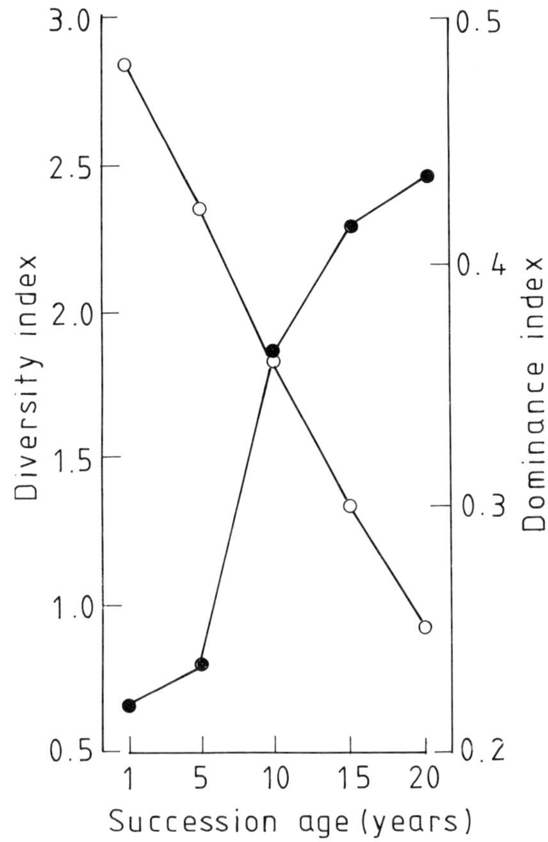

Figure 6.1 Species diversity (●) and species dominance (○) in successional communities up to 20 years at Burnihat in Meghalaya (from Toky and Ramakrishnan, 1983a)

A variety of strategies are adopted by successional species among which stump, root and rhizome sprouts and invasion through seeds are common. The relative importance of different species and their strategies could vary depending upon the length of the jhum cycle, the species composition and structure of the vegetation prior to clearing for cultivation, and the previous land use history (cf Chapter 4). In sites under a 5-year jhum cycle, at Burnihat, *Eupatorium odoratum* is a component of early succession. The production of a large number of seeds, which are light in weight and wind dispersed help it to succeed as an early colonizer. The importance of this was stressed by Hayashi and Numata (1968) and Raynal and Bazzaz (1973). Salisbury (1942) emphasized that species with lightweight and highly mobile seeds often invade highly disturbed ecosystems, whereas species with heavier seeds are often animal disseminated and usually enter the succession at a later stage. While early successional herbs are *r*-strategists, late successional herbs are

Table 6.1 Abundance of predominant species in successional communities of 1, 5, 10, 15 and 20 years of age, developed after jhum at Burnihat in Meghalaya (from Toky and Ramakrishnan, 1983a)

	1			*5*			*10*		
	Frequency (%)	*Density* (m^{-2})	*Basal area* (cm^2m^{-2})	*Frequency* (%)	*Density* (m^{-2})	*Basal area* (cm^2m^{-2})	*Frequency* (%)	*Density* (m^{-2})	*Basal area* (cm^2m^{-2})
Eupatorium odoratum	100	18.7	190	100	25.3	426			
Imperata cylindrica Beauv.	100	22.6	76	100	30.2	141			
Panicum khasianum Munro	48	5.8	6						
Setaria glauca Beauv.	40	2.5	3						
Grewia elastica Royle	12	2.9	9						
Cyperus globosus Allioni	40	6.5	11						
Panicum maximum Wall.	24	15.4	11	60	15.2	12			
Ficus hispida	20	2.2	10						
Arundinella bengalensis (Spreng.) Druce				80	42.2	61			
Callicapra macrophylla Vahl.				20	0.005	0.1			
Dendrocalamus hamiltonii Nees and Arn.				8	0.08	0.7	44	0.009	0.3
Carex cruciata Vahl.				44	8.02	16	90	0.38	13
Bauhinia purpurea				20	0.003	0.1			
Dillenia indica				28	0.008	0.1	64	0.007	1
Anidesma diandrum Roth							60	0.009	0.3
Careya arborea Roxb.							44	0.005	0.5
Phyllanthus emblica Wall.							40	0.007	0.5
Wrightia tomentosa R. and S.							32	0.003	1
Vitex peduncularis Wall.							28	0.003	0.4
Terminalia belerica Roxb.							40	0.005	0.5
Lagerstroemia parviflora Roxb.									
Vitex glabrata Br.									
Dillenia pentagyna Roxb.									
Schima wallichii Chois.									
Cedrela toona Roxb.									

continued

Table 6.1 *continued*

	15			*20*		
	Frequency (%)	*Density* (m^{-2})	*Basal area* (cm^2m^{-2})	*Frequency* (%)	*Density* (m^{-2})	*Basal area* (cm^2m^{-2})
Eupatorium odoratum						
Imperata cylindrica Beauv.						
Panicum khasianum Munro						
Setaria glauca Beauv.						
Grewia elastica Royle						
Cyperus globosus Allioni						
Panicum maximum Wall.						
Ficus hispida						
Arundinella bengalensis (Spreng.) Druce						
Callicapra macrophylla Vahl.						
Dendrocalamus hamiltonii Nees and Arn.				44	0.018	0.4
Carex cruciata Vahl.	96	0.45	32	100	0.56	45
Bauhinia purpurea						
Dillenia indica	60	0.01	1	68	0.02	1
Anidesma diandrum Roth	68	0.02	1	72	0.03	2
Careya arborea Roxb.						
Phyllanthus emblica Wall.	44	0.003	0.8	56	0.02	3
Wrightia tomentosa R. and S.						
Vitex peduncularis Wall.						
Terminalia belerica Roxb.	48	0.007	1	40	0.016	2
Lagerstroemia parviflora Roxb.	60	0.008	2	52	0.008	0.2
Vitex glabrata Br.	48	0.007	1			
Dillenia pentagyna Roxb.	32	0.004	0.4	40	0.009	1
Schima wallichii Chois.	36	0.008	0.7			
Cedrela toona Roxb.	12	0.004	0.4	20	0.006	0.6
				24	0.012	0.2

k-strategists (Kushwaha *et al.*, 1981a,b), *Imperata cylindrica* which is also an early colonizer, has rhizomes which are protected from fire and are difficult to eradicate during cultivation. This strategy, along with the production of lightweight wind-dispersed seeds, even under constant disturbance (Kushwaha *et al.*, 1983b) contribute to its success. These early species which adopt an exploitative strategy (Grime, 1974; Harper and White, 1974; Marks, 1974) are able to attain dominance in abandoned fields that are temporarily enriched with nutrients and radiant energy (cf Chapter 5). The changes in the demography of these weeds have been adequately dealt with in Chapter 4 and therefore this is not considered here. Thus, depending upon the past land use history, five different weed communities were recognized at Burnihat, which were essentially determined by the available propagule source in the soil, prior to farming and the kind of disturbance regime (cf Table 4.1). Where shorter 4- to 6-year jhum cycles occur the 'initial floristic model' of Egler (1954) (cf Table 4.14) was found to operate. In contrast, the 'relay floristics composition' model (Clements, 1916; Odum, 1969) was found to operate only under long jhum cycles of 10–20 years with a comparatively low frequency of fire disturbance.

Although the weedy communities may differ during the first 5 years of fallow regrowth, based upon the past land use history and perturbance regime at lower elevations, they are replaced by the bamboo, *Dendrocalamus hamiltonii* and other woody species during the 10–20-year period (Table 6.1). The 15- and 20-year-old fallows at Burnihat have many shade-intolerant tree species such as *Bauhinia purpurea*, *Careya arborea*, *Cedrela toona*, *Dillenia indica* and *Schima wallichii*. At higher elevations, the Khasi pine (*Pinus kesiya*) is an important tree species (Mishra and Ramakrishnan, 1983a). Further, at higher elevations at Shillong, since the sparsely distributed pine trees are not clear-cut and burnt before jhum, larger trees of this species may be present right from the very early succession following jhum. In older fallows, more pine regeneration may happen, eventually leading to a closed tree canopy (Plate 6.1).

Litter production

The total annual production of litter increased significantly at Burnihat with fallow age reaching 1 kg m^{-2} yr^{-1} in the 20-year-old fallow (Table 6.2). In a *Eupatorium* dominated 5-year-old fallow, this species contributed about 70% of the total litter. Later, the bamboo *Dendrocalamus hamiltonii* accounted for about 45% or more of the total litter in fallows of 10 years or older. Litter from trees increased with successional age and accounted for about half of the total in 10-year-old and older fallows.

The values of litter production for successional communities in the present study fall within the published range for mature and secondary tropical and

Plate 6.1 An early successional jhum fallow at Shillong (1,500 m altitude) in Meghalaya with the exotic weed, *Eupatorium adenophorum*, the native grass *Imperata cylindrica* and the fern *Pteriduim aquilinum*, among many others. Note Khasi pine (*Pinus kesiya*) regeneration

Table 6.2 Annual litterfall ± 95% confidence limits ($g\,m^{-2}\,yr^{-1}$) in successional communities developed after jhum at Burnihat in Meghalaya (from Toky and Ramakrishnan, 1983a)

	Succession (yr)				
Litter category	*1*	*5*	*10*	*15*	*20*
Dicotyledon leaves	80 ± 10	210 ± 30	310 ± 50	350 ± 60	420 ± 70
% of total	67	44	44	45	43
Bamboo leaves	–	10 ± 4	320 ± 50	320 ± 40	430 ± 60
% of total	–	2	45	42	44
Grass leaves	40 ± 6	120 ± 20	–	–	–
% of total	33	25	–	–	–
Dicotyledon twigs	–	140 ± 30	50 ± 10	30 ± 10	36 ± 6
% of total	–	29	7	4	3
Bamboo twigs	–	–	30 ± 8	70 ± 10	90 ± 20
% of total	–	–	4	9	10
Total	120 ± 20	480 ± 80	710 ± 120	770 ± 150	976 ± 180

sub-tropical forests (Toky and Ramakrishnan, 1983a), although Laudelout and Meyer (1954) report higher values for mature forests in Zaire. The relatively high values of litter production in secondary successional communities in the present study as compared to 0.55 $kg\,m^{-2}\,yr^{-1}$ for a 50-year-old one in the same area (Singh and Ramakrishnan, 1982a) is due to the fast

Plate 6.2 A jhum plot at Lailad (296 m altitude) in Meghalaya surrounded by secondary successional fallow regrowth. In the foreground is herbaceous weedy community and in the background is the developing bamboo (*Dendrocalamus hamiltonii*) forest

developing vegetation during the early successional stages and subsequent rapid turnover of biomass. Further, in early fallows, *Dendrocalamus hamiltonii* and a number of tree species are deciduous; in older forests most tree species are evergreen as in a 50-year-old fallow at Lailad (296 m altitude) (Plate 6.2) in Meghalaya (Shukla and Ramakrishnan, 1982b). About 40% of the total leaf litter of 4.3 t ha^{-1} yr^{-1} is reported (Singh and Ramakrishnan, 1982a) to accumulate in the drier period of February–March; in other months leaf-drop is relatively low.

Biomass and production

A large increase in the aboveground biomass with age up to 20 years was noted at Burnihat in Meghalaya, with maximum of 15 kg m^{-2} in 20-year-old fallow (Table 6.3). Herbs accounted for all the biomass in 1-year-old fallow; in subsequent years, their contribution declined sharply. The biomass contribution by *Dendrocalamus hamiltonii* and by shrubs and trees increased with the age of the fallow up to 20 years.

The successional communities studied here showed an increase in production up to 20 years at Burnihat (Table 6.4); this may be due to the growth of exploitative species which utilize resources available after jhum, accelerated

Table 6.3 Changes in standing biomass ± 95% confidence limits ($g\,m^{-2}\,yr^{-1}$) of different categories in successional communities developd after jhum at Burnihat in Meghalaya (from Toky and Ramakrishnan, 1983a)

	Succession (yr)				
Category	*1*	*5*	*10*	*15*	*20*
Bamboo					
Main stem	–	200	2,660	5,090	6,180
Branches	–	20	440	530	960
Leaves	–	10	130	280	350
Total	–	230 ± 50	3,230 ± 540	5,900 ± 920	7,490 ± 1,340
% of total		10	56	57	51
Shrubs and trees					
Main stem	–	600	1,790	3,170	5,330
Branches	–	–	480	1,000	1,520
Leaves	–	20	230	280	400
Total	–	620 ± 110	2,500 ± 400	4,450 ± 900	7,250 ± 1,350
% of total		27	43	43	49
Herbs					
Main stem and branches	400	1,060	10	10	10
Leaves	100	420	10	10	10
Total	500 ± 80	1,480 ± 200	20 ± 3	20 ± 4	20 ± 4
% of total	100	63	0.3	0.2	0.1
Total	500 ± 80	2,330 ± 280	5,750 ± 1,100	10,370 ± 1,700	14,760 ± 2,560

Table 6.4 Changes in rates of accumulation of biomass, litterfall and net primary productivity in successional communities developed after jhum at Burnihat in Meghalaya (from Toky and Ramakrishnan 1983a)

	Successional age (yr)				
	1	*5*	*10*	*15*	*20*
Accumulation in bole and branches ($kg\,m^{-2}\,yr^{-1}$)	0.4	0.4	0.7	0.9	0.8
Litterfall ($kg\,m^{-2}\,yr^{-1}$)	0.1	0.4	0.7	0.8	1.0
Net primary production ($kg\,m^{-2}\,yr^{-1}$)	0.5	0.8	1.4	1.7	1.8
Biomass accumulation quotient (standing biomass/net primary productivity)	0.9	2.5	4.3	5.9	8.3

decomposition of forest floor organic matter, and high radiant energy. The rapid increase in production after 5 years appears to be due to a shift in structure from a predominantly herbaceous community to one with bamboo and other shade-intolerant tree species. Similar high rates of production in early successional stages have been reported by other workers, with a rate of increase in biomass of 1.9–2.7 $kg\,m^{-2}\,yr^{-1}$ (Toky and Ramakrishnan, 1983a). The general pattern at higher elevations at Shillong was the same, but the values reported here are lower (Mishra and Ramakrishnan, 1983a).

Table 6.5 Fine root biomass (kg/ha ± SE at 50-year-old forest fallow at Lailad in Meghalaya (from Ramakrishnan and Singh 1983)

Diameter class (mm)	*Soil horizon* (cm)		
	0–50	*50–100*	*100–150*
< 1	2,017 ± 611	625 ± 176	130 ± 37
1–2	976 ± 296	674 ± 190	155 ± 43
2–4	1,429 ± 414	6,243 ± 171	210 ± 59
4–6	1,396 ± 282	730 ± 206	130 ± 37
6–10	1,736 ± 526	1,196 ± 337	570 ± 161
10–20	3,656 ± 707	2,694 ± 759	726 ± 204
20–50	5,181 ± 818	543 ± 153	351 ± 99
50–100	413 ± 125	2,882 ± 890	0

50-year-old forest fallow at Lailad in Meghalaya

A 50-year-old forest stand at Lailad (296 m altitude) in Meghalaya had tree species such as *Schima wallichii, Castanopsis indica* and *Shorea robusta* as the dominants in its undisturbed core zone (Singh and Ramakrishnan, 1982a). The average biomass values of 137×10^3 kg ha^{-1} yr^{-1} for this stand is lower than that reported by other workers for other tropical forests (Jordan and Kline, 1972) but was within the range of 60–350 t ha^{-1} reported for some temperate and tropical seasonal forests (Whittaker, 1975). The average litterfall value of 5.5 t ha^{-1} yr^{-1} reported for this stand (Singh and Ramakrishnan, 1982b) is lower than 9.7 t ha^{-1} yr^{-1} reported for a 20-year-old fallow at Burnihat, close by (Toky and Ramakrishnan, 1983a). Fast developing vegetation in a younger fallow and rapid changes in species composition, often entire individuals contributing to litter production, result in an over-shoot of litter production. This over-shoot in litter production is responsible for the difference between 20-and 50-year-old fallows (Singh and Ramakrishnan, 1982b).

The net primary productivity of this forest stand at Lailad worked out to be about 16.5 t ha^{-1} yr^{-1} which is also lower than that for a 20-year-old secondary forest at Burnihat (Toky and Ramakrishnan, 1983a). The exploitative nature of the species in the younger stand at Burnihat with fast growth and turnover rates would account for its higher net primary productivity. However, the productivity values for the higher elevation pine stand of a similar age is lower (Mishra and Ramakrishnan, 1983a).

An important characteristic of the biomass distribution below the soil, at this stage of forest development at Lailad, is the occurrence of fine roots in the surface soil layers (Ramakrishnan and Singh, 1983). Complementing the higher fertility in the first 50 cm of the top soil (Ramakrishnan and Toky, 1981; Ramakrishnan and Singh, 1983), about 50% of the fine root biomass is confined to this depth (Table 6.5). This agrees, although less strikingly,

Table 6.6 Soil characteristics of the sacred grove (0–5 cm depth) during the monsoon period at Cherrapunji in Meghalaya (from Khiewtam, 1986)

pH	5.1
Carbon (%)	3.0
Nitrogen (%)	0.32
C/N ratio	9.4
NO_3-N (mg 100 g^{-1})	0.18
PO_4-P (mg 100 g^{-1})	0.02
Potassium (millimole 100 g^{-1})	0.32
Calcium (millimole 100 g^{-1})	1.68
Magnesium (millimole 100 g^{-1})	5.24

with more mature tropical rain forests of Amazonia (Klinge, 1973) and the climax vegetation in the north-eastern India (Khiewtam, 1986). The total root biomass of 34.7 t ha^{-1} in the present study is also comparable with the values for other tropical rain forests (Klinge, 1973). The significance of these observations are obvious. In a 50-year-old stand, as nutrients are released in the surface soil layers and the soil itself is weakly formed and highly leached, a surface-located fine root system is advantageous. As discussed below, as more nutrients become locked up in the living biomass at the climax stage, the root system becomes more surface-feeding and a fine root mat is developed between the litter layer and the mineral soil, for picking up nutrients from the decomposing litter.

Climax vegetation at Cherrapunji

The climax vegetation at higher elevations in Meghalaya, as at Cherrapunji (1,300 m altitude) about 50 km south of Shillong is represented by the sacred grove, a forest protected by the local people for religious and cultural reasons (Khiewtam and Ramakrishnan, 1989). Cherrapunji receives exceptionally high rainfall of 10,372 mm, which may go up to 24,555 mm in an exceptional year such as 1974 (Ramakrishnan, 1985e). Because of a very fragile highly leached soil over limestone formations that have extensive underground tunnels of stalactites and stalagmites (Kratz topography), the soil is highly acidic in spite of calcium (pH 3.9–5.2) and with poor nutrient supply (Table 6.6).

Among the trees in the sacred grove, *Engelhardtia spicata* was the dominant species followed by tree species such as *Syzygium communii Echinocarpus dasycarpus*, *Drimycarpus racemosus* and *Elacocarpus lanceaefolius*. Important shrubs are *Clerodendron nutans*, *Phlogocanthus* spp., *Randia* spp., *Thelypteris* spp., *Strobilanthes coloratus* and *Boehemeria platyphylla*. The vegetation is generally stunted because of extreme environmental conditions, although with a high species diversity. (Table 6.7).

Table 6.7 Shannon index of general diversity (H), index of dominance (C), and evenness index (e), for sacred grove at Cherrapunji in Meghalaya (from Khiewtam, 1986)

	Sacred grove		
Species	*H*	*C*	*e* (%)
Trees and shrubs	2.9	0.08	77.9
Ground layer	3.1	0.07	90.8

Table 6.8 Litterfall pattern in the sacred grove at Cherrapunji in Meghalaya (from Khiewtam, 1986)

Period	*Quantity* ($kg\,ha^{-1}\,yr^{-1}$)	*Rate of fall* ($kg\,ha^{-1}\,yr^{-1}$)
Monsoon season	$6{,}202 \pm 237$	16.9 ± 0.2
Dry season	$5{,}581 \pm 207$	15.3 ± 0.4

Categories : Leaf, 61%; Branch, 20%; Flowers, 1%; Fruits, 12%; Unclassified, 6%

The litterfall in the sacred grove is high, with substantial fall occurring during the monsoon season (Table 6.8). Rainy season leaf fall occurring in overstorey species such as *Englehardtia spicata* and *Drimycarpus racemosus* would help in better light penetration to the forest floor, particularly during this predominantly cloudy period of the year. The annual litter production of 11,783 $kg\,ha^{-1}$ is comparable to mountain rain forests in Columbia (Jenny *et al.*, 1949), lowland rain forest in Ghana (Nye, 1961) or Panama (Hains and Foster, 1977). The fall is comparable with another montane climax forest at Shillong, at an elevation of 1900 m (Boojh and Ramakrishnan, 1982f). However, it is also very close to the values for a 20-year-old fallow at Burnihat at lower elevations of Meghalaya (Toky and Ramakrishnan, 1983a). This is in contrast with similar aged stands at higher elevations of Meghalaya, where lower temperature conditions and highly acidic and nutrient-poor soils affect biomass production and consequently litterfall (Mishra and Ramakrishnan, 1983a).

If one were to calculate the approximate net primary productivity of the sacred grove by multiplying annual litter production by a factor of 3 (Bray and Gorham, 1964), then, the net primary productivity of the sacred grove at Cherrapunji would work out to be 35.3 $t\,ha^{-1}\,yr^{-1}$. However, it may be noted that for a 20-year-old forest fallow at Burnihat, the actual net primary productivity is only about two-times that of annual litterfall; if this calculation is applied here, then the net primary productivity of the sacred grove could be not more than 23.6 $t\,ha^{-1}\,yr^{-1}$ which is only a little higher than the reported values for a 20-year-old fallow at Burnihat (Toky and Ramakrishnan, 1983a), but almost five-times more than that for a 15-year-old forest fallow at Shillong (Mishra and Ramakrishnan, 1983a). However, it is interesting to note here that the net primary productivity of a 22-year-old pure pine plantation at Shillong had a net primary productivity of 20.1 $t\,ha^{-1}\,yr^{-1}$ (Ramakrishnan and Das, 1983; Das and Ramakrishnan, 1987), which is closer to the values for the sacred grove calculated as litterfall × 2.

With a poorly developed and highly leached soil and with Kratz topography, the soil itself has very little nutrients to offer (Table 6.6) to support such a large biomass of the sacred grove at Cherrapunji. The fine root mat development on the surface layers of the soil and that particularly located over the mineral soil are important for supporting the large aboveground biomass and for optimizing cycling of nutrients released through litterfall.

Root growth has been related to environmental conditions, of which soil moisture stress and low soil temperature have often been implicated to limit the growth (Lyr and Hoffman, 1967; Russell, 1977). Our results agree with this. Up to a soil depth of 30 cm, the fine root system below 20 mm diameter was maximum during the wet summer (13,955 $kg\,ha^{-1}$) and minimum during the winter (10,755 $kg\,ha^{-1}$). More than 50% of the fine root biomass was made up of diameter classes less than 6 mm (Fig. 6.2). The data presented for depth-wise distribution of the fine root biomass in the sacred grove show that about 45% of the biomass is located between 0–10 cm soil depth (Table 6.9).

Fine root productivity was looked at for the different diameter classes and at different soil depths (Table 6.10). The productivity was maximum for the finer diameter classes and at 0–5 cm soil depth. This is suggestive of the importance of the active fine root biomass on the surface layers of the soil including that located above the mineral soil. Since fine root biomass in the sacred grove at Cherrapunji did not vary significantly over a 3-year observational period, one could assume that the productivity values broadly are indicative of its annual replacement rate.

The role of a fine root system particularly the surface root mat in direct cycling of nutrients from the decomposing litter is well recognised (Klinge, 1973; Stark and Jordan, 1977; Ramakrishnan and Singh, 1983). Apart from litter nutrients being directly recycled after their release through decomposition, even the nutrient wash-out by rain through direct fall and stemflow may be conserved (Klinge, 1973), through the surface root mat.

Conclusions

The fallow regrowth during the early secondary succession would vary depending upon the earlier land use history – the jhum cycle length, the pre-existing vegetation and the cropping procedures. Apart from these, the ecology of the area is important. Thus, for example, the rate of change is much slower at higher elevations than at lower elevations. This is because

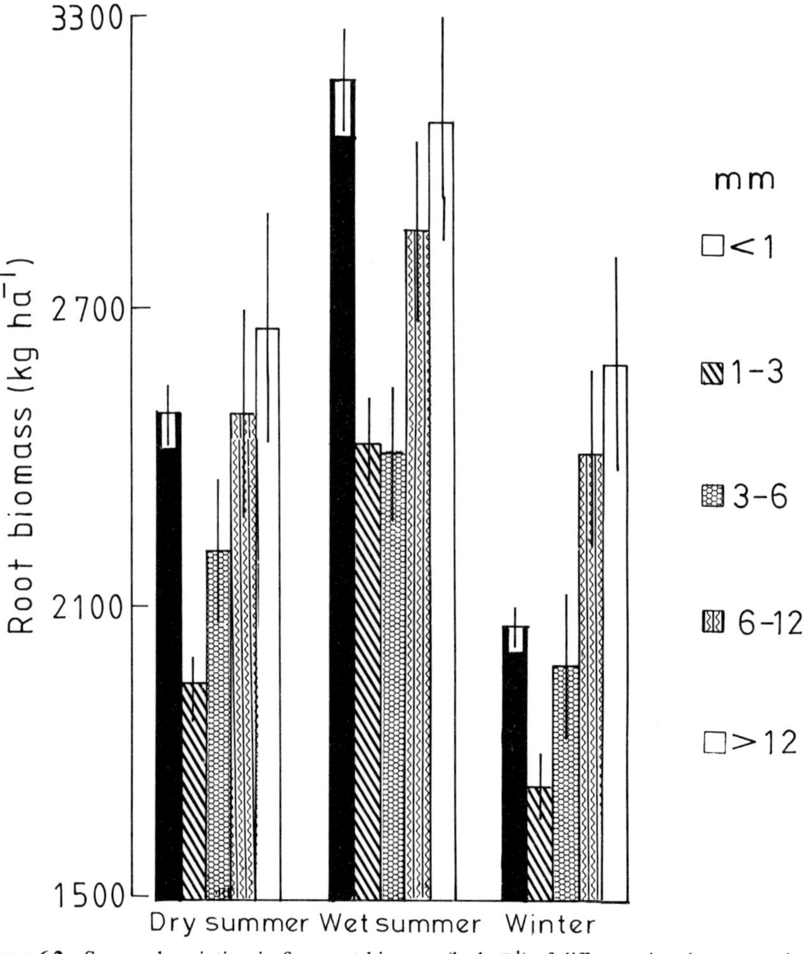

Figure 6.2 Seasonal variation in fine root biomass (kg ha^{-1}) of different size classes ranging below 20 mm diameter class within a soil column of 0–30 cm in the sacred grove at Cherrapungi in Meghalaya. Vertical bars represent SE values (from Khiewtam, 1986)

Table 6.9 Depth-wise distribution of fine root biomass in the soil profle at Cherrapunji in Meghalaya (from Khiewtam, 1986)

Soil depth (cm)	*Biomass* (kg ha^{-1})
0	882
0–5	2,276
5–10	2,912
10–20	3,645
20–30	4,237

Table 6.10 Fine root productivity in the sacred grove at Cherrapunji in Meghalaya (from Khiewtam, 1986)

	Productivity ($kg\,ha^{-1}\,yr^{-1}$)
Diameter class (mm)	
0–1	1,108
1–3	706
3–12	453
> 12	491
Soil depth (cm)	
0 (above mineral soil)	365
0–5	1,093
5–10	738
10–20	325
20–30	679
Total	3,200

cooler temperatures, higher soil acidity and slower mineralization rates under pine forest litter would hamper ecosystem recovery processes. With a thin soil cover under which the system operates, the stability of more mature ecosystem types such as that of a 60-year-old fallow at Lailad in Meghalaya or indeed a climax sacred grove forest is dependent upon efficient nutrient cycling processes being maintained. The development of a large surface-located fine root biomass as has happend in these forest types enables this system to remain stable, although they are highly fragile. The fragility of such a maturely developed ecosystem would be further accentuated if the climatic and edaphic conditions become more unfavourable due to excessive rainfall (reaching up to about 24 m) or 'Kratz' geology (with limestone formations and extensive underground tunnels), as at Cherrapunji in Meghalaya.

ARRESTED SUCCESSION

In north-eastern India, large-scale timber extraction for industrial purposes has been going on during the last few decades (Ramakrishnan, 1985d). Fuelwood extraction for local use and for export outside the region is another forest based activity. According to 1971 census data of the population of the region, this loss through firewood would amount to $189 \times 10^7\,kg\,yr^{-1}$. To top it all, jhum is a major agricultural activity. With increase in population pressure and reduced land area for jhum, partly because of jhum itself and partly because of large-scale timber extraction, the cycle has mostly come down to 5–6 years, with the less frequent cycles of 10–20 years. This has

Plate 6.3 A thicket of impenetrable bamboo forest in pure formation represents an arrested succession

resulted in an arrested succession of weeds or bamboo in the region (Plate 6.3). Vast areas have been converted to these seral communities which tend to remain stabilized in the arrested stage. We shall consider the secondary successional vegetation dynamics that has led to this arrested succession in the region, leading to extensive weed communities and/or bamboo forests.

Arrested succession of weeds

Factors contributing to weed potential under jhum and some of the aspects related to weed potential in early secondary successional fallows including demographic analysis have been discussed at length in Chapter 4. It would be appropriate here just to summarize some of the aspects related to biomass and productivity under an arrested succession of weeds. An outcome of the earlier consideration of weed potential under jhum was that an arrested weed community would result, if a short jhum cycle of 4–5 years is imposed at a site over a period of time. This is because the secondary succession itself is not allowed to proceed beyond the weed stage (Toky and Ramakrishnan, 1983a; Mishra and Ramakrishnan, 1983a) and the seed source of the larger species gets depleted at the expense of a build-up of the weed seed bank in the soil (Saxena and Ramakrishnan, 1984b).

In a 4–5-year-old weed fallow dominanted by *Mikania micrantha* at Lailad in Meghalaya, we have seen how total weed biomass and the net primary productivity of *Mikania micrantha* increased up to a 4–5-year-old fallow (cf Table 4.4). Along with these changes, the contribution of litter by *Mikania micrantha* and other herbs also increased only up to a 4-year-old fallow, *Mikania* alone contributing 58–69% of the litter during this period (Swamy and Ramakrishnan, 1987b).

In an arrested succession of weeds, we have both C_3 and C_4 plant species. The C_3 species with large biomass contributions to early secondary successional communities are all exotics, including the *Eupatorium* species and *Mikania micrantha.* On the other hand, all of the important native species, including *Imperata cylindrica, Saccharum* spp., *Panicum* spp. and *Thysanolaena maxima* have the C_4 photosynthetic pathway. Several less abundant C_3 species occur among the natives.

The biomass of C_3 species is higher than that of C_4 species in early succession (Saxena and Ramakrishnan, 1984c). The rapid growth of a species such as *Eupatorium odoratum* (Fig. 6.3) is achieved partly through a larger light interception surface, and partly through more allocation of resources to the shoot system. Similar results were obtained for *Mikania micrantha* (Swamy and Ramakrishnan, 1987a, 1988b). These results suggest that C_4 species are not always more productive than C_3 species (Caldwell *et al.*, 1977); the productive potential of plants depends on factors in addition to photosynthetic rate per unit leaf area (Black, 1971).

Continuous imposition of short jhum cycles of 4–6 years results in early successional communities monopolized by exotic C_3 species (Saxena and Ramakrishnan, 1984c; Ramakrishnan and Vitousek, 1989). In contrast, long jhum cycles impoverish the soil seed bank owing to more intense fire effect (Saxena and Ramakrishnan, 1984b), thereby affecting the seed-propagated C_4 species. Thus, under an arrested succession resulting as a consequence of constant imposition of short jhum cycles, the weed communities would have domination by exotics. These aspects related to plant strategies are considered in greater detail in Chapter 8.

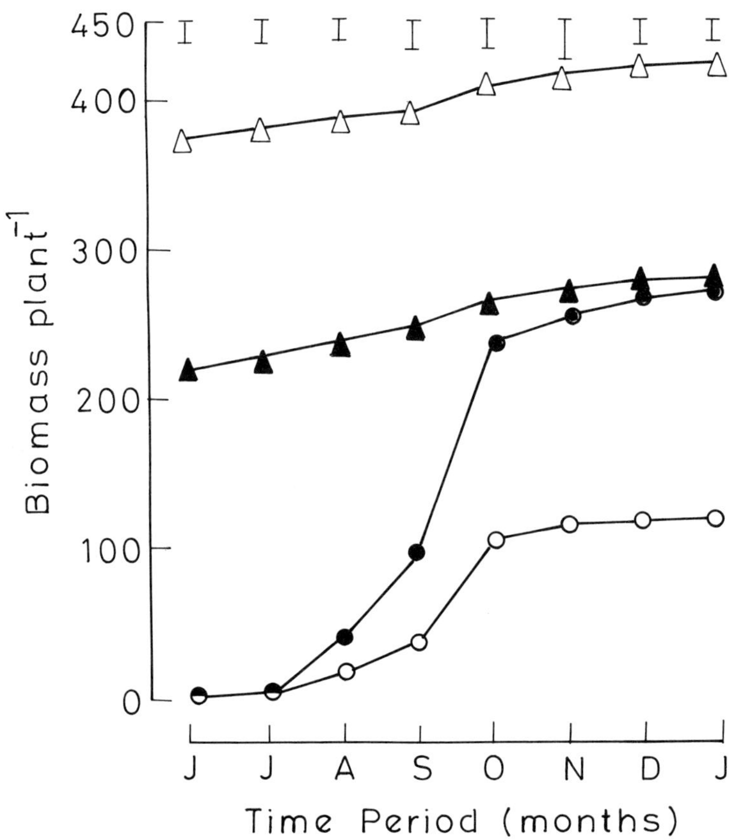

Figure 6.3 Growth curves of *Eupatorium odoratum* in different jhum fallows. ●, recent fallow after the burn and uncropped; ○, 2-yr-old fallow; and ▲, 6-yr-old fallow. Vertical bars represent LSD ($p = 0.05$) from Saxena and Ramakrishnan, 1984a)

Seral grasslands at Cherrapunji – an arrested succession

Cherrapunji (1,313 m altitude) located 50 km south of Shillong in Meghalaya has a typical monsoonic climate with a annual average rainfall of 10,372 mm (mean for 1973), about 96% of which occurs between April and October. The mean monthly maximum and minimum temperatures during the summer are 24°C and 14°C, respectively. Indeed, this area is considered to be one of the wettest places in the world where annual rainfall may be > 24,500 mm. November to February are winter months of which February is relatively dry. Mean monthly maximum temperatures for the winter period are 17°C and 6°C, respectively. The dry summer lasts only 1 transitional month

Table 6.11 Soil characteristics under grasslands at Cherrapunji in Meghalaya (from Ramakrishnan and Ram, 1988)

	Range
Moisture (%)	14–17
pH	5.2–5.8
Organic Carbon (%)	1.4–2.0
Potassium (millimole $100\,g^{-1}$)	0.10–0.20
Calcium (millimole $100\,g^{-1}$)	0.67–1.30
Magnesium (millimole $100\,g^{-1}$)	1.60–2.90
Nitrate – N (mg $100\,g^{-1}$)	0.10–0.29
Phosphate – P (mg $100\,g^{-1}$)	0.08–0.18

(March) between the winter and the rainy seasons, with mean monthly maximum and minimum temperatures of 20°C and 12°C, respectively.

The grasslands at Cherrapunji represent varied levels of site degradation, and are arrested seral communities in common with the rest of the Indian sub-continent. At Cherrapunji, the climax is represented by the sacred groves occuring as a relict vegetation (Ramakrishnan *et al.*, 1981b; Khiewtam and Ramakrishnan, 1989). These climax rain forests are extremely fragile, and disturbance leads to different types of savanna grasslands. Clear-cutting of the forest for jhum in the past possibly has contributed to grassland formation in this high rainfall area. After clear-cutting of the forest, natural recovery processes on steep slopes are retarded because of rapid wash-out of the top soil and nutrients, so that these grasslands are often maintained in an arrested state. Gradual elimination of seed source of trees has led to the disappearance of this component. This was confirmed by a study where we could record tree seedling regeneration occurring in an *Obseckia* type grassland (see below), when the seed source is available near-by. This system is particularly interesting because this is the first study, as far as we know, on grasslands developed under an annual rainfall of over 10,000 mm (Ramakrishnan and Ram, 1988; Ram and Ramakrishnan, 1988a,b).

Vegetation

These grasslands are developed on soils that are generally acidic, highly leached and nutrient poor (Table 6.11). Of the four grassland types recognised, the *Osbeckia* type is developed on deeper soils and is often subject to frequent fire during the dry season. The *Arundinella* type (Plate 6.4) is derived from burning the *Osbeckia* type and would revert to the latter in the following year in the absence of fire. Regular burning at 2-year intervals results in the development of the *Ischaemum* type grassland. On slopes where the soil is very thin, the grasslands remain in a desertified stage (*Eragrostiella* type) even without fire. The soil pH is very low here and the nutrient status is the poorest.

Plate 6.4 *Arundinella* type of grasslands of the desertified Cherrapunji (1,300 m altitude) in Meghalaya developed on soils of a mean depth of about 40 cm are maintained through yearly fire during the summer

Many species are shared between the *Osbeckia* and the *Arundinella* types (Table 6.12). After the burn, some species such as *Gentiana quadrifolia* and *Ischaemum goeblii* appeared, others, for example *Phyllanthus simplex* increased in number, while others such as the shrub *Osbeckia crinita* declined. Species composition of the *Ischaemum* type is similar to that in the *Arundinella* type, except for increases in *Ischaemum goeblii, Fimbristylis complanata* and *Osbeckia capitata.* These species are also in common with the *Eragrostiella* type.

In the *Osbeckia* type, perennial species constituted about 58% of the species in the community. Annuals were more important in the burnt sites, contributing 60% of the species in the *Ischaemum* type, and even more on desertified land (*Eragrostiella* type). It seems reasonable to conclude that fire may cause an increase in annuals on these grasslands (Reynold and Bohning, 1956; Daubenmire, 1968) and a shift from dicot species to grasses (Humphrey, 1958; Sharrow and Wright, 1977).

Soil fertility may also contribute to the shift from perennials to annuals. This trend was evident on the desertified *Eragrostiella* type (Plate 6.5). However, the success of perennial species such as *Arundinella bengalensis, Chrysopogon gryllus* and *Carex cruciata* in infertile soils is because of well-developed rhizomes that help in regeneration after the burn.

Table 6.12 Importance value indices (IVI) in species of grasslands at Cherrapunji in Meghalaya. Species occurring with values < 5% are omitted (from Ramakrishnan and Ram, 1988)

Species	Osbeckia *type*	Arundinella *type*	Ischaemum *type*	Eragrostiella *type*
Grasses and other monocots				
Arundinella bengalensis	24.6	32.8	14.6	21.1
Arundinella khaseana	–	11.2	8.3	3.6
Arundinella nepalensis	13.4	2.3	5.4	–
Axonopus compressus	–	–	5.4	–
Capilipium assimale	2.3	–	5.4	–
Carex cruciata	21.3	14.5	6.5	4.4
Chrysopogon gryllus	24.0	20.3	8.2	21.1
Dimeria fuscescens	–	–	12.1	17.1
Eragrostiella leioptera	–	–	12.7	40.2
Eulalia trispicata	6.4	21.2	9.6	–
Eulalia quadrinervis	10.0	15.1	–	–
Fimbristylis complanata	–	11.1	38.5	26.6
Fimbristylis thomosonii	20.1	–	18.1	2.9
Heteropogon contortus	5.2	–	–	–
Ischaemum goeblii	–	19.1	53.3	21.9
Paspalum arbiculare	10.5	25.8	1.8	10.8
Setaria glauca	14.0	19.9	5.6	–
Forbs				
Anotis oxyphylla	–	–	–	5.3
Borreria articularis	–	–	–	29.5
Gentiana quadrifaria	–	20.9	24.0	13.2
Leucus ciliata	5.7	–	–	–
Osbeckia capitata	10.2	9.6	20.8	18.0
Phyllanthus simplex	11.9	24.3	11.5	13.6
Plectoranthus japonicus	16.3	3.7	3.1	–
Pogostemon quircularis	3.1	10.0	10.7	2.4
Shrubs				
Melastoma malabathricum	1.1	6.0	–	–
Osbeckia crinita	51.5	19.9	4.7	23.0

Biomass

Perennials contributed 71–83% of the total living biomass in these grasslands (Table 6.13). Biomass increased slightly during the 2nd year after burning. Soon after the burn, standing dead biomass and litter declined drastically, but recovered rapidly during subsequent years. It may be noted here that the *Osbeckia* type was last burnt 6 years prior to the initiation of the study; the *Arundinella* type had similar conditions to the *Osbeckia* type, except for an additional burn executed in March, 1 month before the study period. The *Ischaemum* type had 2-yearly regular burns, the last being in March, 1 month before the study period (Ramakrishnan and Ram, 1988).

Plate 6.5 *Eragrostiella* type of grassland of the desertified Cherrapunji (1,300 m altitude) in Meghalaya occurs on highly degraded slopes with partially weathered mineral soil

More than 70% belowground biomass was located in the top 0–10 cm of soil. Under the *Ischaemum* type about 81% of the root biomass was present in this zone. The *Eragrostiella* type had lower belowground biomass compared to others.

The *Ischaemum* type generally had a lower net primary productivity than the *Osbeckia* and *Arundinella* types. The *Eragrostiella* type had the lowest productivity. For the *Ischaemum* and *Arundinella* types, aboveground productivity declined after burning but recovered in the following year. This was also reflected in the total productivity of the system. Belowground productivity however, did not change much.

The decline in aboveground biomass after burning has been observed by others too (Smith, 1960; San Jose and Medina, 1976). However, the accompanied decline in aboveground productivity only in the following years, and not soon after, contrasts with those reported in many studies (Kucera and Ehrenreich, 1962; Hadley and Kieckhefer, 1963) where biomass and productivity in the 1st year after burning exceeded that of unburnt plots. A time lag in the recovery to the original level after the burn, as noted here was also observed by Cook (1939), Smith (1960) and Mott (1982). Extreme climatic conditions with heavy losses of nutrients through water erosion (Ram and Ramakrishnan, 1988b) could be responsible for this time lag in recovery.

Table 6.13 Contribution of plant biomass ($g\,m^{-2} \pm$ SE) and primary productivity $g\,m^{-2}\,yr^{-1}$) in grasslands at Cherrapunji in Meghalaya (from Ramakrishnan and Ram, 1988)

	Osbeckia *type*	Arundinella *type*		Ischaemum *type*		Eragrostiella *type*
	1983	*1983*	*1984*	*1983*	*1984*	*1983*
Aboveground biomass						
Perennials living	334 ± 25.7	261 ± 21.8	266 ± 23.5	205 ± 18.5	231 ± 18.0	146 ± 12.4
Annuals living	67 ± 4.9	79 ± 6.1	79 ± 6.4	76 ± 5.9	82 ± 6.8	60 ± 4.6
Total living	401 ± 18.4	340 ± 24.2	345 ± 27.9	281 ± 7.2	313 ± 23.9	206 ± 11.8
Standing dead	169 ± 15.0	28 ± 1.9	129 ± 6.8	21 ± 0.7	118 ± 6.1	49 ± 4.5
Litter	72 ± 4.8	5 ± 0.3	49 ± 3.2	4 ± 0.2	65 ± 3.0	33 ± 3.0
Total biomass	642 ± 29.2	372 ± 26.5	523 ± 30.2	306 ± 23.2	496 ± 31.2	288 ± 24.2
Belowground biomass						
Soil 0–10 cm	820 ± 65.2	773 ± 47.4	780 ± 49.4	797 ± 62.7	811 ± 54.3	555 ± 39.0
10–20 cm	159 ± 10.2	172 ± 13.0	182 ± 9.6	114 ± 7.7	117 ± 7.0	130 ± 6.3
20–30 cm	109 ± 8.3	111 ± 8.6	115 ± 6.2	76 ± 4.7	78 ± 5.0	49 ± 2.8
Soil Total	1,088 ± 81.4	1,056 ± 41.6	1,077 ± 75.2	987 ± 71.0	1,006 ± 71.5	734 ± 26.6
Primary productivity						
Aboveground	579	420	569	357	485	346
Belowground	990	986	997	932	950	631
Total	1,569	1,406	1,566	1,289	1,435	977

Table 6.14 Contribution of different categories of litter (g m^{-2}) in grasslands at Cherrapunji, in Meghalaya (from Ram and Ramakrishnan, 1988c)

	Site							
	Osbeckia		Arundinella		Ischaemum		Eragrostiella	
Litter category	*type*	%	*type*	%	*type*	%	*type*	%
Dicot leaves	80	16	39	13	8	3	36	17
Dicot twigs	98	19	39	13	8	3	32	14
Grasses	266	52	168	56	188	71	118	55
Sedges and other monocots	66	13	54	18	61	23	29	14
Total	510 ± 46.7		300 ± 29.1		265 ± 24.4		215 ± 19.1	

Litter

Soon after fire, which is an important element in the *Arundinella* and *Ischaemum* type grasslands, litter production also declined and the recovery started only in the 2nd year after the burn (Ram and Ramakrishnan, 1988c); this agrees with the biomass recovery pattern discussed above. This may be related to extreme nutrient deficiency in the soil at Cherrapunji, for where soil fertility is greater litter production recovery would be quicker, occurring soon after the burn.

One of the important consequences of the burn is a decline in the litter contribution through dicot leaves and twigs and an increase in the litter contribution through grasses, sedges and other monocots (Table 6.14) due to a shift in community structure from perennials to annuals, noted earlier. A shift from dicots to monocots, particularly to grasses, may be advantageous after severe perturbation of an already stressed environment because of the better stress tolerance of C_4 grasses. Thus, the physiological advantages conferred by the C_4 photosynthetic pathway in reducing photorespiratory losses and maintaining a high photosynthetic/transpiration ratio, over the C_3 pathway are expected to be maximal following the burn when incident solar radiation and temperture would be higher, imposing a higher water stress (Saxena and Ramakrishnan, 1984c).

A massive transfer from live to dead standing mass (and eventually to the litter compartment) occurred both during the post-monsoon period after maximum herbage production in the rainy season and also in March which is the driest month of the year at Cherrapunji (Ram and Ramakrishnan, 1988c). This trend is similar to that of the monsoonic grassland of Kurukshetra (Singh and Yadava, 1974).

While 100% of the aboveground productivity of grasses and monocots is converted into litter, the conversion of aboveground net primary productivity of dicots into litter is not total. Therefore, an average of 62–88% of the aboveground net primary productivity of these grasslands is transferrred to

Table 6.15 Turnover rates for aboveground (AG) and belowground (BG) components of seral grasslands at Cherrapunji in Meghalaya (from Ram and Ramakrishnan, 1989)

	Turnover rate		*Turnover time*	
Grassland type	*AG*	*BG*	*AG*	*BG*
Osbeckia	0.90	0.91	1.11	1.10
Arundinella	1.13	0.93	0.89	1.07
Ischaemum	1.16	0.94	0.86	1.06
Eragrostiella	1.20	0.86	0.83	1.16

the litter compartment, 62% being under *Eragrostiella* type and 88% under *Osbeckia* type, the other two falling in between; this is comparable with the monsoonal grassland at Kurukshetra (73%). The seral grasslands at Cherrapunji have litter production (215–510 $g\,m^{-2}\,yr^{-1}$) less than that observed in the sacred grove climax forest here (1,178 $g\,m^{-2}\,yr^{-1}$) (Kheiwtam, 1986) and lower than many other temperate and tropical grasslands (Ram and Ramakrishnan, 1988c). Such low values for these grasslands developed on infertile highly leached soils are to be expected, the *Eragrostiella* type on desertified slopes having the least values.

Primary productivity

Total net primary productivity for these grasslands (Table 6.13) is generally lower than the values reported by others under a monsoonic climate (Ramakrishnan and Ram, 1988). However, the more interesting point that emerges is the relative biomass and productivity of aboveground versus belowground parts. Thus the shoot/root ratio in the grassland types subjected to frequent fire is lower, compared to the *Osbeckia* type not subjected to frequent fire. This obviously is an adaptation to withstand stress due to perturbation (Saxena and Ramakrishnan, 1983a). The *Eragrostiella* type on highly nutrient deficient soils is also always under stress and therefore a low shoot/root ratio is to be expected; more root surface helps in nutrient absorption (Chapin, 1980) and vegetative reproduction (Keely and Keeley, 1977).

In most studies, belowground productivity is reported to be almost half that of the aboveground productivity; the reverse is true at Cherrapunji. Higher productivity of belowground parts may partly be due to the need for efficient uptake of nutrients under low soil fertility (Nye and Tinker, 1977) and partly because of drought conditions (Struik and Bray, 1970) due to poor soil moisture retention in spite of high rainfall.

Turnover rate for the aboveground component was minimal under *Osbeckia* type and maximal under *Eragrostiella* type; the latter had a minimal value for the belowground part (Table 6.15). A higher turnover rate for the aboveground component of grasslands under stress would ensure effective recycling of nutrients, while lower turnover rates for belowground parts would ensure survival through storage of resources (Ram and Ramakrishnan, 1989). The annual turnover of biomass of the belowground parts of about 1, is comparable to that of the aboveground parts and agrees with the observation on another monsoon grassland in Kurukshetra, India (Singh and Yadava, 1974). This suggests that belowground biomass is highly dynamic in the tropics and is almost replaced every year, as much as the aboveground compartment. This is unlike the situation in temperate grasslands, where the reported values are often much lower ranging from 0.19 to 0.69 (Sims and Singh, 1978). Obviously, the degraded and arrested grasslands at Cherrapunji which are subjected to a variety of stresses, show ecosystem properties adapted for survival under stress.

Bamboo forests : a case of arrested succession

The entire north-eastern hill region is characterized by extensive bamboo formations, which have considerable socio-economic value for the local communities (Rao and Ramakrishnan, 1987b). These bamboo forests often represent arrested secondary successional stages after jhum in north-eastern India. As already discussed, bamboos appear during secondary succession, after the first 5–6 years of herbaceous weeds and dominate the community during the first 10–25 years as in *Dendrocalamus hamiltonii* an important species in north-eastern India; beyond 25 years or so they decline (Toky and Ramakrishnan 1983a; Rao and Ramakrishnan, 1987a). Others may extend up to about 35 years of fallow regrowth. Because of their specialized growth habit, they form conspicuously large formations. Bamboos flower once in their lifetime, which may be sporadic, once in 30–40 years in *Dendrocalumus hamiltonii* (Canvendish, 1905; Gupta, 1972; Janzen, 1976; Varmah and Bahadur, 1980), or gregariously, once in 35–40 years or so (Rao and Ramakrishnan, 1987a). In the small or large gaps formed due to death of individuals which have already flowered, saplings of trees take over and suppress these light demanding bamboo species, leading to their elimination. In other words, some of the important species of north-eastern India, such as *Dendrocalamus hamiltonii*, *Bambusa tulda*, *Bambusa khasiana* and *Neohouzeoua dulloa* usually dominate the stand up to 25–35 years or so of the fallow regrowth after which they eventually disappear or at least drastically decline. While some species that are strictly light-demanding, such as *Dendrocalamus hamiltonii* are eliminated as light availability in the community declines after a period of about 25 years of fallow regrowth, others which can tolerate shade but prefer a high light environment, such as *Neohouzeoua dulloa*, may continue even in older communities (Rao and

Ramakrishnan, 1987a). However, vast areas of bamboo forests in north-eastern India and elsewhere in the region are a consequence of large-scale disturbance such as jhum (Haig *et al.*, 1958; Drew, 1974; Rao and Ramakrishnan, 1987a). They remain in an arrested successional stage if jhum cycles of 10–30 years are constantly imposed on the same site (Rao and Ramakrishnan, 1987a, 1988a,b), just as an arrested weed stage of succession happens if a short 4- or 5-year jhum cycle is constantly imposed on the same site (Saxena and Ramakrishnan, 1984b).

Population dynamics

This study was done on two bamboo species, at Lailad (200 m altitude) in Meghalaya, with constrasting ecological characteristics – *Dendrocalamus hamiltonii* a deciduous bamboo of the lower elevations up to an altitude of 1,200 m, a strict light demander occurring in fallow regrowths of about 5–30 years, and *Neohouzeoua dulloa*, an evergreen species of similar altitudinal range, a light demander but with some tolerance for shade occuring in similar fallow regrowths but extending into a 60-year-old fallow, as an undergrowth (Rao and Ramakrishnan, 1987a).

In early successional communities of up to 25 years of fallow regrowth *Neohouzeoua dulloa* is vigorous, but exclusively based on clonal reproduction. Gregarious flowering in this species supposed to be at an interval of about 35 years (Janzen, 1976) had not yet occurred as predicted (Rao and Ramakrishnan, 1987a). The clumps established through seedlings here are generally more than 1 m^2 but have vigorous vegetative reproduction through culms established from a close-knit but extensively branched underground rhizome system. The number of culms may be about 50 under more favourable environmental conditions. In older fallows, the species grows as an understorey component, but the clump size is much less, with only half the number of culms per clump. The culms attain a height of 4–5 m and a diameter of 3–6 cm.

Dendrocalamus hamiltonii, with sporadic flowering, observed during the study period has both clonal and seedling regeneration. The number of culms per clump may be up to 60, each culm being 16–18 m long and 12–25 cm in diameter. This species occassionally found in a 60-year-old fallow, always occurred in the disturbed peripheral zone where light penetration is adequate (Singh and Ramakrishnan, 1982a).

Starting with a low recruitment of culms in a 1-year-old fallow, ultimate population size of both species improved up to 20 or 25 years of fallow age (Table 6.16). The population size of the more shade-tolerant *Neohouzeoua dulloa* in a 60-year-old fallow was not very much altered, but that of the less shade-tolerant *Dendrocalamus hamiltonii* declined. The increment in the number of culms per unit area with fallow age up to 20–25 years may be

Table 6.16 Population flux of the culms of *Neohouzoua dulloa* 1–60-year-old successional fallows at Lailad in Meghalaya. Values in parentheses are for *Dendrocalamus hamiltonii* (from Rao and Ramakrishnan, 1987a)

	Fallow age (yr)						
	1	*5*	*10*	*15*	*20*	*25*	*60*
(a) Number of culms ha^{-1} at the beginning of study (April-May 1983)	400 (1,000)	2,000 (2,200)	5,200 (5,500)	4,600 (6,000)	5,000 (6,900)	3,300 (7,300)	3,000 (150)
(b) Number of culms produced ha^{-1} during study period	1,100 (700)	1,600 (1,900)	2,100 (2,900)	2,300 (3,100)	2,300 (4,000)	1,500 (1,800)	1,200 (26)
(c) Number of culms lost ha^{-1} during study period	400 (200)	800 (800)	1,100 (1,400)	700 (1,400)	700 (2,100)	300 (1,100)	300 (16)
(d) Number of culms ha^{-1} at the end of study (April–May 1984)	1,100 (1,500)	2,800 (3,300)	6,200 (7,300)	6,000 (7,700)	6,000 (8,800)	4,600 (8,000)	3,900 (160)
(e) Number of culms ha^{-1} recorded during study period	1,500 (1,700)	3,600 (4,100)	7,300 (8,400)	6,900 (9,100)	7,300 (10,900)	4,800 (9,100)	4,200 (176)
(f) Net change (d − a)	700 (500)	800 (1,100)	1,000 (1,800)	1,400 (1,700)	1,000 (1,900)	1,300 (700)	900 (10)
(g) Rate of change (d/a)	1.8	1.4	1.2	1.3	1.2	1.4	1.3
(h) Percentage annual mortality (c/e × 100)	26.7 (11.8)	22.2 (19.5)	14.1 (16.7)	10.1 (15.4)	9.6 (19.3)	6.3 (12.1)	7.1 (9.1)
(i) Growth index of population (b/a)	2.8 (0.7)	0.8 (0.9)	0.4 (0.5)	0.5 (0.5)	0.5 (0.6)	0.5 (0.3)	0.4 (0.2)

Table 6.17 Recruitment of the culms of *Neohouzoua dulloa* through each monthly cohort in 1–60-year-old successional fallows at Lailad in Meghalaya (expressed as a percentage of the total recruitment and percentage contribution to the population by different cohorts) (from Rao and Ramakrishnan, 1987a)

	Fallow age (yr)						
Monthly cohort	*1*	*5*	*10*	*15*	*20*	*25*	*60*
May							
% of tot. recruitment	27.3	56.3	42.9	39.1	47.8	46.2	55.6
% contribution	27.3	17.9	12.9	8.1	2.9	13.0	12.8
June							
% of tot. recruitment	45.5	25.0	33.3	34.8	30.4	38.5	22.2
% contribution	18.2	17.1	6.5	8.1	8.1	10.4	5.1
July							
% of tot. recruitment	27.3	18.8	14.3	17.4	13.0	15.4	28.2
% contribution	18.2	3.6	3.2	4.8	1.6	4.3	5.1
August							
% of tot. recruitment	0.0	0.0	9.5	8.7	8.7	0.0	0.0
% contribution	0.0	0.0	3.2	1.6	3.2	0.0	0.0

related to the age of the stand, as shown under plantation conditions in Japan (Udea, 1960). While the reduced vigour of bamboo in a 1-year-old fallow may be a consequence of disturbance at the site during jhum operations, the low vigour in a 60-year-old fallow may be a consequence of shade stress to which these light demanding species are subjected to (Rao and Ramakrishnan, 1987a, 1988a).

More monthly cohorts (four) in 10- to 20-year-old fallows as compared to three cohorts only in others, also support the observation regarding greater vigour of the two species at these two sites (Table 6.17). The larger recruitment of culms through the first cohort and the greater contribution of this cohort to the population structure, although only next to the pre-existing population, helps in extending the favourable growth time period for the population as a whole; similar conclusious were made by Doust (1981) for *Ranunculus repens.*

Seedling regeneration in *Dendrocalamus hamiltonii* is very low (Table 6.18). Heavy predation by jungle fowl, other migratory birds and red ants is an important factor. The low seed production during sporadic flowering may not offer sufficient attraction for other large animals reported for gregariously seeding species (Janzen, 1971, 1976). Although it is difficult to predict which of the clumps would flower in any given year, our observations over a 4-year period suggest that those clumps ready to flower conserve the resources within, by not producing new culms for a year, or sometimes for 2 years as in *Dendrocalamus strictus* (Troup, 1921). Flowering resulted in total death of both above-and-belowground organs unlike that reported for *Phyllostachys*

Table 6.18 Population flux of *Dendrocalamus hamiltonii* seedlings in 5–25-year-old successional fallow at Lailad in Meghalaya. Values are shown for plots with and without litter cover (from Rao and Ramakrishnan, 1987a)

	Fallow age (yr)							
	5		*10*		*20*		*25*	
	with litter	*without litter*	*with litter*	*without litter*	*with litter*	*without litter*	*with litter*	*without litter*
(a) Number of seedlings m^{-2} at the begining of study (May 1983)	212	5	346	89	528	34	599	13
(b) Number of seedling recorded m^{-2} (produced) during study period	810	36	1,221	296	2,276	263	2,088	32
(c) Number of seedling lost m^{-2} during study period	804	35	1,208	293	2,251	258	2,071	318
(d) Number of seedlings m^{-2} at the end of study (May 1984)	6	1	13	3	25	5	17	3
(e) Net change (d–a)	−206	−4	−333	−86	−503	−29	−582	−10
(f) Rate of change (d/a)	0.03	0.2	0.04	0.03	0.05	0.15	0.03	0.23
(g) Percentage annual mortality (c/b × 100)	99.26	97.22	98.94	98.99	98.94	98.10	99.19	99.07

Table 6.19 Reproductive output of *Dendrocalamus hamiltonii* in 5–25-year-old successional fallows at Lailad in Meghalaya (from Rao and Ramakrishnan, 1987a)

	Fallow age (yr)			
	5	*10*	*20*	*25*
(a) Clumps ha^{-1}	115	115	116	115
(b) Percentage fertile clumps ha^{-1}	4.39	5.26	3.45	3.95
(c) Number of culms per clump	12	29	31	31
(d) Number of inflorescences per culm	346	461	489	501
(e) Seeds per inflorescence per culm	15	58	68	73
(f) Seeds per culm	5,347	25,632	33,532	36,623
(g) Seed production (kg) per clump	1.6	8.0	10.0	11.0
(h) Seeds ha^{-1}	26,994	155,048	134,195	166,360

bambusoides from Japan (Numata *et al.*, 1974), where the post-flowering phase may have a few slender culms arising out of largely dead rhizomes. Peak seedling recruitment in a 20-year-old fallow suggests peak performance of this species during this secondary successional phase after jhum. Increase in seed production in up to 25 years of fallow regrowth is related to increased culm production and the larger biomass of individual culms (Table 6.19).

Biomass contribution

Three arrested successional bamboo forests at Machipani in the Garo hills of Meghalaya, Lailad in the Khasi Hills of Meghalaya and Chenkii in the Nagaland, at an altitudinal range of 200–300 m were studied with regard to biomass and litter production. All the three sites had *Dendrocalamus hamiltonii*; additionally *Bambusa tulda* was found only in the Garo hills, *Neohouzeoua dulloa* in the Khasi hills and *Bambusa khasiana* in the Naga hills, in successional fallows ranging between 5 and 25 years (Table 6.20). The results presented here are in agreement with our studies at Burnihat (cf Table 6.1 to 6.4). Since these sites consistently have had 5- to 25-year jhum cycles, the vegetation during the fallow phases represent arrested succession (Rao and Ramakrishnan, 1989). The population size of all bamboo species except *Neohouzeoua dulloa* increased up to 25 years of fallow regrowth; this species showed a decline in the 25-year-old fallow.

In these arrested secondary successional fallows, bamboos contributed a larger proportion of biomass than trees and shrubs or herbs (Fig. 6.4), but as woody biomass increased, that of herbs decreased. Despite its numerical inferiority at the Garo hills site (cf Table 6.20), *Dendrocalamus hamiltonii* produced more aboveground biomass than the other bamboo species, at all sites, although the differential varied with fallow age. In the north-eastern

Table 6.20 Density (stems m^{-2}) and basal area ($cm^2\,m^{-2}$) of bamboo species *Dendrocalamus hamiltonii, Bambusa tulda, B. khasiana* and *Neohouzeoua dulloa* under different aged fallows before slashing for jhum in north-eastern India (from Rao and Ramakrishnan, 1989)

	Garo hills		*Khasi hills*		*Naga hills*	
Fallow age	D.hamiltonii	B.tulda	D.hamiltonii	N.dulloa	D.hamiltonii	B.khasiana
5-year						
Density	0.3	0.7	0.2	0.2	0.4	0.2
Basal area	30.0	70.0	28.7	5.8	43.0	35.0
10-year						
Density	0.5	0.7	0.6	0.5	0.8	0.5
Basal area	75.0	75.5	74.3	9.0	90.1	55.0
15-year						
Density	0.7	0.8	0.6	0.5	0.9	0.5
Basal area	80.0	83.0	87.1	11.0	101.2	79.6
20-year						
Density	0.7	0.8	0.7	0.5	0.9	0.5
Basal area	99.7	107.3	112.5	12.4	121.4	91.7
25-year						
Density	0.7	0.7	0.7	0.3	0.9	0.5
Basal area	121.8	105.3	134.2	7.0	135.7	100.2

region as a whole, although many bamboo species are present, *Dendrocalamus hamiltonii* forms an important component of the arrested bamboo succession.

The biomass accumulation rate for *Dendrocalamus hamiltonii* and *Neohouzeoua dulloa* peaked in a 15-year-old fallow suggesting that growth conditions are, perhaps, more favourable here than in others (Rao and Ramakrishnan, 1988a). Further, these studies also show that initially during succession, there was a rapid transfer of resources from below- to aboveground organs and a subsequent retranslocation of resources in older fallows. This strategy helps in quick initial growth of bamboo followed by extensive underground stoloniferous growth, for survival during the following slash-and-burn operation.

Conclusions

Arrested succession of weeds and bamboos operate under short or relatively long jhum cycles. With a 4- to 5-year jhum cycle being predominant in the region, weeds have taken over vast tracts of land. Where the cycles are between 10–30-years, bamboos form important and extensive formations. Both these categories of species (weeds and bamboos) are light demanders with highly efficient seed/vegetative reproduction; they, therefore, ensure their dominance in the community through rapid spread and quick exploitative growth strategy. These early successional category of species are able to

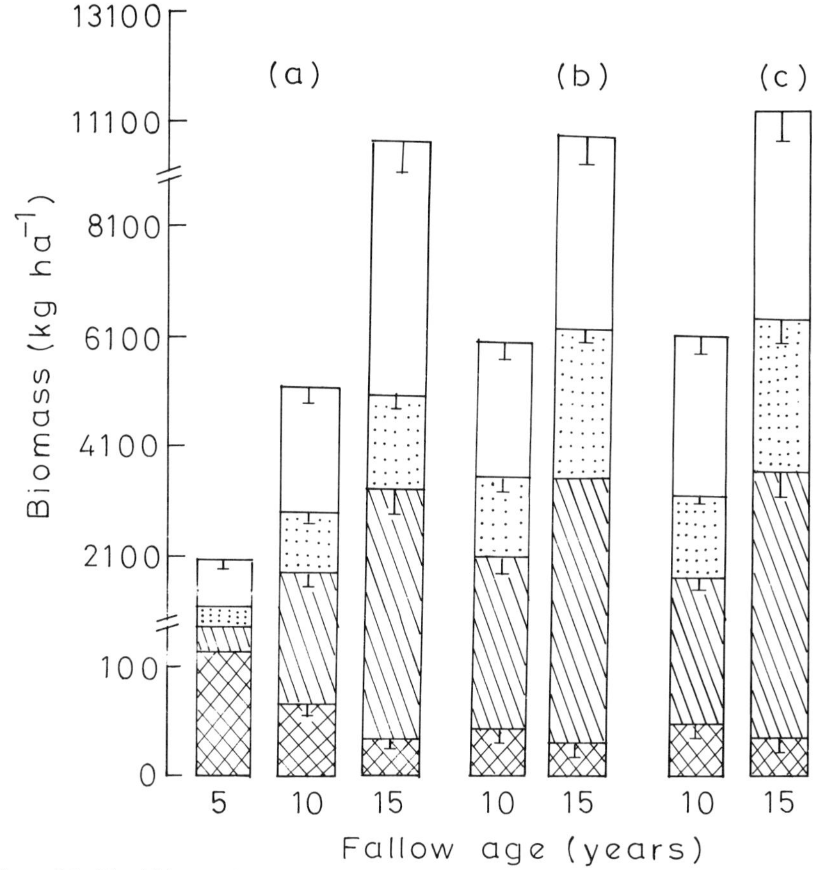

Figure 6.4 Total biomass in secondary successional 5-, 10-, and 15-yr-old fallows in (a) the Garo hills, (b) the Khasi hills and (c) the Naga hills in north-eastern India. Vertical bars represent 95% confidence limits. Double-hatched column, herbs; hatched column, *Dendrocalamus hamiltonii*; stippled column, the other bamboo species (cf text); and open column, shrubs and trees (from Rao and Ramakrishnan, 1989)

capitalize upon the resources of an enriched substratum, which is transient, indeed a useful strategy for any species establishing after slash-and-burn of a forested ecosystem. It is reasonable to expect that they would play an important role in nutrient conservation under jhum as discussed in Chapter 5 and 7. When a given jhum cycle is imposed on vast tracts of land over a period of time, the germ plasm of larger shrubs and trees is eliminated and almost pure formations of weed/bamboo of arrested succession result. The situation becomes further accentuated under extreme edaphic and climatic conditions where degraded grassland formations develop, as at Cherrapunji.

GENERAL CONSIDERATIONS

Summarized here are patterns of invasion, colonization and extinction of species and populations during secondary succession, with emphasis upon changes in vegetation structure, biomass and productivity. The frequency, intensity, the extent of perturbation, and fragility of the forest ecosystem would determine the secondary successional processes, and arrested succession of weeds, grasslands or bamboo forests. Extensive bamboo formations under 10- to 20-year jhum cycles over large areas represent one of the stages in the arrested succession. Land degradation and site desertificaton have two subsequent component stages: arrested weed communities forming extensive formations when the jhum farmer tends to avoid those sites for farming, if he has alternatives, and a bald landscape with very sparse vegetation as under degraded grasslands discussed for the Cherrapunji site in Meghalaya. Recovery and restoration of these desertified sites are difficult through natural processes because a tree cover cannot develop due to the absence of seed source in the vicinity due to elimination of the tree germ plasm over a period of time. Restoring such degraded ecosystems through human effort is complex and expensive.

Short jhum cycles of 4–6 years are important for invasion by exotic weeds that have largely occurred subsequent to the Second World War (Ramakrishnan and Vitousek, 1989). *Eupatorium* spp. and *Mikania micrantha* are now among the most aggresive weeds of the sub-continent. These species having a C_3 photosynthetic pathway, unlike the more important natives, live in apparent coexistence due to mutual avoidance in a heterogenous soil environment. Thus, it is probable that invasion by these species has markedly altered myriad ecosystem properties, including productivity and nutrient budgets, in this succession. However, the success of these biological invaders is inextricably linked with a novel disturbance regime.

Bamboo forests play an important role in the economy of the region (Rao and Ramakrishnan, 1987b), being used as firewood and house construction material, as well as the use of bamboo for handicrafts. Bamboo shoot is an important food source during the rainy season. Bamboo is also exported for industrial purposes. The socio-economic analysis of bamboo carried out for the East Khasi hills District of Meghalaya suggests that there is a severe shortage of this commodity in the region. Our own studies on the biology of bamboo (Rao and Ramakrishnan, 1987a,b, 1988a, 1989) are significant for the management of this important resource. The differential plasticity of some of the bamboo species, their differential vegetative and seedling regeneration strategies, ability to exploit high light regime and/or shade tolerance through architectural growth form, exploitative nutrient utilization and nutrient conservation strategies are all critical ecological attributes that could be exploited for better management of bamboo forests of the secondary successional environment.

In many parts of north-eastern India, mature climax forests have been converted into seral vegetation types so that often the climax vegetation is represented by relict sacred groves maintained by tribal societies, for cultural or religious reasons (Boojh and Ramakrishnan, 1983b; Khiewtam and Ramakrishnan, 1989). Everything within these groves is under protection, and even removal of dead wood or twig is a taboo. Once, these groves were widely distributed with a grove for each village. With the advent of modernity, many of these sacred groves have disappeared and those that exist are highly disturbed. In a study of these groves around Cherrapunji in Meghalaya, we found that their size ranged between 0.2 and 4.0 km^2, under varied degrees of disturbance (Khiewtam and Ramakrishnan, 1989). However, some of these, such as the Mawsmai grove at Cherrapunji and the Mawphlang grove near Shillong, are well preserved with regular ceremonies and rituals performed for the ruling deities. These climax forests degrade rapidly under severe disturbance because of extreme fragility of the system; being relict forests, therefore, their value for conservation ecology is significant (Ramakrishnan, 1984c). Indeed, small-scale disturbances under long jhum cycles of 20 years or more as in the past has contributed to maintenance of biological diversity in the forst ecosystem types of north-eastern India. However, large-scale disturbances of the present, by the industrial man, compounded by the drastically reduced jhum cycles have adversely altered biological diversity in the region.

CHAPTER 7

NUTRIENT CYCLING IN JHUM FALLOWS

INTRODUCTION

The success of jhum under long cycles, in north-eastern India and similar systems elsewhere in the humid tropics, is to a large extent related to nutrient cycling patterns and processes of the forest fallow phase. After clear-cutting and burning of the forest, the ecosystem loses its ability to hold nutrients. Much ash is blown away by strong winds during the dry period just prior to monsoon rains (Toky and Ramakrishnan, 1981b). Large losses of soil carbon and nitrogen also occur through volatilization during the burn; some fire-related loss of soil phosphorus is also reported (Mishra and Ramakrishnan, 1981; Swamy and Ramakrishnan, 1988c). Nutrient depletion in the soil continues during the cropping phase, partly due to losses related to hydrology (Toky and Ramakrishnan, 1981b; Mishra and Ramakrishnan, 1983c), and partly due to crop removal (Swamy and Ramakrishnan, 1988c). This depletion of soil nutrients generally continues through the early fallow phases up to about 10 years of fallow regrowth, due to rapid transfer of nutrients from the soil compartment to the fast developing vegetation. The recovery period for soil nutrients does not start until after about 10 years of fallow regrowth, when build-up in the soil is due to litterfall and decomposition (Ramakrishnan and Toky, 1981; Mishra and Ramakrishnan, 1983d). On the basis of a detailed consideration of these aspects related to soil nutrients, we have concluded in Chapter 5 that jhum cycles shorter than 10 years or so are untenable in north-eastern India. Therefore, we shall not discuss these aspects here, but shall consider only the nutrient cycling patterns.

Rapid regeneration of vegetation following jhum, reduces nutrient loss and allows a return to the steady-state cycling characteristic of mature forest (Toky and Ramakrishnan, 1983b; Mishra and Ramakrishnan, 1983b; Jordan, 1985; Vitousek and Stanford, 1986). Compared with temperate forests, relatively little is known about nutrient cycling in tropical and sub-tropical forests (Greenland and Kowal, 1960; Nye, 1961; Golley *et al.*, 1975;

Grubb and Edwards, 1982; Jordan, 1985). Nevertheless, these studies suggest certain patterns; the annual rate of uptake and return of nutrients may be greater in tropical forests than in temperate forests and a larger proportion of the nutrients appear to be in the vegetation (Rodin and Brazilevich, 1967; Edwards and Grubb, 1982).

There are very few scattered attempts to study the pattern of nutrient cycling in tropical secondary successional fallows after shifting agriculture. Bartholomew *et al.* (1953) found that the storage capacity of leaves and litter is saturated at an early stage in the succession in fallows in the Belgian Congo and, thereafter, the total storage capacity increases more slowly and in the woody material. Nakano and Syahbuddin (1989), on the basis of their studies in Thailand, conclude that soil nutrient depletion continues up to about 4 years of fallow regrowth, and that soil fertility is adequate at the beginning of fallow phase for forest to recover. Based on available information, Trenbath (1989) predicts, using models, the relationships between soil fertility on the one hand and the human carrying capacity and site degradation on the other. In this chapter, we shall consider the changing secondary succession following jhum. While doing so, apart from looking at general patterns and processes during succession, we shall also evaluate as to how nutrient cycling operates under stress situtions as under arrested grassland succession or under desertified situations. The objective here is to look at :

(1) Nutrient accumulation in the soil and vegetation and the rate of transfer of the nutrients between the vegetation, litter and soil in successional communities;

(2) Key species and key processes contributing to nutrient conservation;

(3) The consequences of biological invasion by exotic weeds on nutrient cycling; and

(4) The process of desertification and related changes altering nutrient cycling patterns.

NUTRIENT CYCLING DURING SECONDARY SUCCESSION

General patterns in up to 20 years of fallow regrowth

Low elevation jhum fallows

Under the low elevation jhum system at Burnihat (100 m altitude), the mineral nutrient content in the aboveground vegetation increased during the first 20 years of fallow regrowth (Table 7.1). Herbaceous species contained all the nutrient in the 1st year, 60% after 5 years and a negligible amount in older fallows. The bamboo, *Dendrocalamus hamiltonii* contained

Table 7.1 Nutrient quantity ($g\,m^{-2}$) in various compartments of living aerial shoot biomass during 20 years of succession following jhum at Burnihat in Meghalaya (from Toky and Ramakrishnan, 1983b)

	Fallow age (yr)				
	1	*5*	*10*	*15*	*20*
Trees and shrubs					
N	–	2.6	11.0	18.5	29.2
P	–	0.3	1.3	2.2	3.6
K	–	5.6	21.8	38.7	62.5
Ca	–	3.2	13.1	22.8	36.6
Mg	–	1.3	5.5	9.3	14.8
Bamboo					
N	–	0.6	8.2	15.3	19.6
P	–	0.1	1.2	2.2	2.8
K	–	2.3	32.4	58.9	75.4
Ca	–	0.2	3.2	5.6	7.4
Mg	–	0.2	3.3	6.1	7.8
Herbs					
N	3.1	11.1	0.1	0.1	0.1
P	0.5	1.6	0.02	0.01	0.01
K	3.5	11.0	0.1	0.1	0.04
Ca	1.4	4.5	0.1	0.02	0.02
Mg	1.5	5.0	0.1	0.03	0.02

40–45% of the nitrogen, 44–49% of the phosphorus, 54–60% of the potassium, 17–20% of the calcium and 35–40% of the magnesium in the aerial shoots of the 10–20-year-old fallows. The amounts in trees and shrubs increased between 5 and 20 years, with proportionately more calcium to potassium than bamboo.

Among the cations in the living biomass, the concentration of potassium was especially high in the bamboo, *Dendrocalamus hamiltonii*, which is reflected in the quantity stored in it. The higher concentration of potassium in bamboo is due to its fast accumulation by this species which is an important component of the early succession. This is obvious when the total quantity of potassium in bamboo in a 15–20-year-old fallow is compared with that held in all the other species of trees and shrubs combined. Thus, in early succession studied here, potassium is predominant over calcium in the living biomass whereas other workers (Golley *et al.*, 1975; Grubb and Edwards, 1982) have shown a reverse trend in mature communities.

Fig. 7.1 shows the pattern of distribution of the elements in the boles, branches and leaves of all plants. The boles had the greatest amounts and the leaves least in the fallows of 10–20 years of age. In the 1-year-old fallow, which was devoid of trees, the stem and branches contained much of the

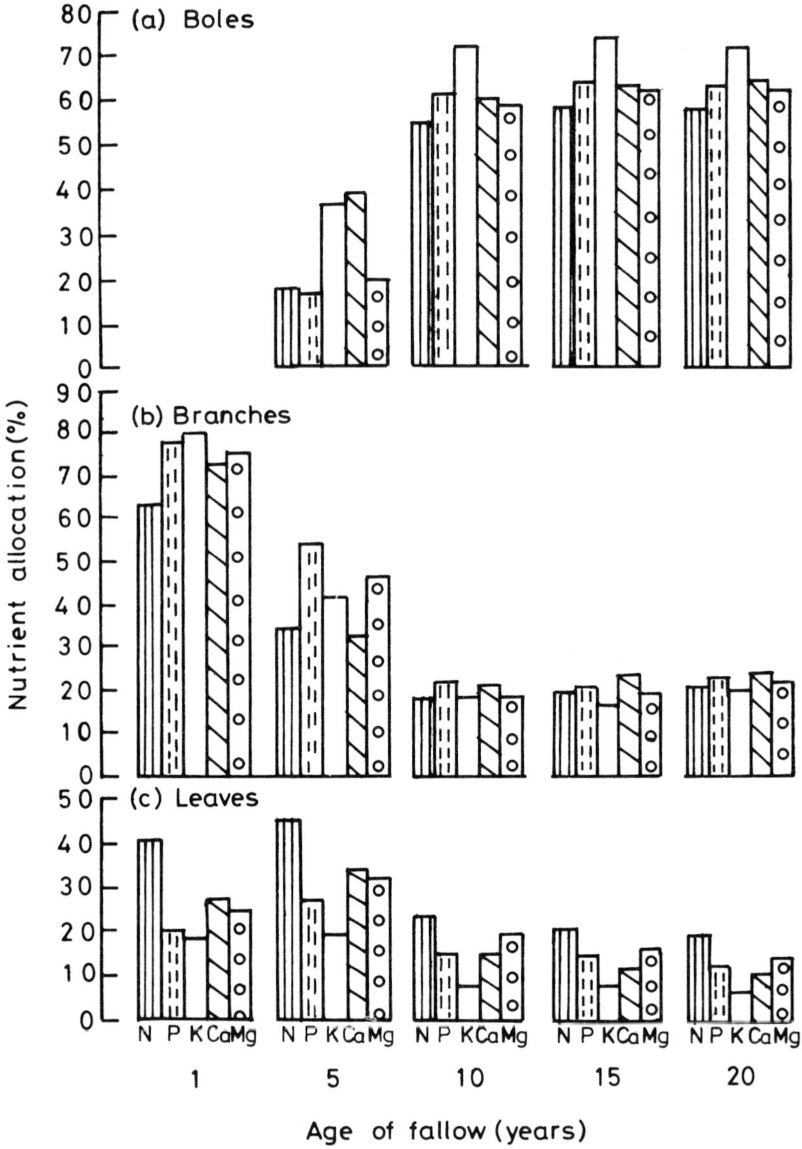

Figure 7.1 Percentage distribution of nutrients in the aboveground vegetation in successional communities up to 20 years old at Burnihat in Meghalaya. (a), boles; (b), branches; and (c), leaves (from Toky and Ramakrishnan, 1983b)

elements. Leaves had a greater proportion of the elements in 1- and 5-year-old fallows than in the older ones. Potassium, calcium, magnesium and phosphorus were mainly in the bole and branches whereas the proportion of nitrogen was greatest in the leaves.

Table 7.2 Rates of uptake of nutrients ($g\,m^{-2}\,yr^{-1}$) in the living aerial shoot biomass (including the annual litterfall) during 20 years succession following jhum at Burnihat in Meghalaya. Values are also given for the enrichment quotients (element held in vegetation/element uptake) of Woodwell, *et al.*, 1975) (from Toky and Ramakrishnan, 1983b)

Year	*N*	*P*	*K*	*Ca*	*Mg*
0–1					
uptake rate	3.1	0.5	3.5	1.5	1.5
enrichment quotient	0.9	1.0	0.9	0.9	1.0
1–5					
uptake rate	5.7	0.5	6.0	4.3	3.0
enrichment quotient	2.5	4.0	3.0	1.8	2.1
5–10					
uptake rate	7.6	0.4	12.0	5.3	3.1
enrichment quotient	2.5	6.2	4.5	3.7	2.8
10–15					
uptake rate	9.2	0.7	13.5	6.3	4.2
enrichment quotient	3.7	6.3	7.2	4.5	3.7
15–20					
uptake rate	10.9	0.9	14.5	7.8	4.8
enrichment quotient	4.5	7.1	9.5	5.6	4.7

The rate of accumulation of nitrogen, phosphorus and magnesium decreased during the first 10 years and then increased sharply during the next 10 years, whereas calcium showed a steady increase during the 20 years; potassium increased for 15 years and then decreased.

The high enrichment quotient (concentration in vegetation/uptake) for potassium in fallows of 10–20 years of age indicates maximum accumulation of this element in the living biomass (Table 7.2). A similar observation was made in early succession in Zaire where *Musanga cecropioides* selectively accumulated potassium (Bartholomew, *et al.*, 1953), although the quantities of potassium was much lower than in a 20-year-old fallow at Burnihat. During the first 10 years of fallow regrowth, when bamboo is not a major component of the vegetation that is largely herbaceous, the enrichment quotient for phosphorus was more than that for other nutrients indicating the selective accumulation of this element. For calcium and magnesium, the annual uptakes and enrichment quotients were much higher in older fallows, probably because of a rapid growth of trees.

Matching our own observations in a 50-year-old fallow at Lailad in Meghalaya (Singh and Ramakrishnan, 1982b), Greenland and Kowal (1960) reported almost three times more calcium than potassium in the aboveground biomass in a 40-year-old forest in Ghana. Recent studies of Grubb and Edwards (1982) showed higher concentrations of calcium and

Table 7.3 Nutrient quantities ($g m^{-2}$) in living aerial shoots, soil and litter on the ground during 20 years of succession following jhum at Burnihat in Meghalaya (from Toky and Ramakrishnan, 1983b)

	Fallow age (yr)				
	1	*5*	*10*	*15*	*20*
Living aerial shoots					
N	3	14	19	34	49
P	1	2	3	4	6
K	4	19	54	98	138
Ca	1	8	16	28	44
Mg	2	7	9	15	23
Soil (0–40 cm)					
N	978	1,110	1,225	1,055	1,050
P	1	1	2	2	3
K	99	57	26	118	120
Ca	136	80	104	109	108
Mg	130	108	72	120	121
Litter on ground					
N	0.3	0.8	1.8	2.3	2.1
P	0.01	0.05	0.1	0.1	0.1
K	0.1	0.1	0.5	0.3	0.7
Ca	0.2	0.5	1.1	0.8	1.4
Mg	0.1	0.4	0.9	0.6	0.8
Total ecosystem					
N	981	1,125	1,246	1,091	1,101
P	2	3	5	6	9
K	103	76	81	216	259
Ca	137	88	121	138	153
Mg	132	115	82	136	144

magnesium than potassium in the trunks of large trees. However, the percentage nutrients in the plant compartment of the aerial shoots for the 20-year-old fallow was : N, 5%; P, 69%, K, 53%; Ca, 28%; Mg, 16% (Table 7.3). A comparison of these results with those on the different rain forest types discussed by Edwards and Grubb (1982) suggests that, in contrast to the conventional idea of a tropical rain forest with about 90% of the nutrient capital in the plants, the fraction in the present case is low. With the development of secondary vegetation, the weight of nutrients in the living biomass increased due to a linear increase in biomass with age (Toky and Ramakrishnan, 1983a; cf Table 6.3).

The concentration and quantity of all elements, except nitrogen, was generally higher in the soil under a 1-year-old fallow than it was later; this followed their release due to slash-and-burn. The rapid decline in nutrient levels in the soil in general as discussed in Chapter 5, is due to depletion by rapidly developing vegetation, with recovery in the soil occurring after

10 years. For soil nitrogen, there is a rapid phase of recovery after fire, due to increased microbial activity, as discussed in Chapter 5; this is because of more favourable conditions for biological nitrogen fixation. After 10 years of fallow regrowth, there is a decline in soil nitrogen levels, probably due to the establishment of a balance between nitrogen fixation and mineralization. Similar observations have been made for the 'Lua' forest fallow systems of shifting agriculture in Thailand (Zinke *et al.*, 1978) and in south-western Nigeria (Aweto, 1981).

The quantity of nitrogen in the whole ecosystem was highest in the 10-year-old fallow, whereas that of phosphorus increased throughout. Quantities of cations that started off at a high level after the burn, declined to a minimum after 5–10 years, and increased again to a maximum in the 20-year-old fallow. The proportion of different nutrients in the living biomass, as compared to the soil gradually increased with fallow age; this was more obvious for phosphorus and potassium,

The return of elements to the soil through litterfall increased with the age of the fallow (Table 7.4). This is a function of the linear increase in the litterfall *per se* with age (Toky and Ramakrishnan, 1983a; cf Chapter 6). The return of nitrogen was greatest, followed by potassium and calcium. In the 1-year-old fallow, nutrient return was exclusively through leaf litter, in the 5-year-old fallow the contribution by twigs of dicots was maximum (18% of the total), and in the 20-year-old fallow, 94% of the nutrient return was through leaf fall. A higher weight of potassium and lower weights of other elements are returned annually, compared with the values reported by Ewel (1976) for fallows up to 14 years old in Gantemala. This is due to the relatively high concentration of potassium in the predominantly bamboo litter, as already discussed. Indeed, where bamboo was a major component in the fallow, its accumulation of potassium and phosphorus in the litter equalled or exceeded that of the other species combined; its accumulation of magnesium was relatively low.

The quotient of the annual return of nutrients through litterfall to the annual uptake by the vegetation expressed as a percentage, increased for 5 years for potassium and for 10 years for other elements; later it tended to stabilize at a slightly lower level (Fig. 7.2). The retention of potassium in the living biomass was high as the proportion returned through the litterfall was low; that of phosphorus was almost as high as potassium whilst that of nitrogen was the lowest.

Fractional annual turnover (nutrient flux expressed as a percentage of the content of that nutrient in the vegetation and soil) (Reiners and Reiners, 1970) of most elements in the soil increased with fallow age (Table 7.5). This is because of the marked increase in uptake by the developing vegetation and increased storage in the living biomass which is proportionately higher than the rate of release through litter. The only exceptions to this are

Table 7.4 Nutrient quantities in various components of litterfall ($g\,m^{-2}\,yr^{-1}$) during 20 years of succession following jhum at Burnihat in Meghalaya (from Toky and Ramakrishnan, 1983b)

	Fallow age (yr)				
	1	*5*	*10*	*15*	*20*
Leaves (excluding bamboo)					
N	1.3	3.7	3.4	3.9	4.6
P	0.1	0.2	0.1	0.2	0.2
K	0.8	2.1	2.2	2.6	3.0
Ca	0.9	2.4	2.7	3.1	3.7
Mg	0.6	1.7	1.6	1.9	2.2
Bamboo leaves					
N	0	0.1	2.6	2.7	3.5
P	0	0.01	0.2	0.2	0.3
K	0	0.1	2.4	2.4	3.2
Ca	0	0.03	0.7	0.7	1.0
Mg	0	0.03	0.9	0.9	1.2
Twigs (excluding bamboo)					
N	0	0.4	0.1	0.1	0.1
P	0	0.04	0.02	0.01	0.01
K	0	0.8	0.3	0.1	0.2
Ca	0	0.7	0.2	0.1	0.2
Mg	0	0.4	0.1	0.1	0.1
Bamboo twigs					
N	0	0	0.1	0.1	0.2
P	0	0	0.01	0.02	0.02
K	0	0	0.2	0.3	0.5
Ca	0	0	0.1	0.1	0.1
Mg	0	0	0.04	0.1	0.1
Total					
N	1.3	4.2	6.2	6.8	8.4
P	0.1	0.25	0.33	0.43	0.53
K	0.8	3.0	5.1	5.4	6.9
Ca	0.9	3.1	3.7	4.0	5.0
Mg	0.6	2.1	2.6	3.0	3.6

phosphorus and potassium which reached their maximum annual turnover in the soil in the 5- to 10-year-old fallow. The high fractional turnover value for phosphorus in the soil in the 5-year-old fallow may be due to a higher uptake by the herbaceous vegetation that predominates then, and that for potassium in the 10-year-old fallow being related to rapid bamboo growth at this time.

In a study of the litter decomposition rates of a few important early successional species coming up during early succession (Toky and Ramakrishnan, 1984), the annual rate of loss of litter varied between 90% in a 10-year-old jhum fallow and 100% in a fresh 0-year-old fallow. The generally

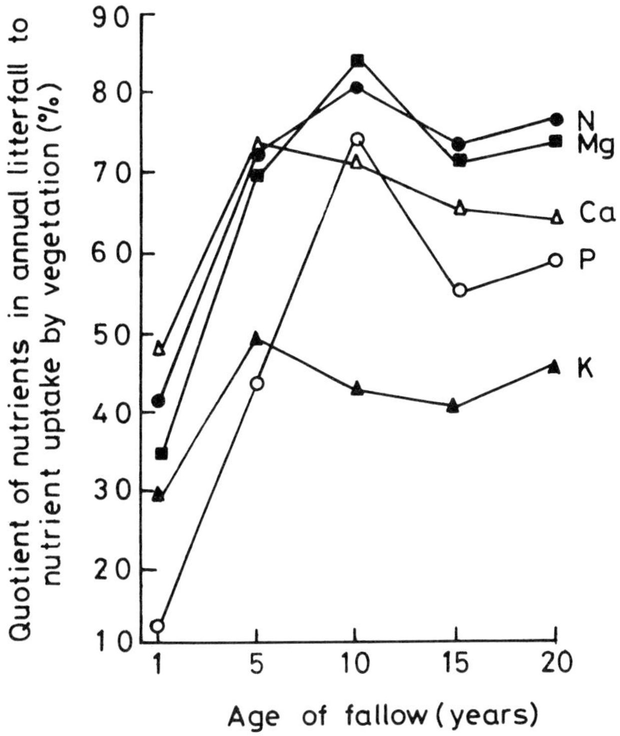

Figure 7.2 The quotients of annual return of nutrients through litterfall to the annual uptake by the vegetation, expressed as percentage in successional communities up to 20 years old at Burnihat in Meghalaya (from Toky and Ramakrishnan, 1983b)

slower rate of release of nutrients from the decomposing litter of the younger fallows, compared to that of the older ones, was related with better retention in the soil sub-system, against losses through hydrology at a time when the vegetation is still developing (Toky and Ramakrishnan, 1981b). On the other hand, quicker release of nutrients from the litter of the older fallows would be conserved in the living biomass of the rapidly developing vegetation.

High elevation jhum fallows

At higher elevations, as at Shillong in Meghalaya (1,500 m), the general pattern of nutrient cycling in early successional fallows is similar. The nutrient inventory for the high elevation system is generally lower (Mishra and Ramakrishnan, 1983d). Nutrient recovery in the soil is also generally low, compared to the lower elevation jhum fallows. With a high fractional turnover rate for potassium in the soil, this labile element is conserved in

Table 7.5 The fractional annual turnover of elements (the weight that left the compartment divided by the weight held in that compartment, expressed as percentage) (Reiners and Reiners, 1970) of soil and vegetation during 20 years of succession following jhum at Burnihat in Meghalaya (from Toky and Ramakrishnan, 1983b)

	Fallow age (yr)				
	1	*5*	*10*	*15*	*20*
N					
Soil	0.3	0.5	0.6	0.8	1.0
Vegetation	42	29	32	20	17
P					
Soil	50	50	20	35	30
Vegetation	10	11	12	10	8
K					
Soil	3	10	46	11	12
Vegetation	20	16	9	6	5
Ca					
Soil	1	5	5	6	7
Vegetation	64	39	23	15	11
Mg					
Soil	1	3	4	3	4
Vegetation	40	32	29	20	16

the total vegetation through an accumulator species like *Dendrocalamus hamiltonii* which is absent at this altitude.

A relatively higher rate of nutrient cycling under the high elevation jhum fallows is due to the presence of large pine trees (*Pinus kesiya*) right from the beginning of the fallow regrowth phase. As discussed in Chapter 1, under the modified version of jhum at this altitude, the sparsely distributed pine trees are left undisturbed during jhum, and only some of the lower branches of the pine trees are lopped for the slash-burn. These pine trees with a fast growth rate and rapid turnover of needles that are produced in three flushes have an exploitative growth strategy (Das and Ramakrishnan, 1986; cf Chapter 9). Apart from accumulating large quantites of nutrients in their living biomass (Das and Ramakrishnan, 1987), these nutrients are recycled rather rapidly through litterfall (Das and Ramakrishnan, 1985). With an annual litterfall of about 9,000 kg ha^{-1} in a 22-year-old pine forest floor through litterfall, the mass present on the forest floor is substantial (Table 7.6).

The rate of pine litter decomposition was evaluated as dry weight loss (%) along with decomposition constant *k*, as well as half time ($t_{0.50}$) and time to 95% loss of dry weight ($t_{0.95}$). Considering the rates over 2 years, the rate of decomposition was fast for *Pinus kesiya* (Table 7.7), and indeed

Table 7.6 Nutrients ($kg\,ha^{-1}$) returned through litterfall and forest floor levels in a 22-year-old pine (*Pinus kesiya*) plantation at Shillong in Meghalaya (Das and Ramakrishnan, 1985)

Nutrient	*Litterfall*	*Forest floor*
Nitrogen	64.7	276.9
Phosphorus	15.4	38.3
Potassium	40.1	55.4
Calcium	49.2	87.5
Magnesium	24.2	40.7

Table 7.7 Dry weight loss and decomposition parameters of *Pinus kesiya* needle litter at Shillong in Meghalaya

Decomposition time (months)	*Dry weight loss* (%)	*k*[a]	$t_{0.50}$	$t_{0.95}$
12	37.5	0.46	1.50	6.52
24	79.2	0.78	0.88	3.84

[a] annual decomposition rate

Table 7.8 Residence time and fractional annual turnover of nutrients on the forest floor of a 22-year-old pine stand at Shillong in Meghalaya (from Das and Ramakrishnan, 1985)

Nutrient	*Residence time* (yr)	*Fractional annual turnover* (%)
Nitrogen	4.31	23.22
Phosphorus	2.49	40.24
Potassium	1.38	72.38
Calcium	1.78	56.26
Magnesium	1.68	59.38

higher than that for other pine forests elsewhere (Das and Ramakrishnan, 1985). Temperature and moisture conditions being the two important abiotic factors controlling the rate of decomposition under natural conditions, they are relatively favourable considering the altitude at Shillong. Thus, the rate of litter loss, which is higher during the monsoon months, as supported by the study on CO_2 evolution, is indicative of favourable conditions for decomposition during this time of the year (Das and Ramakrishnan, 1985). This is also reflected in the residence time or turnover of nutrients and fractional annual turnover values for different nutrients in the 22-year-old pine stand (Table 7.8). Residence time for nitrogen and phosphorus was much higher and it was least for potassium; correspondingly, the annual turnover rates of the former two were the lowest.

Even though nutrient cycling rates during the early forest fallow phase at higher elevations are generally low, *Pinus kesiya*, as an important

Table 7.9 Nutrient input through water in a 50-year-old fallow at Lailad in Meghalaya (from Singh and Ramakrishnan, 1982c)

	Nutrients ($kg\,ha^{-1}$)				
Input	*N*	*P*	*K*	*Ca*	*Mg*
Stem flow	0.17	0.02	0.71	0.84	0.10
Through-fall	8.39	0.89	31.28	35.19	5.03
Rain water	4.33	0.43	7.80	9.96	4.77

component of the early fallow phases up to about 60–100 years, plays an important role in rapid turnover of nutrients, under the given climatic conditions. The success of this species on soils of low fertility is due to the rapid turnover rate of nutrients. The frequency of flushing during the year, the shorter life span of the needles (Ramakrishnan and Das, 1983; Das and Ramakrishnan, 1986) and the consequent high litter production, along with faster release of nutrients, all contribute towards the success of *Pinus kesiya* as an early colonizer on comparatively infertile soils.

50-year-old fallow at Lailad in Meghalaya

In the 50-year-old fallow, we could only follow nutrient flow patterns through rain water and through litterfall and not a total budget analysis as was done for fallows up to 20 years old (Singh and Ramakrishnan, 1982b,c). The nutrient input into the system through water is substantial (Table 7.9), with higher values for the through-fall. Amongst the cations, potassium and calcium inputs are high. On the other hand, the loss that occurred through run-off and percolation water was minimal compared with the early successional fallows discussed in Chapter 5. However, the loss of calcium and potassium from a 50-year-old fallow was higher compared to other nutrients, and this may be due to greater release of nutrients from the vegetation through litterfall and decomposition. Since the storage of calcium in the living biomass would be higher in an older forest stand (Grubb and Edwards, 1982), this would be reflected in the litterfall too. With substantial bamboo growth (*Dendrocalamus hamiltonii*) in the disturbed peripheral open-canopy zone of the forest (Singh and Ramakrishnan, 1982a), substantial release of potassium through litterfall could be expected.

With an annual average litterfall of $5.5\,t\,ha^{-1}\,yr^{-1}$ in this forest stand, the nutrient release through litter is rapid (Singh and Ramakrishnan, 1982b) with comparable values reported for mature rain forests as in New Guinea (Edwards, 1977). Among the nutrients returned to the forest floor, the proportion of calcium was the highest. In spite of the high levels of nutrients released through litterfall, much for it is captured by the fine root system

Table 7.10 Nutrient quantity in the fine-root biomass in a 50-year-old forest at Lailad in Meghalaya (from Ramakrishnan and Singh, 1983)

	Nutrients (kg ha^{-1})				
Soil depth (cm)	*N*	*P*	*K*	*Ca*	*Mg*
0–50	148.1	5.4	21.0	21.2	16.5
50–100	85.1	4.3	11.2	14.9	18.9
100–150	0.9	1.2	1.3	1.6	0.9

in the surface soil layers (Ramakrishnan and Singh, 1983). Indeed, the fine root system in the surface soil layers itself has substantial quantities of nutrients stored in it (Table 7.10). This is slowly released during the rapid turnover of the root system with a decay rate which could be as high as one quarter of the total root biomass in a year (Khiewtam, 1986). In the case of the 50-year-old forest stand, this fine root biomass (< 1–100 mm diameter) at a given time is as high as 16,804 kg ha^{-1}, up to a soil depth of 50 cm (cf Table 6.5).

Sacred grove climax forest at Cherrapunji

The climax forest represented by the relict sacred grove at Cherrapunji is an extremely fragile ecosystem. Similar sacred groves also occur elsewhere in Meghalaya as at Mawphlong near Shillong (Plate 7.1) The fragility of this ecosystem is essentially related to the tight cycling of nutrients through the litter. The soil itself has poor nutrient status (cf Table 7.6) and much of the nutrient input into the system is either through rain water or through that cycled through litter.

The canopy, being a totally closed one, eliminates direct fall input as such but substantial input, much more than that observed in a 50-year-old fallow at Lailad (cf Table 7.9), occurs by through-fall and stem flow (Table 7.11). Amongst the cations, calcium followed by potassium predominate. With an exceptionally high annual average rainfall of 10,372 mm of a monsoonic kind, much of this input happens during the wet season (Khiewtam, 1986). Indeed, with such a high rainfall on steep slopes of 30–40° angle, the losses through run-off and infiltration water could be substantial. It is in this context, rapid uptake of nutrients released through litterfall by the living biomass becomes significant.

With an annual litterfall of 11,783 kg ha^{-1}, the nutrient return through it is substantial (Table 7.12). Indeed, the release through litter other than leaf was also significant. The rate of release is also very rapid because of a high rate of decomposition of the litter, with annual percentage weight loss of up to 98% for the leaf litter and up to 68% for the branch litter (Khiewtam, 1986). Amongst the cations, calcium release was high; this

Plate 7.1 The rain forest, a fragile ecosystem, represented by the relict sacred grove forest at Mawphlong (1,500 m altitude) protected by the tribals for religious/cultural reasons. In the foreground is the totally desertified landscape where the forest is unable to advance in spite of the rich seed source of the sacred grove. Note the large monoliths which are symbolic of the tribal belief

Table 7.11 Nutrient input/output (kg ha^{-1}) through water in the sacred grove at Cherrapunji in Meghalaya (from Khiewtam, 1986)

	Input		*Output*	
Nutrient	*Through-fall*	*Stem flow*	*Run-off*	*Percolation*
NO_3–N	10.4	4.0	8.4	2.1
PO_4–P	0.8	0.5	1.0	0.3
K	48.6	38.6	46.1	15.0
Ca	100.6	35.9	65.0	23.9
Mg	39.6	19.1	37.2	12.2

agrees with our observations on the 50-year-old forest fallow at Lailad, discussed earlier (cf Singh and Ramakrishnan, 1982b). The fine root biomass of 13,952 kg ha^{-1} located in the 30 cm layer of the soil, of which 882 kg ha^{-1} is located above the mineral soil (cf Table 6.9) effectively recycles the nutrients released through litterfall and puts it back into the living biomass compartment (Plate 7.2). This fine root system ranging up to 20 mm in diameter, has a productivity of 3,200 kg ha^{-1}, and an annual death rate equal to its productivity. Therefore, it also recycles one quarter

Table 7.12 Annual return of nutrient (kg ha^{-1}) ($\pm$ SE) from different litter fractions in the sacred grove at Cherrapunji in Meghalaya (from Khiewtam, 1986)

Litter fractions	*N*	*P*	*K*	*Ca*	*Mg*
Leaves	101.0 ± 8.2	22.7 ± 0.9	65.0 ± 2.9	110.6 ± 6.3	56.9 ± 1.8
Branches	22.6 ± 1.9	4.9 ± 0.32	10.4 ± 0.6	19.5 ± 1.2	8.9 ± 0.5
Flowers	2.3 ± 0.14	0.2 ± 0.01	0.5 ± 0.03	0.8 ± 0.03	0.6 ± 0.02
Fruits	14.2 ± 1.5	2.3 ± 0.2	17.3 ± 0.8	8.8 ± 0.6	6.7 ± 0.5
Miscellaneous	7.2 ± 0.4	0.9 ± 0.05	5.5 ± 0.4	5.3 ± 0.3	3.0 ± 0.2
Total	151.3 ± 12.1	31.0 ± 1.5	98.7 ± 4.7	145.0 ± 8.4	75.8 ± 3.0

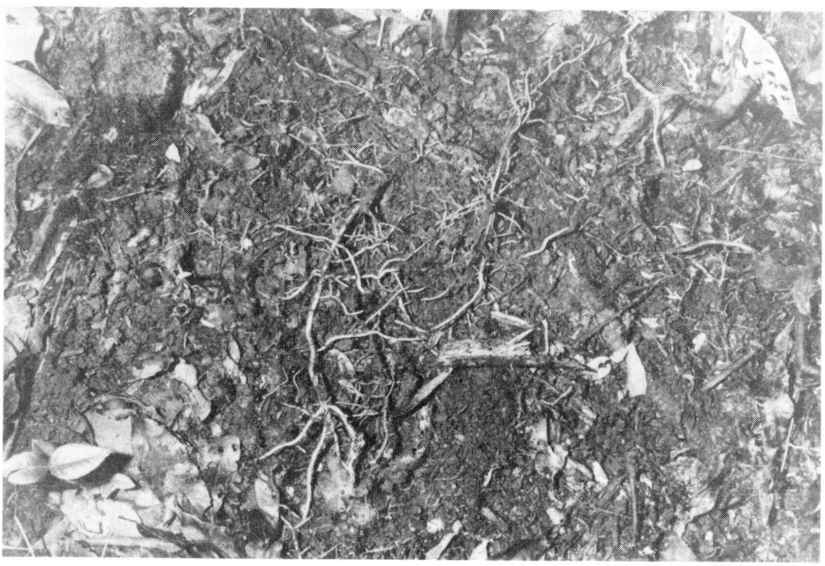

Plate 7.2 The extensive soil surface fine-root mat which could be up to 15 t ha^{-1} found in the sacred grove climax forests of Meghalaya

of its own nutrient load through gradual decay (cf Chapter 6). This could be a substantial addition to the soil recycling pool (Table 7.13).

As also discussed earlier, the fine root mat present in the surface layers of the soil and that particularly located over the mineral soil is important for supporting the large biomass developed over the weak soil. Since nutrients are to be recycled as soon as litter starts decomposing (or else it could be lost through rain water, under the high rainfall situation of Cherrapunji) the presence of a larger proportion of the finer roots compared to coarser ones above the mineral soil is helpful; the reverse is true within the soil (Khiewtam, 1986). Further, greater fine root production during the

Table 7.13 Nutrient quantity in the fine-root biomass up to a soil depth of 30 cm at Cherrapunji in Meghalaya (from Khiewtam, 1986)

Nutrient	*Quantity* ($kg\,ha^{-1}$)
N	209.9 ± 4.9
P	5.2 ± 0.2
K	54.8 ± 1.1
Ca	35.5 ± 0.8
Mg	32.6 ± 1.4

rainy season helps in more efficient uptake from the rapidly decomposing litter at this time of the year. Apart from the short-circuiting of litter nutrients by the fine root mat present on the surface of the mineral soil before it enters the mineral soil, the rest is mopped up by the fine root system present in the 30 cm soil column (Klinge, 1973; Stark and Jordan, 1977; Khiewtam, 1986). Abundantly available mycorrhizae (Stark, 1971; Ramakrishnan and Swamy, 1987) could augument the efficiency of nutrient cycling in this delicately balanced ecosystem type.

Conclusions

Soon after the slash-and-burn operation, the soil nutrient status would differ, both quantitatively and qualitatively, between the short and long jhum cycles. The quantitative differences are directly related to the biomass build-up before the burn (Ramakrishnan and Toky, 1981; Mishra and Ramakrishnan, 1983d). Herbaceous communities up to about 5 years of age generally have more phosphorus, whereas the bamboo forests of *Dendrocalamus hamiltonii* of 10–25 years of age have more potassium (Toky and Ramakrishnan, 1983b; Rao and Ramakrishnan, 1989), so that different proportions of the elements are released in ash, under short and long jhum cycles. These qualitative differences in the nutrients after slash-and-burn, obviously are related to the vegetation structure of the fallow phase. Indeed, under a very long jhum cycle of 60 years, as studied by us in Arunachal Pradesh in north-eastern India, calcium and magnesium would be proportionately higher because of greater accumulation of these elements in the bole and branches of large trees. Here in lies an important linkage between the cropping and fallow phases of the jhum system.

This linkage between the agroecosystem and the forest fallow operating through the process of nutrient cycling could be fragile because of highly leached nutrient-deficient soils on which vegetation develops, when short jhum cycles are constantly imposed in the same area. Otherwise, under longer cycles, after the destruction of nutrient cycling during jhum, the system would recover rapidly because of rapid regeneration of vegetation and through a quick succession. The nutrient losses from the system are gradually minimized, thus promoting a return to steady-state cycling. A reasonably stable cycling is established during the first 20 years of fallow regrowth and a 50-year-old stand is sufficiently stable, with a rich species diversity characteristic of mature forest.

THE CONCEPT OF KEY SPECIES IN NUTRIENT CONSERVATION AND CYCLING

The secondary successional processes in the jhum fallow system such as those operating in north-eastern India are characterized by certain unique features that set them apart from the secondary succession that would operate in any forest ecosystem, where normal gap phase regeneration operates randomly. Here we have perturbation following a given cycle, which also determines its intensity in terms of the fire event. There is a cropping phase following fire both of which determine the available plant propagules at the site depending upon the disturbance regime. The steep slopes of 30–40° angle on which all these events occur creates considerable soil heterogeneity, both biological and chemical. It is in the context of this secondary successional process determined by cyclical events imposed upon a stress environment (steep topography, leached nutrient-deficient soil under high rainfall) that the role of key species and their role in determining critical nutrient cycling processes, become significant.

Exotic versus native weeds and nutrient cycling properties

Weeds form the chief component of the early successional jhum fallows up to about 5 years old (cf Chapter 6). With shortening of the jhum cycle, weeds become more abundant during the cropping phase (cf Chapter 4) and the subsequent early fallow phase. This increase in weeds under short jhum cycles is due to the continuous presence of the same weed species in the cultivated plots as well as in the 4- to 6-year-old fallow plots. This allows a build-up of the soil bank of propagules (Saxena and Ramakrishnan, 1984b). As already discussed, exotic weeds are important pioneers, with *Eupatorium odoratum* at lower altitudes, *E. adenophorum* and *E. riparium* at higher altitudes (Dev and Ramakrishnan, 1987a,b), and *Mikania micrantha* (Swamy and Ramakrishnan, 1987a); all are native to Latin American tropics introduced to India towards the turn of this century. In contrast to these important C_3 exotic species, the C_4 native grasses also form important components of the early succession in the region. Amongst these grasses, *Imperata cylindrica, Thysanolaena maxima* and *Saccharum arundinaceum* are important at Burnihat (100 m altitude) in Meghalaya (Saxena and

Table 7.14 Mean concentration of three nutrients (± SE) during the growing season in leaf tissue of C_3 versus C_4 species of early successional jhum fallow at Burnihat in Meghalaya (from Saxena and Ramakrishnan, 1983a)

	Nutrient concentration (%)		
Species	*Nitrogen*	*Phosphorus*	*Potassium*
Eupatorium odoratum	2.92 ± 0.12	0.28 ± 0.06	3.60 ± 0.32
Grewia elastica	1.70 ± 0.14	0.21 ± 0.04	1.32 ± 0.24
Imperata cylindrica	0.85 ± 0.05	0.11 ± 0.02	1.46 ± 0.24
Thysanolaena maxima	1.40 ± 0.09	0.19 ± 0.04	1.54 ± 0.20

Ramakrishnan, 1984c). Several less abundant C_3 species, however, also occur among the natives.

Exotic species contribute a major fraction of the total biomass of early successional communities and hence they play a significant role in reducing losses of nutrients in run-off after cropped plots are abandoned (cf Chapter 5). The extent to which this function of exotic weeds would be replaced by natives if the former were not available, is a moot point. However, with their C_3 photosynthetic pathway the exotics do have a systematically higher nutrient uptake requirement (Saxena and Ramakrishnan, 1983a; Swamy and Ramakrishnan, 1988a), and therefore they accumulate larger quantities of nutrients in the plant tissue (Table 7.14). For this reason, under the present shortened jhum cycles, it is likely that nutrient losses would be accelerated in the absence of exotic species.

Nitrogen is volatilized from the soil during fires, causing a decline in soil nitrogen status, and the nitrification process in the soil itself is inhibited in proportion to fire intensity (Saxena and Ramakrishnan, 1986). Production of nitrate-nitrogen which is largely utilized in hot and humid climates with low acidity (Haynes and Goh, 1978), as is the case under jhum in north-eastern India, declines after the burn. However, nitrate-nitrogen recovery starts very soon. It may be anticipated that high efficiency of nutrient use in the C_4 species (Fig. 7.3) may allow them to grow more effectively under land levels of nitrogen during the early part of the growing season when climatic conditions are favourable. In contrast, C_3 species with low efficiency of nutrient use may grow successfully during the later part of the growing season when nutrient availability is improved by detritus input (Ramakrishnan and Toky, 1981). Furthermore, both C_3 and C_4 species also coexist in the same site. The soil nutrient levels, particularly nitrogen are in a state of flux, soon after burn, with high soil heterogenicity. Nitrogen accumulation during the short fallow period is inadequate to replace the amounts removed in crop harvest, volatilized during fires or leached (Mishra and Ramakrishnan, 1984). Further, heavy run-off losses also occur. All these result in heterogeneous distribution of soil nutrients on the steep slope. The

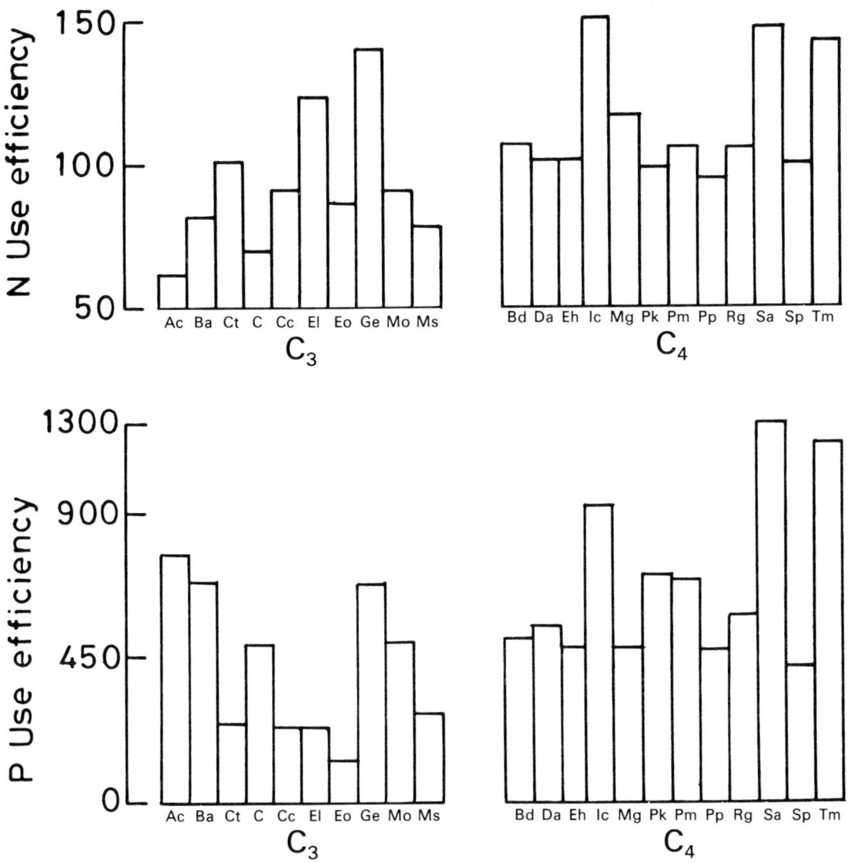

Figure 7.3 Nutrient-use efficiency (expressed as mg dry matter production per mg nutrient absorbed) of different species coming after jhum at Burnihat in Meghalaya. Different columns from left to right are: C_3 species: *Ageratum conyzoides, Borreria articularis, Cassia tora, Costus* spp., *Crossocephalum crepidioides, Erigeron linifolius, Eupatorium odoratum, Grewia elasica, Mimosa pudica, Mollugo stricta*; C_4 species: *Braclinaria distachya, Digitaria adscendens, Euphorbia hirta, Imperata cylindrica, Manisuris granularis, Panicum khasianum, P. maximum, Paspalidium punctatum, Rottboelia goalparensis, Saccharum arundinaceum, Setaria palmifolia, Thysanolaena maxima.* Vertical bars represent standard error of the mean (from Saxena and Ramakrishnan, 1984c)

C_3 species with large utilization of nutrients for a given dry-matter production are suited to occupy nutrient-rich microsites, while the C_4 species with a higher nutrient-use efficiency can successfully colonize nutrient-poor microsites. Indeed, the high efficiency in nutrient-use of C_4 species might have evolved alongside C_4 photosynthesis under conditions of poor nutrient status of the soil in tropical environments where efficient use of nutrients could be important for survival (Brown, 1978).

Table 7.15 Enrichment quotients (elements held in vegetation/element uptake) for *Mikania micrantha* and the vegetation in successional fallows developed after jhum at Lailad in Meghalaya. Values are also shown for total vegetation (from Swamy and Ramakrishnan, 1987c)

Year	*N*	*P*	*K*	*Ca*	*Mg*
0–1					
Enrichment quotient	0.85	0.97	1.00	0.85	0.88
Total vegetation	1.14	1.25	1.32	1.08	1.17
1–2					
Enrichment quotient	1.03	1.12	1.18	1.03	0.98
Total vegetation	1.42	1.74	1.60	1.33	1.29
2–4					
Enrichment quotient	1.19	1.30	1.61	1.22	1.07
Total vegetation	2.12	2.67	2.74	1.90	2.41
4–8					
Enrichment quotient	1.85	1.94	2.55	2.44	1.46
Total vegetation	2.63	5.62	4.01	2.64	3.20
8–12					
Enrichment quotient	1.93	2.19	2.81	2.47	1.52
Total vegetation	3.08	4.64	6.15	3.58	3.34

The manner in which an exotic species can specifically affect nutrient retention in the ecosystem is exemplified by the effect of *Mikania micrantha* on potassium cycling (Swamy and Ramakrishnan, 1987c). This species, being an important component of early successional fallows, generally has a high rate of nutrient uptake, and higher accumulation because of its large biomass, which occurred only in up to 4-year-old fallow at Lailad (296 m altitude) in Meghalaya. The subsequent drastic decline in uptake is related to the decline in population vigour of this species in older fallows, although the uptake by the vegetation as a whole increased. Potassium in particular, is accumulated rapidly by the vegetation up to 4 years into the fallow period, *Mikania micrantha* alone accumulating 10.5 $g\,m^{-2}$ of potassium compared with 3.62 $g\,m^{-2}$ by other herbs and 6.8 $g\,m^{-2}$ by shrubs, trees and bamboos in a 4-year-old fallow (Swamy and Ramakrishnan, 1987c). Indeed, the quotients of annual return of nutrients through litterfall to the annual uptake by *Mikania micrantha* increased with age of the fallow (Table 7.15). The enrichment quotient for potassium was the highest observed.

During the early phases of secondary succession up to 4 years, *Mikania micrantha* maintains a high rate of nutrient cycling through its rapid growth, total death of the aboveground parts during the winter months, and the consequent high litter production. Greater release of nutrients through the litter of *Mikania micrantha* is reflected in the increase in annual turnover through this species with fallow age (Table 7.16). Thus, the nutrient cycling

Table 7.16 Fractional annual turnover (%) of various elements in soil and vegetation compartment and the contribution by *Mikania micrantha* in successional fallows developed after jhum at Lailad in Meghalaya (from Swamy and Ramakrishnan 1987c)

	Age of fallow (yr)				
	1	*2*	*4*	*8*	*12*
N					
Soil	0.48	0.65	0.60	0.63	0.87
Mikania micrantha	29.30	21.05	29.72	45.88	54.38
Total vegetation	16.14	27.221	26.46	33.67	23.23
P					
Soil	61.76	75.49	99.89	50.0	44.98
Mikania micrantha	14.33	16.43	26.61	35.71	33.33
Total vegetation	7.59	15.67	18.72	21.21	12.93
K					
Soil	3.43	8.00	15.81	16.09	10.62
Mikania micrantha	11.96	9.90	18.55	19.26	23.33
Total vegetation	7.79	14.45	15.50	12.25	8.64
Ca					
Soil	1.95	3.49	6.56	6.98	6.35
Mikania micrantha	37.70	29.41	45.08	61.81	67.50
Total vegetation	18.86	29.67	30.55	28.46	19.29
Mg					
Soil	1.90	3.47	3.14	2.46	4.20
Mikania micrantha	31.63	20.59	23.75	43.47	56.25
Total vegetation	19.73	23.01	19.72	30.82	21.43

pattern in *Mikania micrantha* which dominated early successional fallow are somewhat different from those observed in others where this species was not present. This feature along with the role of this species in conserving a labile element such as potassium in the ecosystem through rapid uptake and storage provides a key role to this exotic in nutrient cycling process of early successional fallows, particularly those developed under short jhum cycles where nutrients could be limiting.

Nutrient conservation through bamboos

Bamboos being an important component of the secondary succession in north-eastern India, have a large biomass contribution (cf Fig. 7.4) to jhum fallows of up to about 30 years. In a comparative study of three locations in the north-eastern India–the Garo, the Khasi and the Naga hills, the role of four bamboo species, namely, *Dendrocalamus hamiltonii* (a common and dominant species to all locations), *Bambusa tulda* (Garo hills), *Neohouzeoua dulloa* (Khasi hills) and *Bambusa khasiana* (Naga hills) were evaluated for

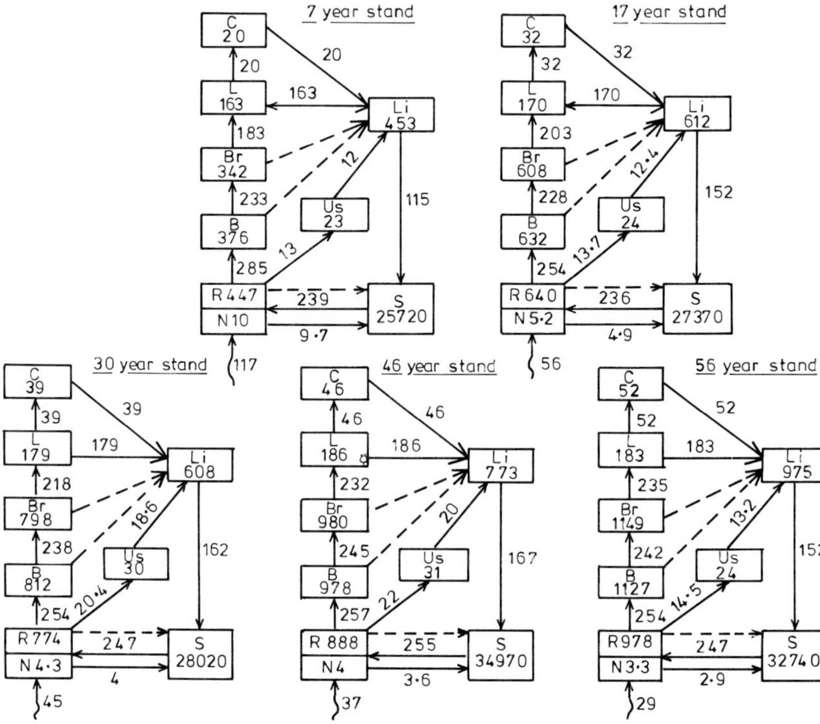

Figure 7.4 Distribution of nitrogen and flow rates in the components of the five different ages of the Nepalese alder (*Alnus nepalensis*) plantation stands. Units are kg ha^{-1} for compartments and kg ha^{-1} yr^{-1} for flows. Arrow with twisted end shows flow through biological nitrogen fixation. Broken lines indicate that values are not estimated. B, bole; Br, branches; C, catkins; L, twig and leaf; Li, litter; N, root nodules; R, roots; S, soil (m^{-1} depth) and Us, understorey (from Sharma and Ambasht, 1988)

their role in nutrient conservation (Rao and Ramakrishnan, 1989). With the role of herbs in nutrient cycling declining after 5 years of fallow regrowth, the contribution by the bamboo increases up to about 25–30 years. It is therefore reasonable to expect a key role for bamboos in nutrient cycling and nutrient conservation, an aspect that is little known, except for our own studies from the region (Toky and Ramakrishnan, 1982b, 1983b; Rao and Ramakrishnan, 1989).

The four bamboo species, under consideration here, always had higher concentrations of nitrogen, phosphorus and potassium than shrubs and trees, but lower concentrations of calcium and magnesium (Table 7.17). Consequently, the quantities of these elements in the bamboo component of the vegetation in 5-, 10- and 15-year-old fallows would be higher and would increase with fallow age because of the biomass increase (cf Fig. 6.4); the quantities released through litter, obviously, would also follow a similar pattern (Table 7.18).

Table 7.17 Mean percentage contents (mean ± 95% confidence limit) of elements in the boles, branches and leaves of bamboos and trees and shrubs in 10-year-old jhum fallows in north-eastern India (from Rao and Ramakrishnan, 1989)

	Bole	*Branch*	*Leaf*
Nitrogen			
Dendrocalamus hamiltonii	0.44 ± 0.05	0.51 ± 0.09	3.20 ± 0.47
Other bamboos[1]	0.41 ± 0.26	0.47 ± 0.06	2.63 ± 0.44
Trees, shrubs	0.43 ± 0.02	0.52 ± 0.02	1.42 ± 0.21
Phosphorus			
Dendrocalamus hamiltonii	0.15 ± 0.03	0.18 ± 0.03	0.47 ± 0.09
Other bamboos	0.12 ± 0.02	0.18 ± 0.03	0.44 ± 0.07
Trees, shrubs	0.07 ± 0.004	0.10 ± 0.005	0.13 ± 0.02
Potassium			
Dendrocalamus hamiltonii	2.0 ± 0.32	2.5 ± 0.37	3.1 ± 0.34
Other bamboos	1.2 ± 0.25	1.8 ± 0.24	2.4 ± 0.29
Trees, shrubs	1.1 ± 0.29	1.3 ± 0.31	1.5 ± 0.32
Calcium			
Dendrocalamus hamiltonii	0.19 ± 0.03	0.23 ± 0.01	0.35 ± 0.02
Other bamboos[1]	0.20 ± 0.03	0.26 ± 0.03	0.38 ± 0.10
Trees, shrubs	0.46 ± 0.03	0.62 ± 0.05	0.92 ± 0.06
Magnesium			
Dendrocalamus hamiltonii	0.14 ± 0.03	0.16 ± 0.03	0.35 ± 0.04
Other bamboos[1]	0.11 ± 0.03	0.13 ± 0.03	0.33 ± 0.04
Trees, shrubs	0.26 ± 0.04	0.26 ± 0.04	0.45 ± 0.04

[1]*Bambusa tulda*, *Neohouzeoua dulloa* and *Bambusa khasiana* in the Garo, Khasi and Naga hills, respectively

Biomass growth of tropical forests can be severely limited by nutrients; so shortages and losses are of concern to the practice of jhum. When slashed forest is burnt, nitrogen losses by volatilization are large (cf Chapter 5). Ewel *et al.* (1981) showed losses of 500 kg ha^{-1} of nitrogen in Costa Rica. Mishra and Ramkrishnan (1984) and Saxena and Ramakrishnan (1986) showed that the losses of about 600 kg ha^{-1} of nitrogen occurring during one cropping season could be replenished only in about 15 years of secondary succession. Phosphorus is in short supply in tropical soils because of its fixation into unavailable forms in the clay fractions rich in iron and aluminium sesquioxides (Gebhardt and Coleman, 1974; Tinker, 1977). Potassium is more readily lost than divalent cations through run-off and percolation water after fire (Toky and Ramakrishnan, 1981b; Mishra and Ramakrishnan, 1983c; Ram and Ramakrishnan, 1988b). Therefore, the rapid accumulation of these elements by bamboos, particularly of potassium, is significant in view of the losses during jhum operations.

Although element storage in the biomass and recycling through litter increased in older fallows, nutrient turnover rates varied with the species composition and site (Table 7.18). However, the fractional turnover rate

Table 7.18 Inventory of elements in the total vegetation (Tv) and (in parentheses) in the bamboo component (Bc), (mean ± 95% confidence limit) slashed in different jhum fallows in north-eastern India (from Swamy and Ramakrishnan, 1989)

Element/ fallow age (yr)	*Aboveground biomass* (kg ha^{-1})		*Release through litter* (kg ha^{-1})		*Annual turnover* (%)		*Annual nutrient accumulation*		*Enrichment ratio*	
	Tv	*Bc*	*Tv*	*Bc*	*Tv*	*Bc*	*Tv*	*Bc*	*Tv*	*Bc*
Nitrogen										
5	75 ± 5.8	49 ± 4.0	37 ± 3.8	17 ± 2.0	49 ± 4.2	35 ± 2.9	12 ± 1.3	6 ± 0.5	6 ± 0.5	8 ± 0.8
10	306 ± 21.5	184 ± 14.6	85 ± 4.0	57 ± 5.2	28 ± 0.9	31 ± 1.73	25 ± 1.92	14 ± 0.85	13 ± 0.18	13 ± 0.18
15	439 ± 44.3	261 ± 21.9	110 ± 11.2	91 ± 8.4	25 ± 1.1	35 ± 0.30	21 ± 2.30	8 ± 1.52	21 ± 1.28	33 ± 4.74
Phosphorus										
5	31 ± 3.3	16 ± 1.3	17 ± 1.8	3 ± 0.2	53 ± 4.6	21 ± 2.0	1 ± 0.1	1 ± 0.1	45 ± 4.1	62 ± 7.0
10	77 ± 6.1	51 ± 6.1	14 ± 0.6	9 ± 1.6	18 ± 0.1	17 ± 1.6	6 ± 0.2	5 ± 0.3	14 ± 0.6	11 ± 1.1
15	115 ± 5.9	79 ± 8.1	26 ± 1.5	13 ± 1.1	23 ± 1.3	17 ± 0.4	7 ± 0.3	5 ± 0.8	16 ± 0.3	18 ± 1.8
Potassium										
5	281 ± 18.9	175 ± 15.6	45 ± 4.0	20 ± 1.8	16 ± 1.2	12 ± 0.6	54 ± 5.2	31 ± 3.0	5 ± 0.5	6 ± 0.5
10	883 ± 68.3	527 ± 34.4	78 ± 7.8	61 ± 5.3	8 ± 0.6	12 ± 0.4	98 ± 12.4	51 ± 3.1	9 ± 1.3	11 ± 0.8
15	1,359 ± 60.8	812 ± 37.2	113 ± 91.8	14 ± 1.7	8 ± 0.8	11 ± 0.5	110 ± 51.1	13 ± 0.6	13 ± 1.9	16 ± 0.8
Calcium										
5	113 ± 5.8	18 ± 1.4	39 ± 3.4	3 ± 0.3	35 ± 3.0	14 ± 1.7	4 ± 0.5	1 ± 0.1	30 ± 3.1	58 ± 5.2
10	381 ± 20.0	59 ± 6.9	45 ± 4.1	9 ± 0.8	12 ± 1.3	15 ± 1.4	24 ± 3.3	6 ± 0.4	16 ± 1.9	10 ± 0.7
15	527 ± 67.5	106 ± 18.9	51 ± 4.9	15 ± 1.2	10 ± 2.1	15 ± 1.9	34 ± 3.7	8 ± 2.7	16 ± 2.0	16 ± 2.8
Magnesium										
5	78 ± 6.6	13 ± 1.0	21 ± 2.4	3 ± 0.2	28 ± 2.5	2 ± 0.1	2 ± 0.2	1 ± 0.1	50 ± 4.8	63 ± 5.6
10	239 ± 44.2	45 ± 4.0	28 ± 3.5	8 ± 0.8	12 ± 0.8	18 ± 1.7	15 ± 6.3	3 ± 1.3	30 ± 4.1	37 ± 5.0
15	292 ± 47.6	69 ± 5.6	32 ± 4.0	14 ± 1.0	11 ± 1.1	20 ± 1.8	13 ± 5.6	2 ± 0.8	38 ± 5.7	51 ± 4.6

for potassium was lower because of its rapid accumulation in the more dominant bamboo, *Dendrocalamus hamiltonii* (Toky and Ramakrishnan, 1982b, 1983b) and slower release through litterfall. This was reflected in the higher annual accumulation rate for potassium. Higher turnover rates of nitrogen and phosphorus, in spite of their larger storage in the biomass, may be because of much higher concentrations in the leaves which determine turnover rates. With bamboos contributing 40–58% of the total slash biomass at the three sites and 49–73% of nitrogen, phosphorus and potassium, they play a significant role in nutrient cycling and conservation under the jhum system in north-eastern India.

Nitrogen accretion through Nepalese alder (*Alnus nepalensis*)

The Nepalese alder (*Alnus nepalensis*) is an important early successional tree species in the north-eastern hill region above an altitude of 500 m. In the altitudinal range of 900–1,900 m, this species grows very vigorously. Apart from the fact that its species is a successful colonizer of landslide affected or freshly exposed rocky and eroded slopes throughout the Himalayan region, particularly the eastern Himalayas (Sharma and Ambasht, 1988), this species is important as an early colonizer of jhum fallows in north-eastern hill region (Boojh and Ramakrishnan, 1981c, 1982a). This species with nodulated roots, colonized by *Frankia* occurring as an endophyte, is effective in biological nitrogen fixation (Sharma and Ambasht, 1988) like many other *Alnus* species (Akkermans and Van Dijk, 1981).

In a study of the energetics of nitrogen accretion in five different ages of *Alnus nepalensis* plantations, Sharma and Ambasht (1988) showed that the percentage contribution of biological nitrogen fixation to annual total nitrogen uptake were 33%, 19%, 15%, 13% and 11% in 7- 17-, 30-, 46- and 56-year-old plantations, respectively (Fig. 7.4). Annual total nitrogen uptake in the 7-year-old stand was 1.3 times that of the 56-year-old stand. Nitrogen accretion through fixation reduced with decrease in nitrogen demand as trees aged (117 kg $ha^{-1} yr^{-1}$ in 7-year-old stand to 29 kg $ha^{-1} yr^{-1}$ in 56-year-old stand); these values compare favourably with the values reported for other species of *Alnus* (Binkley, 1981). However, total nitrogen uptake remained the same in 17-, 30- and 46-year-old plantations.

Under short jhum cycles of 5–6 years, we have already seen that about 600 kg ha^{-1} of soil nitrogen is lost during one cropping season and it takes a minimum of 10 years of natural fallow regrowth to recover all this nitrogen back into the system (Mishra and Ramakrishnan, 1984; cf Chapter 4). It is in this context of putting back the 600 kg ha^{-1} of nitrogen over a period of 4–5 years of the fallow phase that an early successional species such as the Nepalese alder becomes important. Through our studies on the jhum system, we have shown that such a recovery of nitrogen is possible using this key species (Ramakrishnan, 1987d,e, 1988a,c), Apart from nitrogen fixation (Sharma and Ambasht, 1988), the production of nitrogen-rich litter and mineralization also contributes to a biological build-up of soil fertility. Indeed, the Nepalese alder and many legumes that come up during early succession are critical for replenishing nitrogen in the jhum system and in the nitrogen-poor denuded hill slopes.

Earthworms in the forest fallows and their contribution to nutrient cycling

Much of the information available on the ecology of earthworms in forested ecosystems is from temperate regions of the world (Edwards, 1983). These studies show that earthworms play a key role in determining nutrient cycling patterns through their role in litter decomposition and soil turnover (Syers and Springett, 1984). However, information on the ecology and role of earthworms in nutrient cycling is limited (Bhadauria and Ramakrishnan, 1989). The tropical species are largely surface dwellers (endoges of Bouche, 1977) unlike the deep burrowing Lumbricids, (aniques of Bouche, 1977) of the temperate zone. Therefore in the tropical ecosystems their role is in concentrating nutrients in the surface soil layers. To this extent, the functional role of earthworms in the tropical systems could differ from their temperate counterparts. In this section, the role of earthworms in the early successional communities up to 15 years of fallow regrowth is compared with a 35-year-old pine forest and a climax sacred grove forest of mixed broad-leaved species (cf Chapter 6 for details on vegetation), at a higher elevation at Shillong (1,500 m) in Meghalaya (Bhadauria and Ramakrishnan, 1989, 1990).

Earthworms in early successional fallows up to 15 years

Of the five species found during the first 15 years of secondary succession, *Nelloscolex strigosus* and *Drawida assamensis* were restricted to 5- and 15-year-old fallows, respectively, (Fig 7.5). The other three species, namely *Amynthes diffringens*, *Eutyphoeus festivus* and *Tonoscolex horaii* were found in 5-, 10- and 15-year-old fallows. Restriction of *Drawida assamensis* to a 15-year-old fallow, but with low frequency, may suggest its sensitivity to perturbation on the one hand, and to pine litter on the other (the latter aspect is suggested by its absence in a 35-year-old pine forest but its reappearance in the sacred grove mixed broad-leaved forest, as seen from the subsequent discussion). *Tonoscolex horaii* was dominant throughout and it even occurred in a 35-year-old pine forest and in the climax sacred grove forest of the region (Bhadauria and Ramakrishnan, 1990). The

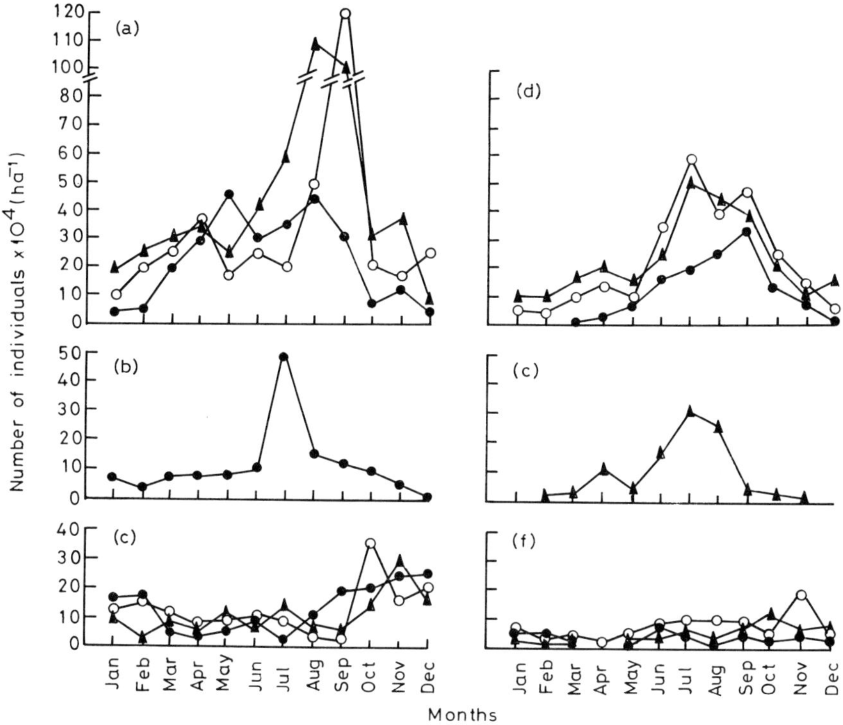

Figure 7.5 Population fluctuations of earthworm species in 5-yr- (●), 10-yr- (○) and 15-yr- (▲) old jhum fallows at Shillong in Meghalaya (a), juveniles; (b), *Nelloscolex strigosus*; (c) *Amynthas diffringens*; (d) *Tonoscolex horaii*; (e) *Drawida assamensis*; (f) *Eutyphoeus festivus* (from Bhadauria and Ramakrishnan, 1989)

population density of earthworms in the fallows was up to 50 individuals (matures) m^{-2}, higher than that reported for temperate fallow plots (Edwards and Lofty, 1977).

The cast production in the fallows improved from 1,507 t ha^{-1} yr^{-1} in 5-year-old fallows to 40 t ha^{-1} yr^{-1} in 15-year-old fallows. These values are much higher than the 8.6 t ha^{-1} yr^{-1} of cast material reported for temperate arable fallow lands (Watanabe and Ruaysoongnern, 1984). Apart from the fact that cast soil pH was higher than that of the normal soil (Bhaduria and Ramakrishnan, 1989), the nutrient concentration in the cast soil was also higher (Fig. 7.6). The elemental concentration was generally higher in the worm casts during July. The increase in concentration of nutrients in the cast soil was more pronounced in the older fallows. The quantities of all nutrients in the worm casts (except calcium) and dead tissue (except nitrogen and phosphours) increased with fallow age (Table 7.19). Nitrogen alone was present in appreciable amounts in the mucus, and increased sharply with fallow age.

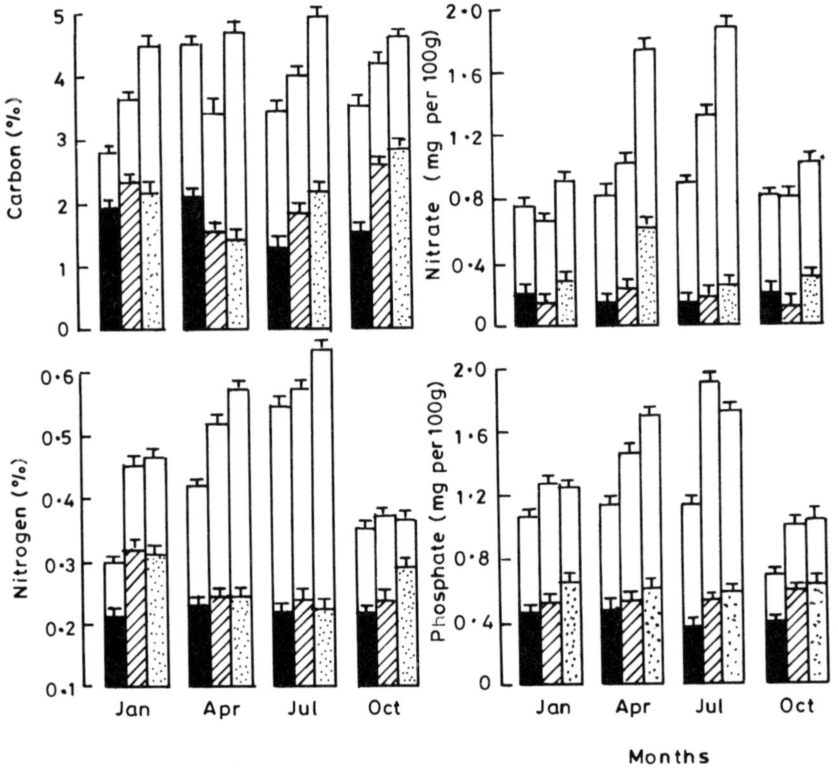

Figure 7.6 Variation in nutrient concentration in earthworm casts and in surface soil (0–10 cm) in 5-yr- (■), 10-yr- (▨) and 15-yr- (▩) old jhum fallows at Shillong in Meghalaya. Vertical bars denote SE (from Bhadauria and Ramakrishnan, 1989)

A general increase in population density and activity with increasing fallow age may be due to more favourable temperature, moisture and humidity conditions with increased organic matter and nutrient accumulation. All the species are restricted to 0–14 cm of the soil profile. This implies that these tropical species are subjected to greater fluctuations in the micro-environment than some of the temperate Lumbricidae which can migrate vertically to avoid environmental fluctuations in the surface layers (Wallwork, 1983). Thus, except for *Amynthes diffringens* which showed an inverse correlation between population size and ambient air temperature peaking in winter months (Fig. 7.5), all others showed positive correlation between population size and earthworm activity on the one hand, with temperature, soil moisture or organic matter on the other (Bhadauria and Ramakrishnan, 1989). Thus, earthworms play a key role in nutrient cycling in the early successional fallow phase following jhum.

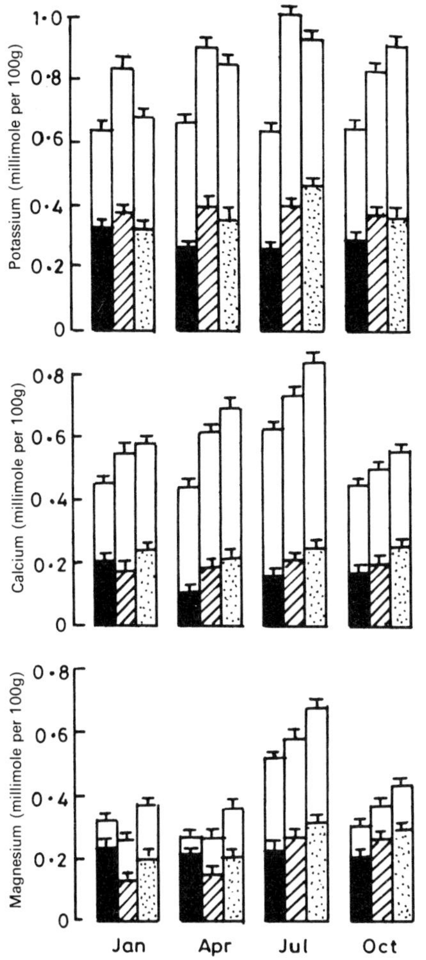

Figure 7.6 *Continued*

Earthworms in seral pine forest and climax sacred grove

Of the five species of earthworms found in the 35-year-old pine forest and the climax sacred grove, *Tonoscolex horaii* occurred in large numbers in both sites (Fig. 7.7). *Amynthas diffringens* and *Eutyphoeus festivus* were confined to pine forest alone, whilst *Perionyx* spp. and *Drawida assamensis* occurred only in the sacred grove. These distributional patterns seem to be related to the soil characteristics and litter quality differences between the two forest types (Das and Ramakrishnan, 1985; Khiewtam, 1986). Edwards and Lofty (1977) suggest that earthworms generally have a narrow range

Table 7.19 Nutrients (kg ha^{-1} yr^{-1}) returned to the soil in dead worm tissue, mucus and worm casts by earthworms in 5, 10 and 15-yr-old fallows developed after jhum at Shillong in Meghalaya. Mucus N refers to nitrogen contributed through mucus production, ND = no data (from Bhadauria and Ramkrishnan, 1989)

	Dead tissue (± *SE*)			*Worm casts* (± *SE*)		
	5-yr fallow	*10-yr fallow*	*15-yr fallow*	*5-yr fallow*	*10-yr fallow*	*15-yr fallow*
Organic C	ND	ND	ND	383 ± 3.9	698.9 ± 5.4	869.8 ± 6.4
Total N	14.5 ± 0.5	15.0 ± 0.6	12.0 ± 0.4	21.3 ± 0.4	44.2 ± 0.4	103.5 ± 1.7
Mucus N	114.5 ± 1.8	118.1 ± 1.8	139.8 ± 2.1			
NO_3-N	ND	ND	ND	0.1 ± 0.01	0.3 ± 0.01	0.4 ± 0.1
P	1.0 ± 0.01	1.1 ± 0.3	0.9 ± 0.03	0.1 ± 0.02	0.2 ± 0.01	0.4 ± 0.02
K	0.7 ± 0.03	0.7 ± 0.1	1.4 ± 0.1	1.2 ± 0.07	9.0 ± 0.6	9.7 ± 0.6
Ca	1.2 ± 0.10	1.3 ± 0.03	11.7 ± 0.8	34.9 ± 1.6	37.8 ± 1.0	13.0 ± 0.1
Mg	0.7 ± 0.02	0.7 ± 0.04	0.8 ± 0.02	0.4 ± 0.03	12.9 ± 1.0	14.5 ± 1.0

Table 7.20 Nutrients (kg ha^{-1} yr^{-1}) (± SE) returned to the soil in dead worm tissue, mucus and worm casts by earthworms in *Pinus kesiya* stands and a sacred grove at Shillong in Meghalaya. Mucus nitrogen refers to nitrogen contributed through mucus production, ND = no data (from Bhadauria and Ramakrishnan, 1990)

	35-year-old pine		*Sacred grove*	
	Dead tissue	*Cast*	*Dead tissue*	*Cast*
Organic carbon	ND	541.7 ± 74	ND	831.7 ± 88
Total nitrogen	5.5 ± 0.3	199.1 ± 13.2	7.1 ± 0.5	233.6 ± 17.0
Mucus nitrogen	31.4 ± 7.0			
Phosphorus	0.4 ± 0.02	0.1 ± 0.02	0.5 ± 0.03	0.1 ± 0.004
Potassium	2.4 ± 0.1	5.9 ± 0.4	3.0 ± 0.2	3.7 ± 0.04
Calcium	0.5 ± 0.03	34.2 ± 2.8	0.6 ± 0.03	21.5 ± 1.8
Magnesium	0.4 ± 0.02	2.2 ± 0.06	0.3 ± 0.02	1.9 ± 0.1

of tolerance for pH. However, with a soil pH of 4.5 for the sacred grove the population size per m^2 is higher than that reported for temperate forests.

The cast production in the two forest types closely followed population numbers discussed above, with maximum activity during the monsoon season and least activity during the relatively dry winter months. However, the cast production in the sacred grove was much higher (41,000 kg ha^{-1} yr^{-1}) than that for the 35-year-old pine forest (32,300 kg ha^{-1} yr^{-1}) (Bhadauria and Ramakrishnan, 1990). These values are higher than the reported values for many temperate broad-leaved forests (5.7–20 t ha^{-1}) (Lee, 1985), although lower than the values reported from Ghana (50 t ha^{-1} yr^{-1}) for a low elevation mixed tropical forest (Nye, 1955).

Carbon and nitrogen recycled through earthworms increased sharply in the sacred grove compared to the pine forest; however, the cations were higher in the pine forest (Table 7.20).

It is well known that the significance of these key soil organisms lies in their ability to rapidly mineralize detritus and to locally concentrate nutrients in the surface worm casts (Nye, 1955; Lee, 1985). With a highly leached forest soil, we have seen how the surface soil layers have an exceptionally high fine-root biomass (cf Chapter 6). The early successional pine (*Pinus kesiya*) trees have a larger root biomass in the surface layers of the soil (Das and Ramakrishnan, 1986) like all other early successional trees (Shukla and Ramakrishnan, 1984b). We have already seen (cf Chapter 6; Khiewtam, 1986) the important role of fine-root biomass in the climax sacred grove for rapid recycling of nutrients; this fine-root biomass in the surface soil layers may be as much as about 14 t ha^{-1} yr^{-1}. In these situations particularly, earthworms have an important role in improving soil physical and chemical properties in the surface soil layers.

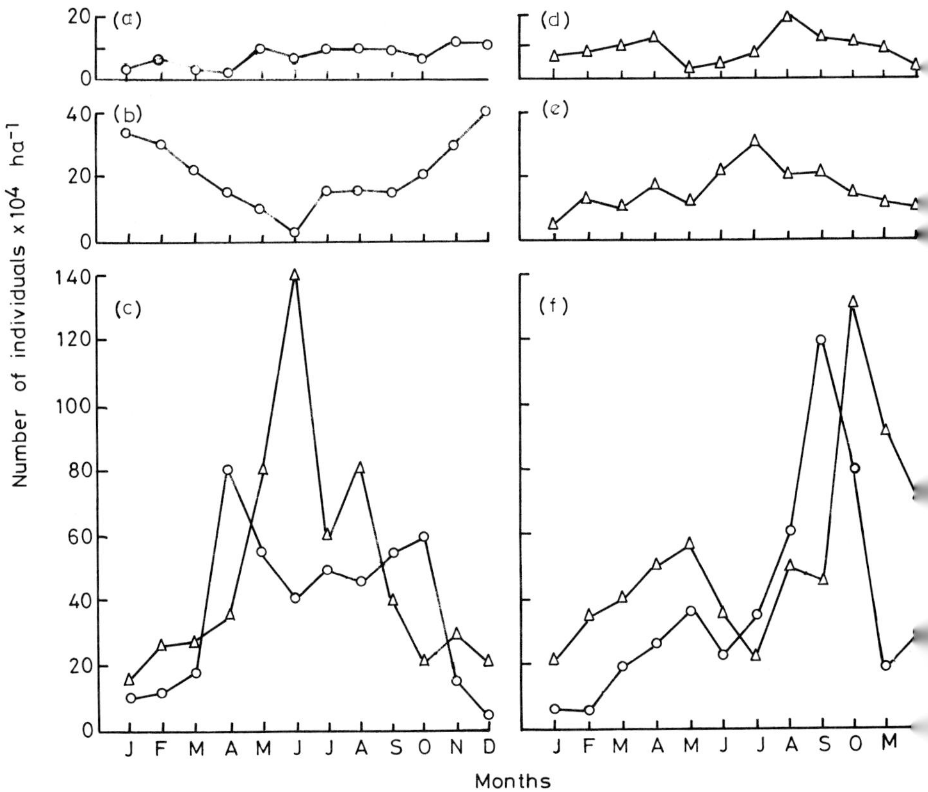

Figure 7.7 Population fluctuations of earthworm species in 35-yr-old pine forest (○) and sacred grove (△) at Shillong in Meghalaya. (a) *Eutyphoeus festivus*; (b), *Amynthas diffringens*; (c) *Tonoscolex horaii*; (d), *Drawida assamensis*; (e) *Perionyx* spp. and (f) juveniles (from Bhadauria and Ramakrishnan, 1990)

Conclusions

Key species during succession are those that either determine the vegetation structure, biomass and productivity of a given seral/climax ecosystem type or would determine an important ecological process such as nutrient cycling, considered in the foregoing pages. The two groups of species, namely the exotic weeds and the bamboos, alter both structural and functional properties of the seral ecosystems as a whole during early succession. Earthworms, on the other hand, have a somewhat limited role mainly affecting soil ecosystem properties right through succession. They have a more critical role in concentrating nutrients in the surface soil layers and indeed increasing the rate of nutrient cycling itself. This may have indirect consequences for the whole ecosystem.

The example of biological invasion by Latin American exotic species illustrates very well both the potential significance of biological invasion into secondary succession and the analytical difficulties inherent in demonstrating these effects. Short jhum cycles of 4–6 years have enabled large-scale spread of a few species, namely, *Eupatorium* spp and *Mikania micrantha*, these invasions having largely occurred subsequent to the Second World War. These exotics with C_3 photosynthetic pathways, high reproductive potential and high nutrient uptake have no doubt markedly altered myriad ecosystem properties, including biomass and nutrient budgets, in this succession. However, the success of these biological invaders is inextricably linked with a novel disturbance regime (short jhum cycles), and it is difficult to separate the effects of the invaders *per se* from those of the disturbance regime.

A non-leguminous species such as the Nepalese alder (*Alnus nepalensis*), an early successional tree species up to about 60 years of fallow regrowth, plays a key role in nitrogen conservation. Many legumes that come up in early succession could also be important. The Nepalese alder is critical for conserving a labile element such as nitrogen and for stabilizing jhum at a 4- to 5-year cycle length, which otherwise would need a minimum jhum cycle length of 10 years.

Bamboos, partly because of their large biomass and dominance in seral communities and partly due to their nutrient uptake capabilities, accumulate large quantities of important nutrients like nitrogen, phosphorus and potassium. *Dendrocalamus hamiltonii*, the dominant bamboo species, by accumulating potassium in particular, conserves this labile element which is very easily lost from the system. Because this and other bamboo species gain dominance in the fallow regrowth 10 years after jhum and are important for up to about 30 years, they play an important role in nutrient conservation under long jhum cycles. What *Dendrocalamus hamiltonii* does to conserve potassium under long cycles, is done by *Mikania micrantha* under short cycles of 4–5 years, since the latter species is dominant in fallow regrowths for up to 4–5 years.

As already mentioned, earthworms have a more limited, although important role in nutrient conservation. In the soil sub-system, they may influence the soil biota. Worm casts are local concentrations of nutrients. In highly leached soils of the humid tropics, earthworm activity is beneficial because of rapid incorporation of detritus into the mineral soil. Use of earthworms through vermiculture could be an important approach for biological maintenance of soil fertility under the jhum system and for sustainable hill agriculture. As seen here, different earthworm species have different ranges of tolerance to a changing successional environment. *Tonoscolex horaii* is a species with a wide range of tolerance and with a high activity level; it occurs right through the cropping phase and all through the fallow phases under jhum, extending up to the climax phase at higher elevations of Meghalaya. A species such as this has a high potential for vermiculture and for use as a key species from amongst this group of key organisms.

NUTRIENT CYCLING UNDER ARRESTED SERAL GRASSLANDS AT CHERRAPUNJI

The mixed broad-leaved forest of Cherrapunji representing the climax vegetation would rapidly degrade into grasslands if the forests are clear-cut for any reason, because of the highly leached nutrient deficient soil with 'Karst topography', developed under exceptionally high rainfall (annual average of 1,037 cm) (cf Chapter 6). Recovery of these degraded grasslands is difficult because of extreme stress conditions related to the climate and soil. An important ecological attribute that is immediately affected after perturbution is nutrient cycling. In a comparative analysis of four grassland types of this area, namely, the *Osbeckia* type on slopes with good soil and burnt 6 years prior to the study; the *Arundinella* type that had similar habitat features as the first, but for an additional burn, 1 month prior to the study period; the *Ischaemum* type that had similar conditions to the other two except for a regular 2-yearly burn, the last being 1 month prior to the study period; and finally the *Eragrostiella* type developed on highly degraded slopes that are gravelly with partially weathered mineral soil, we have evaluated (Ram, 1986) their tendency to accumulate nutrients against the forces that operate leading to losses from the system.

With generally very low concentrations of nutrients in the soil under *Eragrostiella* type and maximal values under the *Osbeckia* type (cf Tables 6.11), the *Arundinella* and *Ischaemum* types had nutrient levels inversely related to fire frequency. Thus the soil nutrient pool under the *Ischaemum* type was lower than under the *Arundinella* type. In these two types, there was a net loss of nitrogen through fire due to volatilization, as also discussed earlier (cf Chapter 5; Ramakrishnan and Toky, 1981; Mishra and Ramakrishnan, 1983d, 1984), and this loss was more under the *Arundinella* type due to more biomass burnt.

The very slow nutrient recovery rate is exemplified for nitrogen through a comparative study for nitrification potential of the sacred grove soil and three grassland types (referred to as 1,2 and 3 for increasing levels of disturbance (Ramakrishnan and Saxena, 1984). The nitrification rate was markedly higher in the forest soil; the grassland soils had very low rates, the highly disturbed site having the lowest level of nitrification (Fig. 7.8).

Some loss of phosphorus was also noted (Ram, 1986; Mishra and Ramakrishnan, 1983d). The cause for this is suggested to be through convective losses of particulates, although there is no obvious mechanism for phosphorus volatilization (Ashton, 1976).

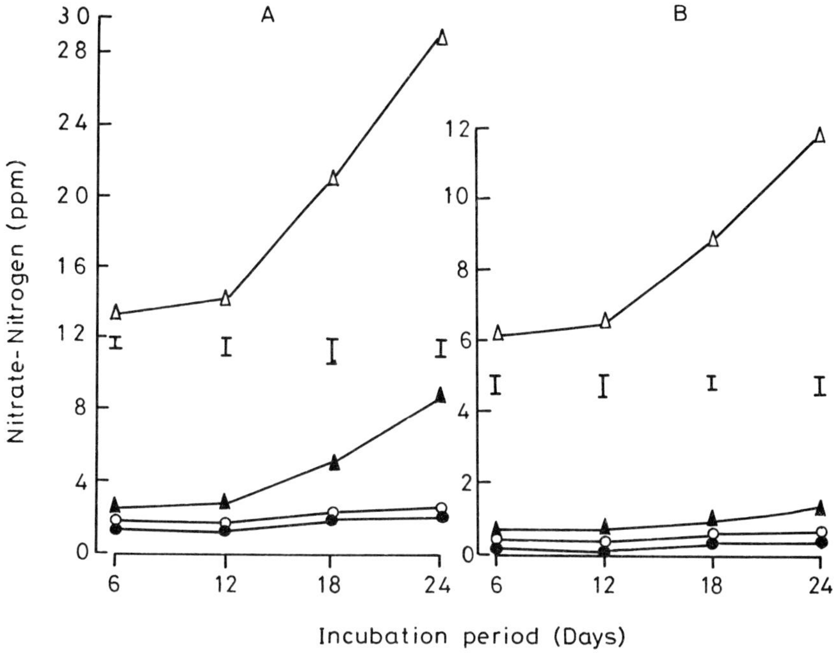

Figure 7.8 Nitrification potential expressed as increase in nitrate-nitrogen concentration during incubation of soils from forest (△) and grassland types: 1, ▲; 2, ○; 3, ●; the numbers representing increasing levels of disturbance. A. dry period (February); B, wet period (June). Vertical bars represent LSD ($p = 0.05$) (from Ramakrishnan and Saxena, 1984)

Substantial quantities of nutrients are lost through run-off and percolation water (Table 7.21) during the monsoon; even the totally desertified site under *Eragrostiella* type loses large quantities every year so that nutrient build-up in the soil becomes curtailed. This kind of a loss under high rainfall conditions would be a major factor for the arrested vegetation (Ram and Ramakrishnan, 1988b). Frequent burning of plots as under the *Ischaemum* type is expected to increase the losses from the system as seen from this study; potassium and nitrogen losses are heavy.

The differences observed between the four grassland types for the nutrients held in the biomass are simply reflections of different biomass values rather than radically different tissue concentrations (Ram, 1986). Thus, with higher biomass for the *Osbeckia* type and very low biomass for the *Eragrostiella* type, the nutrient level in the living biomass and that returned to the grassland floor through litter (Table 7.22) was maximum for the former and minimum for the latter. The decay rate for the litter is very high with a k value of 3.06, which represents the annual decay constant (Ram and Ramakrishnan, 1988c), for the monsoon season when maximum litter decomposition occurs, but with an annual average k value of 1.98. This is a much higher rate than that for temperate grasslands and higher than

Table 7.21 Annual loss of water (cm) and nutrients (g per m^2) through run-off and percolation from different grasslands at Cherrapunji in Meghalaya. Total rainfall in 1982 was 1,038 cm (from Ram and Ramakrishnan, 1988b)

	Vegetation type		
	Osbeckia *type*	Ischaemum *type*	Eragrostiella *type*
Water (cm)			
Run-off	719.7	818.2	856.5
Percolation	136.4	126.6	130.7
NO_3-N			
Run-off	1.86	4.28	1.25
Percolation	0.70	0.58	0.3
PO_4-P			
Run-off	0.18	0.22	0.17
Percolation	0.03	0.03	0.036
K			
Run-off	4.14	7.53	3.38
Percolation	0.83	0.44	0.30
Ca			
Run-off	11.4	8.41	7.44
Percolation	1.73	1.63	1.42
Mg			
Run-off	3.92	3.76	2.46
Percolation	0.44	0.64	0.26

Table 7.22 Nutrients $\pm$ 95% confidence limit (g m^{-2}) returned to the grassland floor through litterfall at Cherrapunji in Meghalaya (Ram, 1986)

	Vegetation type			
Nutrient	Osbeckia *type*	Arundinella *type*	Ischaemum *type*	Eragrostiella *type*
N	2.9 ± 0.3	1.8 ± 0.2	1.0 ± 0.1	1.1 ± 0.1
P	0.7 ± 0.1	0.3 ± 0.01	0.3 ± 0.07	0.2 ± 0.09
K	5.6 ± 0.7	3.5 ± 0.4	2.4 ± 0.3	1.8 ± 0.2
Ca	3.1 ± 0.5	1.7 ± 0.2	1.0 ± 0.2	1.6 ± 0.2
Mg	1.4 ± 0.2	1.1 ± 0.1	0.7 ± 0.1	0.6 ± 0.1

that for other Indian grasslands (Ram and Ramakrishnan, 1988c). Exceptionally high rainfall and high humidity are responsible for this high rate.

In conclusion, it is seen that the *Osbeckia* type alone had a net gain of all the nutrients at the end of the year; others had a net loss (Table 7.23). Even when site quality is comparable to the *Osbeckia* type, disturbance due to fire upsets the nutrient budget under *Arundinella* and *Ischaemum* types; the *Eragrostiella* type of course is highly desertified and has little ability to

Table 7.23 Net changes of elements ($g\,m^{-2}$) in the soil pool over a 1-year period for *Arundinella* and *Ischaemum* types in March/preburn of year 1 and then in the March of the subsequent year at Cherrapunji in Meghalaya (Ram, 1986)

	Osbeckia *type*	Arundinella *type*	Ischaemum *type*	Eragrostiella *type*
Soil pool year 1				
N	932.9	886.0	737.8	549.0
P	0.44	0.42	0.40	0.37
K	19.6	19.3	19.6	18.0
Ca	75.2	63.7	51.3	48.2
Mg	75.2	63.5	50.7	48.2
Soil pool year 2				
N	934.0	875.0	728.0	543.1
P	0.48	0.38	0.39	0.35
K	20.2	18.8	18.8	17.3
Ca	80.1	60.0	46.6	47.3
Mg	75.6	60.0	46.6	47.3
Net changes (gain/loss)				
N	+1.1	−11.0	−9.8	−5.9
P	+0.04	−0.04	−0.01	−0.02
K	+0.6	−0.5	−0.8	−0.7
Ca	+4.9	−3.7	−4.7	−0.9
Mg	+0.4	−3.5	−4.1	−0.9

hold nutrients. In the fire affected grasslands, the nutrient cycling recovery process is, to a certain extent, determined by a well-developed underground biomass compartment which has substantial quantities of nutrients stored in it (Fig. 7.9). The turnover rate for this underground biomass compartment in these tropical grasslands could be five times more than in grasslands of the temperate regions of the world (Ramakrishnan and Ram, 1988; Ram and Ramakrishnan, 1988a, 1989). Frequent perturbations after some time would adversely alter nutrient cycling properties resulting in extreme site degradation as seen under the *Eragrostiella* type of grassland.

Conclusions

The grasslands at Cherrapunji are in an arrested stage due to non-availability of propagules over large areas, due to large-scale deforestation and also, perhaps, partly related to jhum done once upon a time in the distant past. There is evidence of it in the neighbourhood of Cherrapunji (Khiewtam and Ramakrishnan, 1989), although this land use practice is now banned in the Cherrapunji region itself, a decision made by the local community. With a fragile mature forest ecosystem as exemplified by the sacred grove, once the forest was cut down, successive spells of heavy

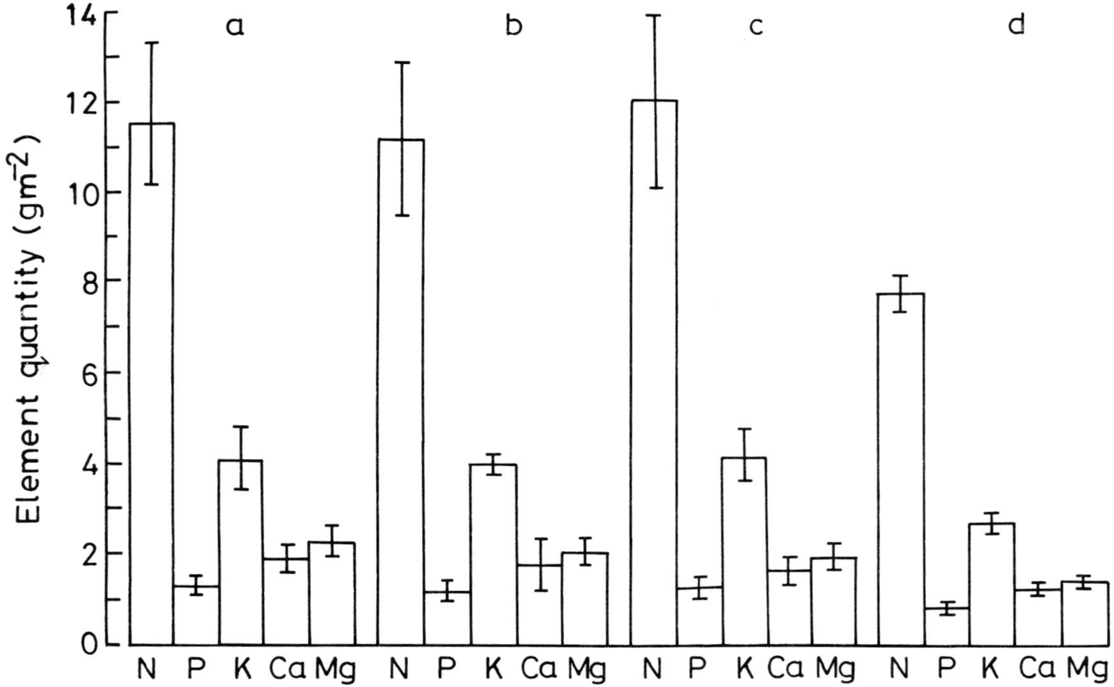

Figure 7.9 Nutrient quantity in belowground biomass at peak growth under *Osbeckia* (a), *Arundinella* (b), *Ischaemum* (c) and *Eragrostiella* (d) types of grasslands at Cherrapunji in Meghalaya (from Ram, 1986)

rainfall would wash out the top soil and along with it a heavy nutrient load. These degraded grassland ecosystem types once established are then not able to revert back to the climax forest vegetation of the area. The problem is accentuated by depletion of tree seed source. Grasslands developed on relatively well-formed soil profile are maintained at varied levels of degradation by manipulated fire of varied frequency and intensity. Extreme degradation results in stunted grasses sticking out of a largely stony and gravelly soil with xerophytic stunted grasses; *Eragrostiella* type is an example. Once that stage is reached, fire is of no consequence and the system is maintained in that state merely through stress created by high rainfall which does not allow soil build-up.

GENERAL CONSIDERATIONS

The nutrient cycling patterns of the jhum system in north-eastern India reveal some important points of applied value. If the perturbance regime is infrequent, the recovery of the system through secondary succession, and stabilization of nutrient cycling properties is rapid. The rapid shift in nutrient cycling patterns during the first 20 years or so of the fallow regrowth is due to : rapid shifts in nutrient pool between the soil and the vegetation compartments when the community structure changes from herbaceous to bamboo/pine dominated and, finally and more slowly, to a mixed broad-leaved forest. These shifts are also partly due to the role played by some of the key communities such as herb communities conserving phosphorus, bamboos conserving nitrogen, phosphorus and potassium and key species such as *Mikania micrantha* and *Dendrocalamus hamiltonii* accumulating potassium in large quantities in the biomass. Some of these key species could have value for agroforestry. Thus the Apatanis use bamboo as an agroforestry species around their agricultural plots. The value of an early successional species such as the Nepalese alder (*Alnus nepalensis*) for recovering nitrogen in the jhum system under very short jhum cycles is particularly significant because the jhum farmer traditionally protects this species in his jhum system because of his intuitive experience that it does some good to the soil. Therefore, agroforestory input using this species is more readily acceptable to him than many other species that may otherwise be biologically sound. Indeed, there exists a close parallelism between ecologically important keystone species and socially relevant keystone species as perceived by these traditional societies.

From a nutrient cycling view point, a 10-year jhum cycle is the minimum for cropping because, among other reasons, soil fertility build-up is adequate at this stage. Large-scale deforestation and highly shortened jhum cycles constantly imposed on a site often has led to extensive desertification of sites, as illustrated through the Cherrapunji case study; this is because of the extreme fragility of the forest ecosystem in the region. Under desertification, nutrient cycling processes are drastically altered and the tendency of a natural and stable ecosystem to accumulate nutrients, against the forces that operate leading to losses from the system, is upset.

CHAPTER 8

EARLY SUCCESSIONAL WEEDS AND THEIR STRATEGIES

INTRODUCTION

Amongst the physical constraints determining primary productivity are two important factors, material and energy, the availability of which has both spatial and temporal dimensions. Conversion of solar energy into chemical energy leads to accumulation of plant biomass. During this process, the plants take up mineral nutrients, apart from water and carbon dioxide. The spatial and temporal distribution of species are determined by adaptation to micro-environmental differences in space and by phenological attributes governed by time. It is in this context that ecophysiological attributes of species and populations become significant for an understanding of community diversity and forest ecosystem function (Ramakrishnan, 1978; Saxena and Ramakrishnan, 1988).

The scope of adaptive strategy analysis (Grime, 1974; Noble and Slatyer, 1980) and ecophysiology of individual species (Bazzaz, 1979; Bazzaz and Pickett, 1980; Tilman, 1985) in understanding the vegetation structure and processes have been emphasized in recent years. However, the information on these aspects from the tropics is limited. Ecophysiological attributes such as regeneration strategies, resource uptake and utilization for growth and reproduction, and the manner in which resources of various kinds are partitioned for major life processes, are all important in evaluating the relative success or failure of the given species in a given environment. If in this analysis, we are dealing with a transient environment such as that during succession, then these strategy differences would determine the relative success of the species as a colonizer/invader, its stability and extinction patterns.

In north-eastern India, the early successional herbaceous communities developed after jhum are characterized by an environment that is subject to considerable stress because of slash-and-burn of the forest and cropping on steep slopes before it is left aside for fallow regrowth. The early successional environment here is characterized by high insulation, low moisture retention in the soil, generally lowered soil fertility but with the available nutrients present in the surface soil layers, and high heterogeneity in soil chemical properties on the steep slopes (cf Chapter 5 and 7). During succession, edaphic and micro-environmental changes occur. Mineral, nutrient and soil moisture status change rapidly and changes in the soil pH occur as litter accumulates and decomposition develops. An open community changes into a closed one with a more complete canopy and increased potential evapotranspiration. The early successional herbaceous weeds which form the major component in such seral communities are exposed to these changes during establishment and development. If the jhum cycle is relatively long, then fire events and the subsequent perturbations during the cropping phase are infrequent. On the other hand, short jhum cycles of 4–5 years, now common in the region, increases the frequency of perturbation. Therefore, the strategies of early successional weeds have to be such as to cope up with varied disturbance regimes, establish themselves in adequate numbers and make effective growth. We shall consider here some of the strategy-related aspects of weed potential under jhum and early fallow phases.

RESOURCE ALLOCATION PATTERNS AND THEIR ADAPTIVE SIGNIFICANCE

Since organisms have limited quantities of resources to expend for various life processes such as growth, maintenance and reproduction, the manner in which accumulated resources are partitioned between these activities would reflect the fitness of an organism in a given environment (Cody, 1966). For an organism, the ecological implications of such partitioning of resources would be critical when the given resource is limited and plant components being considered for allocation represent different alternatives, so that an increase in proportion allocated to one component would be linked with a decrease to the other/others (Harper, 1977). The identification of biomass as a crucial resource has always been a critical assumption in the evaluation of adaptive strategies in plants. Thompson and Stewart (1981) suggest that since reproduction also requires mineral nutrients and the reproductive structures can make no contribution in this regard, mineral allocation may be even more crucial. In this analysis of the herbaceous weeds during early succession following jhum in north-eastern India, we have considered both biomass and nutrient elements as currencies in our evaluation of the allocation strategies (Saxena and Ramakrishnan, 1988).

Early successional environment after jhum at lower elevations

Secondary succession following jhum is characterized by weedy communities during the first 5–6 years of fallow regrowth (Toky and Ramakrishnan, 1983a; Saxena and Ramakrishnan,1984b). We have seen earlier how the population structure of the weedy communities is determined by the jhum cycle at a given site and the patterns and processes involved in community dynamics during early succession (Chapters 4 and 6). We have seen how population numbers of different species such as that of *Imperata cylindrica* and *Mikania micrantha* are determined through fire events and how the demography of weed populations change during early secondary succession, eventually leading to their extinction when a closed canopy cover of large shrubs, bamboos and trees develops.

Differential regeneration patterns of weeds

A useful strategy of any species arising after slash-and-burn of a forested ecosystem for jhum as in the north-eastern hill region of India, should be to capitalize upon the resources of an enriched substratum, which is transient, by rapid multiplication followed by quick growth. A number of species such as *Imperata cylindrica, Thysanolaena maxima* and *Saccharum* spp. have soil-insulated underground vegetative organs that sprout (sprouting strategy) following the burning/cropping phase (Saxena and Ramakrishnan, 1982b, 1983a, 1988). On the other hand, reproduction may occur through viable sexually produced propagules (seeds) present in the soil seed bank (Saxena and Ramakrishnan, 1984b) or that arrive from neighbouring areas. A large number of species belong to this non-sprouting category (Saxena and Ramakrishnan, 1982a). Many, however, have a combination of both strategies. Thus, *Eupatorium* spp. apart from having heavy seed production (Kushwaha *et al.*, 1981a; Ramakrishnan and Mishra, 1981) also sprout through a root-stock. *Imperata cylindrica*, although largely reproducing through extensive underground rhizomes also produces seeds under frequent perturbations (Kushwaha, *et al.*, 1983b; Sarma, 1985). *Mikania micrantha* is another noxious weed that multiplies rapidly through vegetative means by rooting at nodes (through ramets) and at the same time has heavy seed production which is fire-regulated (Swamy and Ramakrishnan, 1987a, 1987d).

In a comparative analysis of early successional and late successional herbaceous species of Burnihat (100 m altitude) in Meghalaya the leaf area ratio (cm^2 leaf area per unit of biomass) was related to seed reproductive effort (Saxena and Ramakrishnan, 1982a). Leaf area ratio is an important parameter because it expresses the proportion of the assimilatory surface to respiratory mass (Evans, 1972). Although a certain degree of species-specific variation was noted for the different categories of species considered here, the late successional species generally had higher leaf area ratios and lower reproductive effort than early successionals. Within the early successionals themselves, the sprouting species generally had lower leaf area ratios and somewhat reduced sexual reproductive effort (with some exceptions), and this may be due to the preferential allocation of photosynthates to the underground vegetative organs of reproduction. The reduced sexual reproductive effort of the sprouting species might have evolved due to the failure of their regeneration through seedlings (Wilson, 1971). On the other hand, if the annuals and perennials of the early successional non-sprouting category of species are considered separately, certain differences are apparent. Whilst reproductive effort is generally dependent upon biomass production efficiency, it was positively correlated with absolute leaf area in the case of perennials alone, indicating that the vigour of the plant is less critical for the annuals having a single possibility of flowering during their life span.

If one were to look at the three kinds of strategies, namely, *r*- and *k*-strategies of MacArthur and Wilson (1967) and the additional competitive strategy occupying an intermediate position between *r*- and *k*-strategies (Grime, 1974), then in the present scheme of things under jhum in north-eastern India, one is able to see the early successional non-sprouting species generally as having *r*-strategy with heavy seed production and/or vegetative reproduction, and the late successionals generally as following *k*-strategy (Saxena and Ramakrishnan, 1982a; Kushwaha *et al.*, 1981b). The typical sprouting species such as *Imperata cylindrica, Thysanolaena maxima, Panicum khasianum* and *Saccharum arundinaceum* with vegetative reproduction are looked upon merely as having growth in a horizontal plane (Harper, 1977) and directing the resources for competition and stress tolerance, respectively by economizing on reproductive growth. The extreme of such an economy in seed production is reflected in *Imperata cylindrica* (Kushwaha *et al.*, 1983b; Sarma, 1985) and *Mikania micrantha* (Swamy and Ramakrishnan, 1987d, 1988a) where perturbation events such as fire determine seed production (cf. Chapter 4). Indeed, in a comparative study of the competitive relationships of plants of *Imperata cylindrica* established from rhizomes and seeds, the plants raised through rhizomes proved to be more aggressive than those raised through seeds (Kushwaha *et al.*, 1983a).

Such a differential strategy of the early successional sprouting and non-sprouting species on the one hand, and that of the late successional species on the other is also reflected in the allocation patterns of biomass and nutrients (Saxena and Ramakrishnan, 1982b). Non-sprouting early successionals allocated a higher proportion of resources to stem and seed components whereas late successional populations, as well as the sprouting early successional populations, allocated more to the underground organs (Fig 8.1). These differences are understandable; high allocation to seed is a

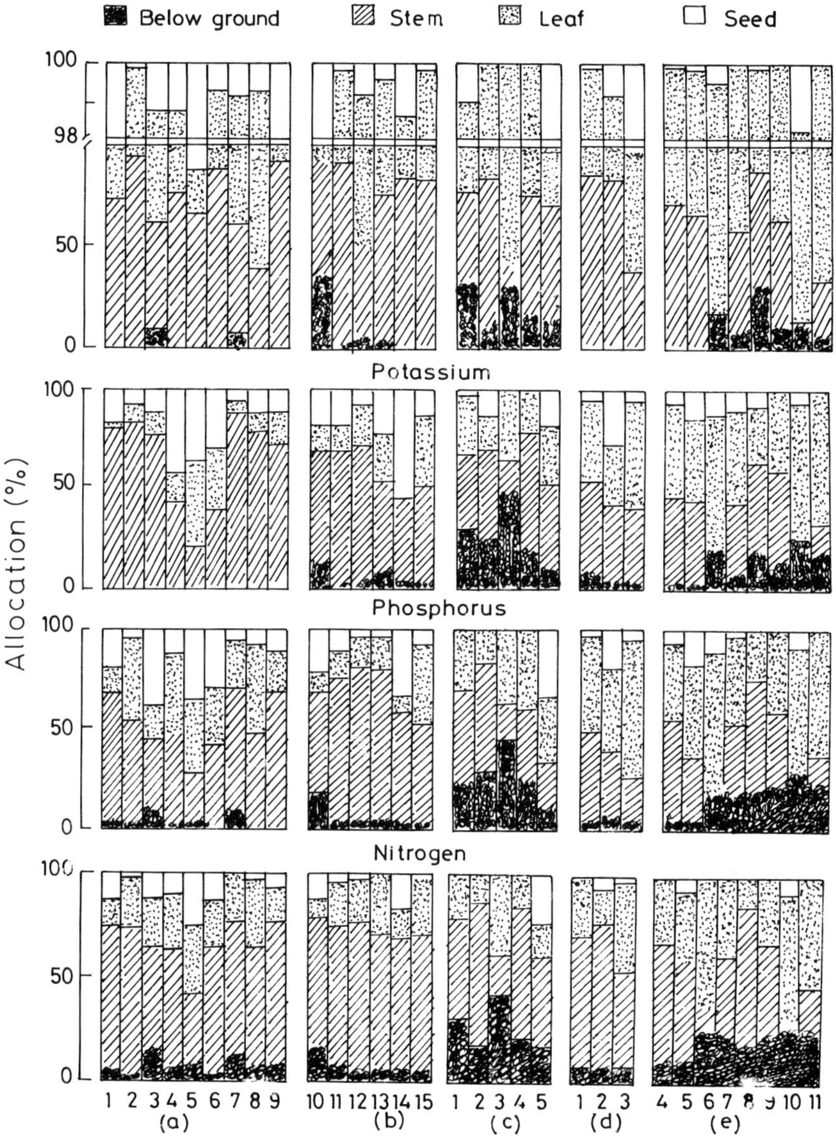

Figure 8.1 Allocation of biomass (%) to different components of early versus late successional species. Early successional non-sprouting annuals (a); and perennials (b); sprouting perennials (c); and late successional non-sprouting annuals (d); and sprouting perennials (e). Each bar represents a different species out of a total of 31 important species studied (from Saxena and Ramakrishnan, 1982b)

strategy of ruderals to exploit a temporarily nutrient-rich soil, whereas higher allocation to underground organs is a strategy for survival under stress due to perturbation/competition.

The contrasting regeneration strategies, namely, sprouting versus non-sprouting strategies of postfire early successionals have been a matter of discussion by different workers. Keeley (1981), based on his study of the Chapparel ecosystem, suggested that the non-sprouting strategy may be more successful in situations of frequent fires than the sprouting strategy and that non-sprouting species are evolved from their sprouting ancestors owing to the loss of sprouting ability under frequent fire regimes (Stebbins, 1974). These conclusions obviously do not agree with the situation under jhum in north-eastern India. With fire regimes being followed by perturbations during cropping under jhum, rapid multiplication and growth by efficiently utilizing nutrient resources from a heterogeneous soil of a steep slope is critical. Both sprouting and non-sprouting strategies could be useful in such a complex situation.

Differential allocation pattern for sprouting versus non-sprouting species

The success of an organism in a given environment is attained through a favourable allocation of the limited available resources to the diverse life purposes such as maintenance, growth and reproduction (Abrahamson and Gadgil, 1973). A useful strategy for any species under the jhum system where fire is used as a tool to release nutrients on to the soil surface in a single flush would be to make quick growth by appropriately allocating biomass and nutrients between the different components such as the root, stem, branches, leaves and reproductive organs of the organism.

Four dominant annual weeds of the early successional fallow of the jhum system, at Burnihat in Meghalaya namely, *Borreria articularis*, *Cassia tora*, *Ageratum conyzoides* and *Erigeron linifolius* were evaluated with respect to biomass allocation versus nutrient allocation strategies (Saxena and Ramakrishnan, 1983b). A striking feature of the allocation pattern was the higher allocation of nitrogen and phosphorus, when compared with biomass and potassium, for seed reproduction by these annuals (Fig. 8.2). This result parallels the findings of Van Andel and Vera (1977) and Williams and Bell (1981). Thus, the measurement of reproductive effort in terms of biomass alone may be erroneous and mineral allocation in addition to biomass allocation may be useful (Thompson and Stewart, 1981).

Allocation strategy analysis may also have another deficiency when the evaluation is done purely on the basis of biomass allocation alone. Thus a species such as *Ageratum conyzoides* showed an exceptionally low allocation of the total dry matter to fruits (1.9%). However, considering the number of seeds produced in relation to photosynthetic production, *Ageratum*

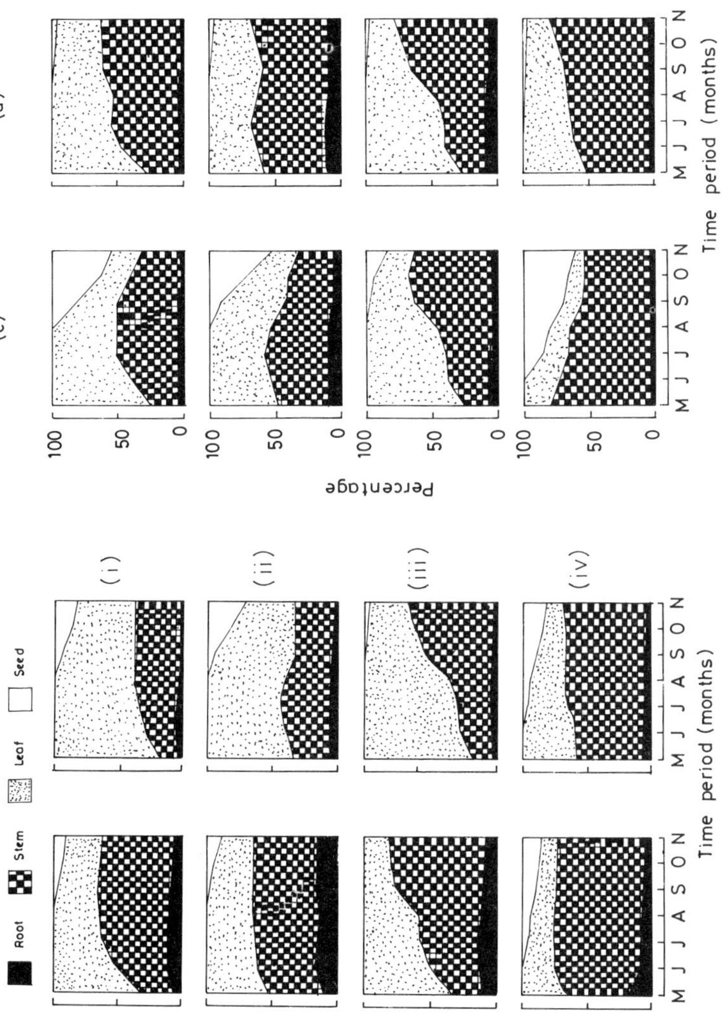

Figure 8.2 Allocation of biomass and nutrients (%) to different components during growth by (i) *Borreria articularis*; (ii) *Cassia tora*; (iii) *Ageratum conyzodies*; (iv) *Erigeron linifolus*. (a), biomass; (b), nitrogen; (c), phosphorus; (d), potassium (from Saxena and Ramakrishnan, 1983b)

Table 8.1 Growth characteristics of four early successional perennials of jhum fallows at Burnihat in Meghalaya (from Saxena and Ramakrishnan, 1983a)

Growth parameters	Eupatorium odoratum	Grewia elastica	Imperata cylindrica	Thysanolaena maxima
Aboveground biomass (g)	218.8 ± 12.8	193.1 ± 11.2	43.4 ± 3.2	43.5 ± 4.8
Belowground biomass (g)	48.0 ± 5.6	55.6 ± 8.5	31.6 ± 4.9	12.3 ± 2.1
Aboveground/ belowground ratio	4.6	3.5	1.4	3.5
Relative growth rate ($mg\,mg^{-1}\,day^{-1}$)	0.036 ± 0.015	0.008 ± 0.003	0.012 ± 0.004	0.010 ± 0.004
Net assimilation rate ($mg(cm^2)^{-1}\,day^{-1}$)	0.302 ± 0.115	0.065 ± 0.014	0.133 ± 0.043	0.108 ± 0.020
Leaf area ratio ($cm^2\,mg^{-1}$)	0.122 ± 0.031	0.121 ± 0.040	0.098 ± 0.016	0.093 ± 0.018

conyzoides showed substantially higher reproductive efficiency than *Borreria articularis* or *Cassia tora*, in spite of the much higher reproductive allocation of biomass in the latter two cases. Apparently the strategy in *Ageratum conyzoides* is to partition the limited reproductive resources into a higher number of propagules instead of increasing the total cost of production. Thus, distinct compromises may occur within, depending upon the species with a particular type of life history. Indeed, it is important to evaluate reproductive strategies both in terms of the number of seeds produced (Ramakrishnan and Mishra, 1981; Kushwaha *et al.*, 1981a; Saxena and Ramakrishnan, 1983b) in addition to the bulk allocation of biomass for reproduction.

In another study at Burnihat, growth and allocation patterns of dry matter and nutrients (nitrogen, phosphorus and potassium) in four important perennial weeds of early succession, namely, *Eupatorium odoratum, Grewia elastica, Imperata cylindrica* and *Thysanolaena maxima* were compared (Saxena and Ramakrishnan, 1983a). *Eupatorium odoratum*, exclusively dependent upon the sexual mode of reproduction, had the highest values for relative growth rate and net assimilation rate (Table 8.1). Compared to other species, this species was more vigorous in growth and had a high aboveground/belowground ratio.

Very few studies have examined the comparative growth and allocation strategies of sprouting and non-sprouting weedy species in communities. Keeley and Keeley (1977) made such an effort to assess the biomass allocation patterns in sprouting and non-sprouting species of *Arctostaphylos* in California Chapparral. However, their study was limited in scope in that it considered only the proportion of biomass allocated to sexual reproduction compared with terminal growth; i.e., allocation to secondary growth and underground organs were ignored. In addition inorganic nutrient allocation was not examined. In the study of the four perennials at Burnihat in Meghalaya, we have considered life time allocation of biomass and nutrients, rather than resource partitioning at a single point in the life cycle. For comparison of perennials (Saxena and Ramakrishnan, 1983a) with annuals (Saxena and Ramakrishnan, 1983b), a one-time estimate is inappropriate because of the retention of tissues over several growing seasons in the former category of species.

The allocation patterns of biomass and different nutrients, expressed as a percentage of the total pool are shown in Fig. 8.3. The survival of a sexually reproducing species such as *Eupatorium odoratum* is ensured through rapid growth rates and the allocation of many resources to the shoot system. On the other hand, the slow growth rates in the other three sprouting species may result from initial allocation of much biomass and nutrients to belowground parts followed by transportation of these resources for shoot growth during early stages of regrowth (Schier and Zasda, 1973). The transfer of underground reserves for shoot growth was evident from an initial decline in belowground allocation in *Thysanolaena maxima* and *Grewia elastica.* However, such a prominent decline was not observed in *Imperata cylindrica.* This may be due to its rapid growth and multiplication through underground rhizomes, simultaneously with translocation for shoot growth.

Since there are larger allocations of available resources to the underground organs of regeneration, the sprouting species are expected to have less allocation to sexual reproduction when compared with non-sprouters. Thus, *Thysanolaena maxima* had a much lower allocation of biomass and nutrients (except potassium which is of less importance for reproduction compared with nitrogen and phosphorus) to reproduction than *Eupatorium odoratum.* As already discussed earlier for the annuals (Saxena and Ramakrishnan, 1983b), the perennials considered here also had high allocation of nitrogen and phosphorus to reproduction compared with total biomass (Saxena and Ramakrishnan, 1981, 1983a).

It has often been shown that vegetative reproductive effort of the colonizing species is more plastic towards various environmental stresses, whilst sexual reproductive allocation remains nearly fixed in terms of percentage allocation regardless of the stress (Harper, 1977). In a comparative study of the allocation strategies of *Eupatorium odoratum* and *Imperata cylindrica* (Saxena and Ramakrishnan, 1981) we find that this was reflected in the greater plasticity in allocation to the rhizome in *Imperata cylindrica* in contrast to the relatively more stable sexual reproductive allocation pattern of *Eupatorium odoratum.* Such results, however, should not be generalized too far in view of some degree of plasticity shown in seed

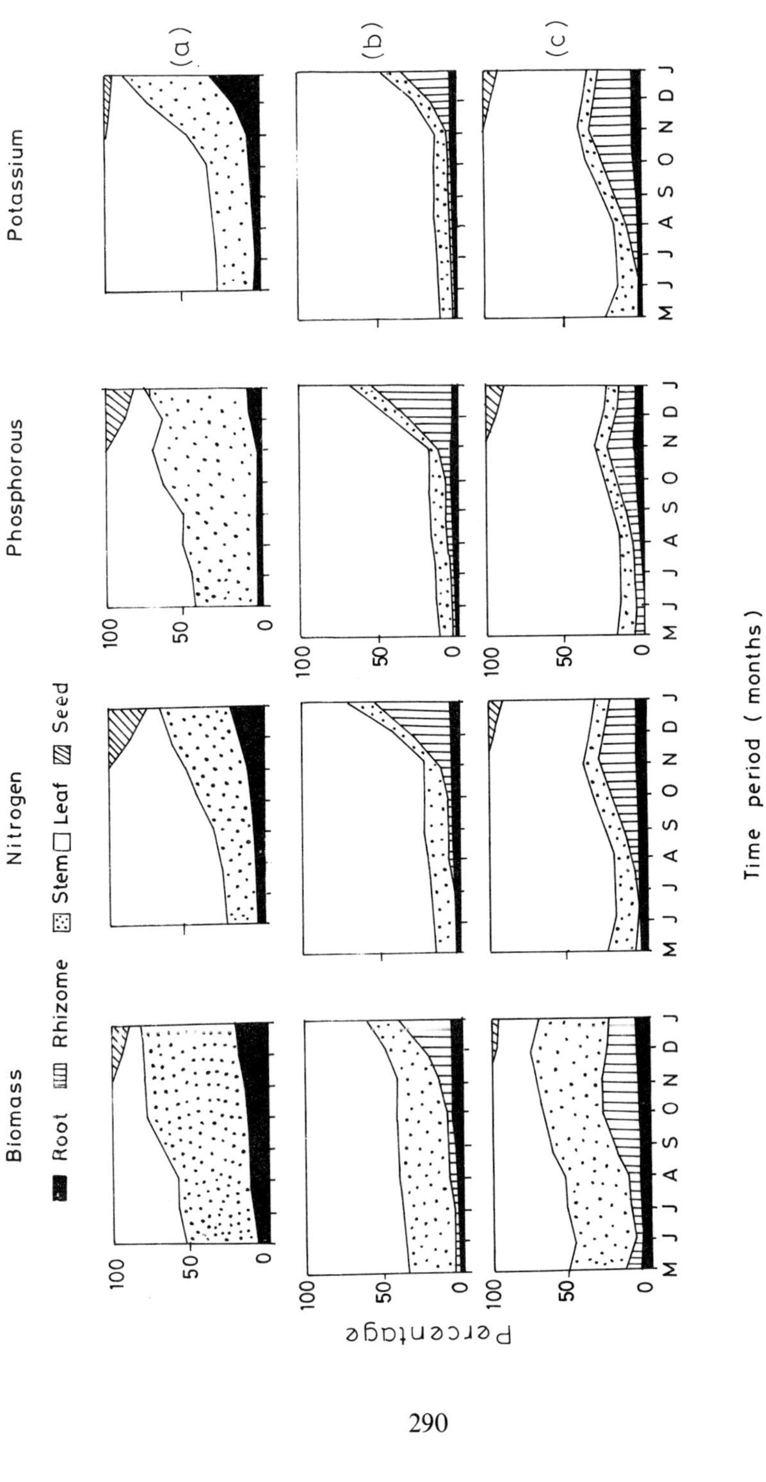

Figure 8.3 Allocation of biomass and nutrients to various compartments (expressed as % of the total pool) during growth of *Eupatorium odoratum* (a), *Imperata cylindrica* (b), and *Thysanolaena maxima* (c), after slash-and-burn of a 20-yr-old jhum fallow (from Saxena and Ramakrishnan, 1983a)

production in some of the earlier studies (Ramakrishnan, 1972; Ramakrishnan and Kumar, 1971; Ramakrishnan and Jeet, 1972; Hickman, 1975).

Species such as *Imperata cylindrica* flower only as a consequence to perturbation (cf Chapter 4). In the study done at Burnihat on the four perennials (Saxena and Ramakrishnan, 1983a), *Imperata cylindrica* and *Grewia elastica* did not flower in the first postfire year of observation. Gill (1975) reported a similar result for a number of postfire shrubs in Australia. Such a difference between the sprouting and non-sprouting strategies supports the prediction of Wilson (1971) that sexual reproduction will seldom be attempted when on averege it fails. However, in species that regenerate primarily by vegetative means, the small numbers of new recruits which may arise from seed are extremely important in maintaining the genetic diversity of the population. Although Keeley (1981) suggests that the non-spouting regenerative strategy is superior under frequent fire regimes, our results on perennials where both strategies are evidently successful, does not seem to support this contention.

Allocation strategy changes over a successional gradient

The early successional environment is dominated by herbaceous weeds up to about 5–6 years of fallow regrowth (Chapter 4 and 6). Weed extinction occurs after this period and is completed during secondary succession over a time period of 10–12 years. Over this time period, the physical and the biological environments are altered. Changes in micro-environmental conditions such as temperature, humidity and light, occur at the ground level. Soil physical and chemical properties are drastically changed (cf Chapter 5). At the population level, demographic changes occur in population structure, as already discussed earlier in Chapter 4, leading to species extinction during succession. An aspect that is important for survival versus extinction of species during secondary succession is the pattern of change in allocation strategies of species. This has been evaluated with respect to some of the important species such as *Eupatorium odoratum* and *Mikania micrantha* (Saxena and Ramakrishnan, 1984a; Swamy and Ramakrishnan, 1987a).

A comparative study of populations of *Eupatorium odoratum* from seral environments of fallows that were just abandoned after jhum (0-year-old) and up to 6-year-old fallows at Burnihat (100 m altitude) in Meghalaya (Saxena and Ramakrishnan, 1984a) revealed sharp plasticity changes. The rapid accumulation of biomass during the growth period until October was more pronounced in the recently fallowed jhum plots (Fig. 8.4). The increase in biomass with fallow age was because of the age of this perennial plant in older fallows.

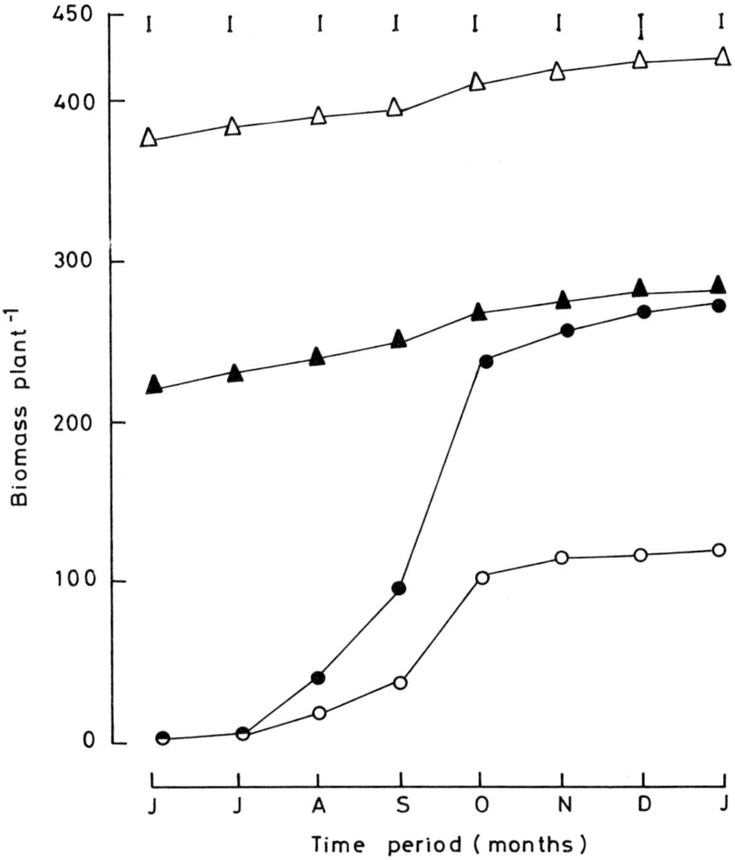

Figure 8.4 Growth curves of *Eupatorium odoratum* in different jhum fallows at Burnihat in Meghalaya. ●, recent fallow but uncropped; ○, recent fallow but cropped; ▲, 2-yr-old fallow; △, 6-yr-old fallow. Vertical bars represent LSD ($p = 0.05$) (from Saxena and Ramakrishnan, 1984a)

Allocation of a high proportion of the available resources to the supporting organs but a lower proportion to the photosynthetic and reproductive structures in the older fallows than in the recent ones (Fig. 8.5) may be because the plants tended to grow taller and avoid shading in the more mature habitats where light could be limiting. Similar shifts in the allocation of the biomass have been shown during seral development for individual species (Roos and Quinn, 1977) and also for the whole community (Newell and Tramer, 1978; Abrahamson, 1979; Saxena and Ramakrishnan, 1982b). It has been shown specifically for colonizing species that vegetative reproductive allocation is often 'plastic' depending upon various environmental stresses whereas sexual reproductive allocation

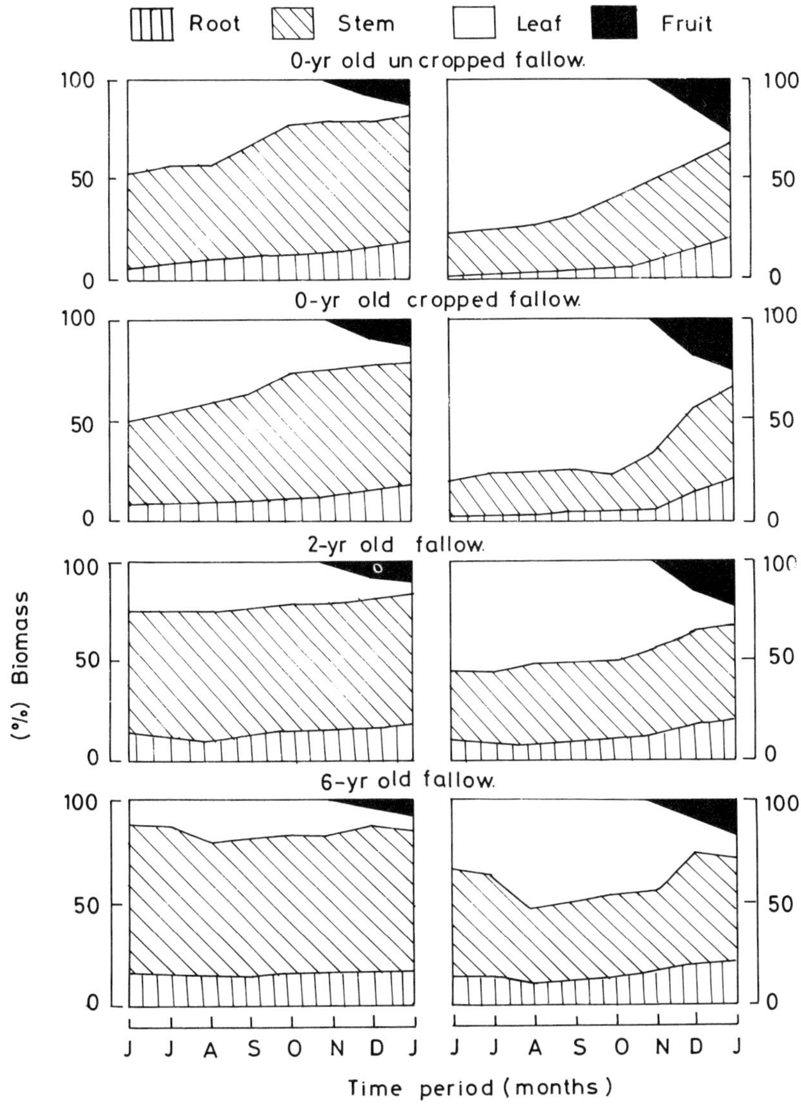

Figure 8.5 Allocation of biomass and nutrients to different component organs (expressed as percentage of total biomass or nutrients) in *Eupatorium odoratum* in different jhum fallows (from Saxena and Ramakrishnan, 1984a)

remains nearly fixed regardless of the imposed stress (Ogden, 1974; Abrahamson, 1980). While such a strategy may operate in the perennials possessing both sexual and vegetative multiplication alternatives, it may not be equally applicable for the species exclusively dependent upon sexual reproduction. In *Eupatorium odoratum*, plasticity in sexual reproductive allocation also occurred during secondary succession. The proportion of all resources received by the roots increased with increasing age of the fallow, although it was most pronounced for potassium. Such an increased allocation to the underground organs in older jhum fallows would be a useful strategy for survival and growth through another slash-and-burn cycle, as roots spout.

In a study of the biomass allocation strategy changes during early succession at Lailad, a species that has a highly developed clonal propagation strategy also showed similar results (Swamy and Ramakrishnan, 1987a; cf Fig. 5.4). Greater allocation of biomass to the rosette root in the older fallows is helpful for survival through another jhum cycle as seen for *Eupatorium odoratum* discussed above. Such a strategy also helps to exploit deeper soil layers for nutrients particularly since the surface soil layers would already be depleted of nutrients by the developing vegetation.

Another measure that is useful to evaluate allocation strategy is the 'reproductive effort' expressed as the percentage of net primary production or nutrient absorbed during the current growing season and allocated to reproductive units. This was done for *Eupatorium odoratum* (Saxena and Ramakrishnan, 1984a). Quantification of the reproductive cost in the perennials is complicated because here, the production of the previous year is added to that of the next year (Ogden, 1974; Hickman, 1975). It may be worth noting that in field studies, reproductive allocation has always been measured as a proportion of the existing biomass of 'new growth' (where the current season's growth is largely distinguished from the growth of the previous years, e.g., overwintering of shoots every year). Moreover, the importance of the underground organs in the resource dynamics is often ignored (Mooney and Billings, 1960; Chapin, 1980; Saxena and Ramakrishnan, 1983a). Such a quantification of reproductive allocation may be erroneous because some of the perennial biomass used in the 'total resource capital' is not produced during the current growing season. Cost of reproduction to the plant in perennials may be calculated more accurately as a proportion of the current increment in the available resources devoted to reproductive units as suggested by Harper and Ogden (1970). Such a measure of reproductive allocation (cf Harper and Ogden, 1970; net reproductive effort expressed as the ratio of reproductive biomass to net primary production) was substantially higher in older fallow plots than in the recent ones (Table 8.2) suggesting that reproductive development in *Eupatorium odoratum* is emphasized at the cost of vegetative development under the stressfull conditions of the former. Further, reproductive allocation of nitrogen and phosphorus (nitrogen in the 2-year-old fallow and both nitrogen and phosphorus in the 6-year-old fallow) were high (Saxena and

Table 8.2 Reproductive effort (calculated as allocation to fruits as the percentage of net primary production or nutrient uptake during the current growing season) of *Eupatorium odoratum* in different jhum fallows at Burnihat in Meghalaya (from Saxena and Ramakrishnan, 1984a)

	Reproductive efforts			
Jhum fallow	*Biomass*	*Nitrogen*	*Phosphorus*	*Potassium*
0-yr-old, uncropped	11.87	26.63	17.28	2.26
0-yr-old, cropped	11.75	23.22	23.40	2.08
2-yr-old	34.65	427.38	69.33	25.97
6-yr-old	46.82	477.67	100.22	13.91
LSD ($p = 0.01$)	4.21	11.56	15.20	6.45

LSD = least significant difference

Table 8.3 Nutrient uptake (expressed as mg nutrient absorbed per g root biomass) by *Eupatorium odoratum* in different jhum fallows at Burnihat in Meghalaya (from Saxena and Ramakrishnan, 1984a)

	Nutrient uptake		
Fallow field	*Nitrogen*	*Phosphorus*	*Potassium*
F_1	71.02	9.21	72.40
F_2	57.60	6.30	60.80
2-year-old	1.82	1.30	5.74
6-year-old	1.40	0.72	4.70
LSD ($p = 0.01$)	9.48	2.26	9.20

LSD = least significant difference

Ramakrishnan, 1983a,b) and exceeded the uptake during the current growing season (Saxena and Ramakrishnan, 1984a), indicating an extreme ruderal strategy in this species. A number of experimental studies have shown the tendency of ruderals to concentrate upon reproductive growth even when subjected to severe stresses (Hickman, 1975; Raynal and Bazzaz, 1973; Van Andel and Vera, 1977).

Lower nutrient uptake observed in the older jhum fallows (Table 8.3) may be due to the poor availability of nutrients in the soil here (cf Chapter 5; Ramakrishnan and Toky, 1981). Such nutrient stress may partly be responsible for low increments in individual biomass in older jhum fallows compared to the recent ones (Fig. 8.4), apart from intense competition from other species. Thus, stress of a different nature and magnitude, as secondary succession progresses, results in gradual replacement of *Eupatorium odoratum* after 5–6 years of seral development (Toky and Ramakrishnan, 1983a). In the cropped jhum fallows, the nutrient status would be lower than in the uncropped plots because of greater perturbations. Therefore, it is reasonable to find low nutrient uptake and low dry matter produced under the cropped fallow plots than under the uncropped ones.

Table 8.4 Mean (± SE) values of relative growth rate (RGR), net assimilation rate (NAR) and leaf are ratio (LAR) growth functions of *Mikania micrantha* in burnt and unburnt early successional fallows at Lailad in Meghalaya (from Swamy and Ramakrishnan, 1988a)

Growth function	*Fallow age* (yr)	*Burnt*	*Unburnt*
RGR ($mg\,mg^{-1}\,day^{-1}$)	2	0.0235 ± 0.0005	0.0148 ± 0.0003
	4	0.0247 ± 0.0004	0.0153 ± 0.0002
	8	0.0257 ± 0.0004	0.0070 ± 0.0004
NAR ($(mg\,cm^{-2})^{-1}\,day^{-1}$)	2	0.171 ± 0.003	0.154 ± 0.004
	4	0.184 ± 0.003	0.163 ± 0.003
	8	0.197 ± 0.006	0.088 ± 0.001
LAR ($cm^2\,mg^{-1}$)	2	0.195 ± 0.006	0.139 ± 0.005
	4	0.254 ± 0.008	0.214 ± 0.005
	8	0.293 ± 0.009	0.010 ± 0.005

Fire effects on seed allocation strategy in Mikania micrantha

The example of *Mikania micrantha* illustrates elegantly the effect of fire on plant strategies under jhum at Lailad (290 m altitude) in Megalaya. This study examined the fire dependence and decline in vigour in older fallow plots of this exotic weed (Swamy and Ramakrishnan, 1988a). Its fire dependence is shown by higher mean values of relative growth rate (RGR), net assimilation rate (NAR) and leaf area ratio (LAR) (Hughes and Freeman, 1967) in the burnt successional fallows of all ages than in unburnt fallows (Table 8.4). This difference was greater in 8-year-old fallow, suggesting that the species rapidly declines in vigour without disturbance. This also corresponds with our observations that seedlings establish only after fire (Swamy and Ramakrishnan, 1987a,d; cf Chapter 4).

The ruderal nature and fire dependence of *Mikania micrantha* is reflected in its allocation strategy, it being able to flower better and at a lower nutrient cost in burnt than unburnt 8-year-old fallow (Table 8.5). A comparative analysis of burnt and unburnt plots of an 8-year-old jhum fallow showed that biomass allocation in *Mikania micrantha* to the inflorenscence was 26% of the total resource in the burnt plot compared to only 12% of the total resource in the unburnt plot (Swamy and Ramakrishnan, 1987a). This would lead to eventual weed suppression, an aspect discussed in much detail in Chapter 4 and 6 (Toky and Ramakrishnan, 1983a; Saxena and Ramakrishnan, 1984b).

Nutrient uptake efficiency of *Mikania micrantha* was higher in the burnt plots perhaps because of increased availablity (Table 8.6). With a decreased nutrient pool in an unburnt 8-year-old fallow, which is related to rapid transfer of nutrients from the soil to the developing vegetation (Ramakrishnan and Toky, 1981; Toky and Ramakrishnan 1983b), it is reasonable to expect reduced uptake efficiency. A higher use efficiency in sites of lower

Table 8.5 Reproductive effort (calculated as biomass or nutrient allocation to the seed as the percentage of net primary production or nutrient uptake during the growing season) of *Mikania micrantha* in burnt and unburnt successional fallows at Lailad in Meghalaya (from Swamy and Ramakrishnan, 1988a)

Reproductive effort	*Fallow age* (yr)	*Burnt*	*Unburnt*
Biomass	2	24	23
	4	25	22
	8	27	22
LSD at 5% level		5.2	6.4
Nitrogen	2	57	53
	4	60	61
	8	64	88
LSD at 5% level		8.4	9.2
Phosphorus	2	51	46
	4	53	49
	8	56	80
LSD at 5% level		7.4	7.3
Potassium	2	31	33
	4	31	39
	8	33	83
LSD at 5% level		9.2	8.4

LSD = least significant difference

uptake efficiency may be considered as a compensatory mechanism for survival.

Competition and coexistence

An early successional environment where weed colonization occurs during the first 5–6 years of fallow regrowth after jhum is characterized by rapid colonization by species that could result in intense intra- and interspecific competition. Intraspecific competition would result in self-thinning and/or plastic responses in order to adjust the population size (Ramakrishnan, 1972). The presence of different species in the same area raises the question as to how continued cohabitation of the two species is possible without one succeeding at the expense of the other. Such a coexistence implies that the species concerned possess divergent ecological features that enable them to avoid each other (Ramakrishnan and Jeet, 1972; Ramakrishnan and Gupta, 1972a). In the north-eastern situation, the weedy colonizers, perhaps, manage to coexist through a variety of strategies. We shall consider some of these possibilities using a few examples.

Table 8.6 Mean uptake efficiency (mg nutrient absorbed per g root biomass) and use efficiency (mg dry matter produced per mg nutrient absorbed) of nitrogen, phosphorus and potassium by *Mikania micrantha* in burnt and unburnt successional fallows (from Swamy and Ramakrishnan, 1988a)

		Burnt		*Unburnt*	
Nutrient-uptake efficiency	*Fallow age* (yr)	*Mean uptake efficiency*	*Use efficiency*	*Mean uptake efficiency*	*Use efficiency*
Nitrogen	2	108	78	85	98
	4	121	76	98	100
	8	128	69	45	175
LSD at 5% level		11	8	7.3	8
Phosphorus	2	12	699	9.5	896
	4	14	672	9.3	1,010
	8	14	583	4.4	2,030
LSD at 5% level		1.3	25	1.2	157
Potassium	2	134	60	115	74
	4	155	56	121	77
	8	166	51	81	98
LSD at 5% level		15	6	11	5

In a study on the growth requirements of *Eupatorium odoratum* and *Imperata cylindrica* at Burnihat (100 m altitude) in Meghalaya, it was found that *Imperata cylindrica* would allocate more biomass to the belowground rhizomes under stress (Kushwaha and Ramakrishnan, 1982; Sarma, 1985). Further, we have already seen that *Imperata cylindrica* resorts to sexual reproduction only when subjected to constant perburbation such as fire (Kushwaha *et al.*, 1983b). Our competition studies, however, suggested that introduced seedlings of *Imperata cylindrica* have very little chance of survival when compared to tillers (Fig. 8.6). Only about 20% of the seedlings originating from introduced seeds survived in the freshly abandoned jhum fallows (0-year-old) at Burnihat in Meghalaya. This is in spite of the fact that over 70% germination of the sown seeds occurred in the 0-year-old fallow. These observations tend to support the prediction of Willson (1971) that sexual reproduction will seldom be attempted when on an average it fails.

It is in this context that the relative advantage of competition between individuals of *Imperata cylindrica* established through rhizomes and seeds was evaluated (Kushwaha *et al.*, 1983a). In pure stands, the reaction of both types of plants to density stress was found to be strikingly similar. This implies that although the plants from rhizomes had a better start, their response to high density stress in pure populations was no better than those arising from seeds. However, in mixtures, the plants arising from

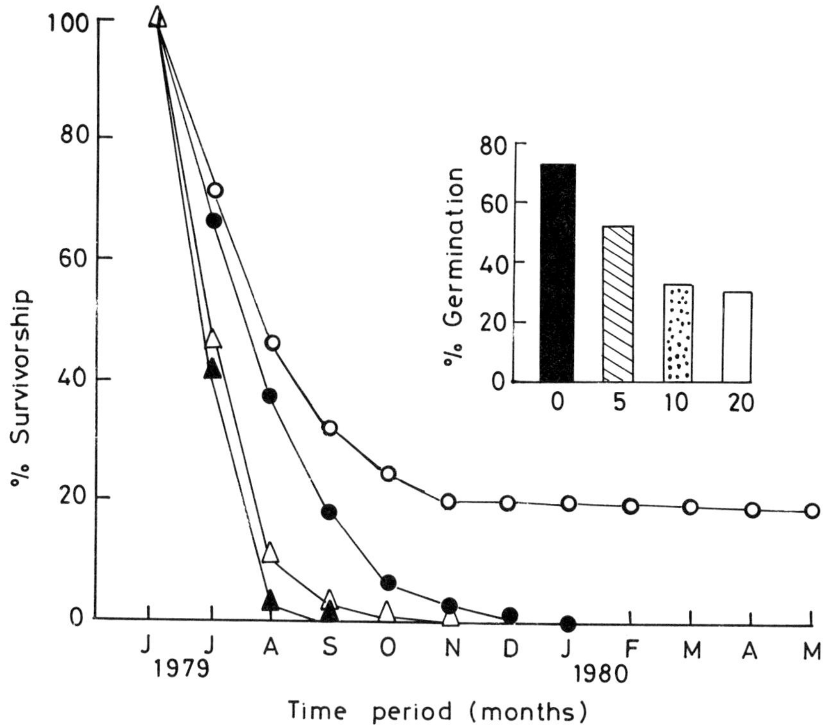

Figure 8.6 Survivorship of introduced seedlings in different jhum fallows at Burnihat in Meghalaya. ○, 0-yr-old; ●, 5-yr-old; △, 10-yr-old and ▲, 20-yr-old fallows. Inset Figure is for seed germination (from Kushwaha *et al.*, 1983b)

Table 8.7 Relative yield total (RYT), relative yield quotient (RYQ) and relative replacement rate (RRR) at three harvests of *Imperata cylindrica* raised from rhizome and seeds (20 individuals in a replacement series in pot culture) from Kushwaha *et al.*, 1983a)

	Relative yield				*RRR*
Harvests	*Rhizome* (r_1)	*Seed* (r_2)	*RYT* ($r_1 + r_2$)	*RYO* (r_1/r_2)	*'Rhizome' with respect to 'Seed'*
H_1	0.69	0.31	1.00	2.22	
H_2	0.53	0.47	1.00	1.12	0.51
H_3	0.51	0.49	1.00	1.04	0.92

rhizomes had a distinct advantage over those arising from seeds as seen from the higher relative yield for the former (Table 8.7). However, this initial advantage gradually tends to level off over a time period as seen from the decline in relative yield between the first and the third harvest.

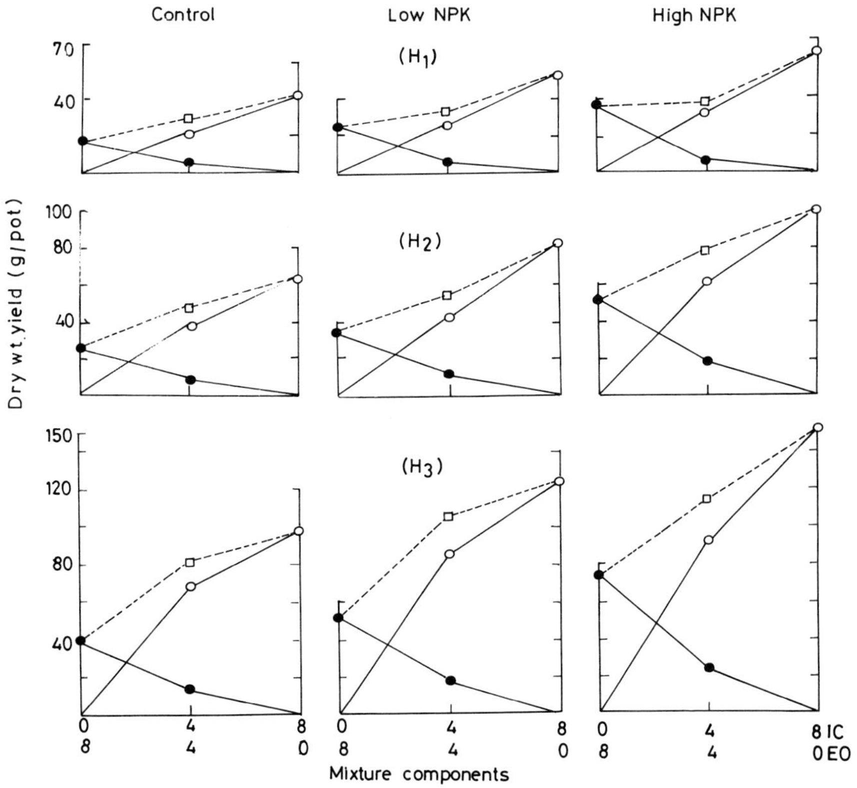

Figure 8.7 Replacement diagrams for dry weight yield (g) at three harvests (H_1, H_2 and H_3 taken after 8, 16 and 24 weeks) and at three nutrient levels (control : unaltered soil; low NPK: 1.2g $NaNO_3$, 2g NaH_2PO_4 and 0.88g K_2SO_4 in one dose; and high NPK: same on two occassions) involving *Imperata cylndrica*, IC (○) and *Eupatorium odoratum*, EO (●), with total yield (□) in mixtures (from Kushwaha *et al.*, 1983c)

The unit relative yield total (RYT) values suggest that the two categories of population compete for the same space (Van den Bergh, 1968).

In the early successional environment where colonization and establishment of species and populations is continuously happening, interspecific competition is likely to be severe, particularly if two of the species have to share common space or resource. The competitive relationships between *Imperata cylindrica* and *Eupatorium odoratum*, two important weeds of the early successional environment of the jhum fallows, may explain their adjustment to each other under conditions of fluctuating nutrient conditions (Kushwaha *et al.*, 1983c). In mixtures, in pots, where the set-up was based on de Wit's replacement series (de Wit, 1960), with three nutrient levels and three harvests (Fig 8.7), *Imperata cylndrica* was more aggressive than the other, as seen from the relative convexity of the replacement curves. These

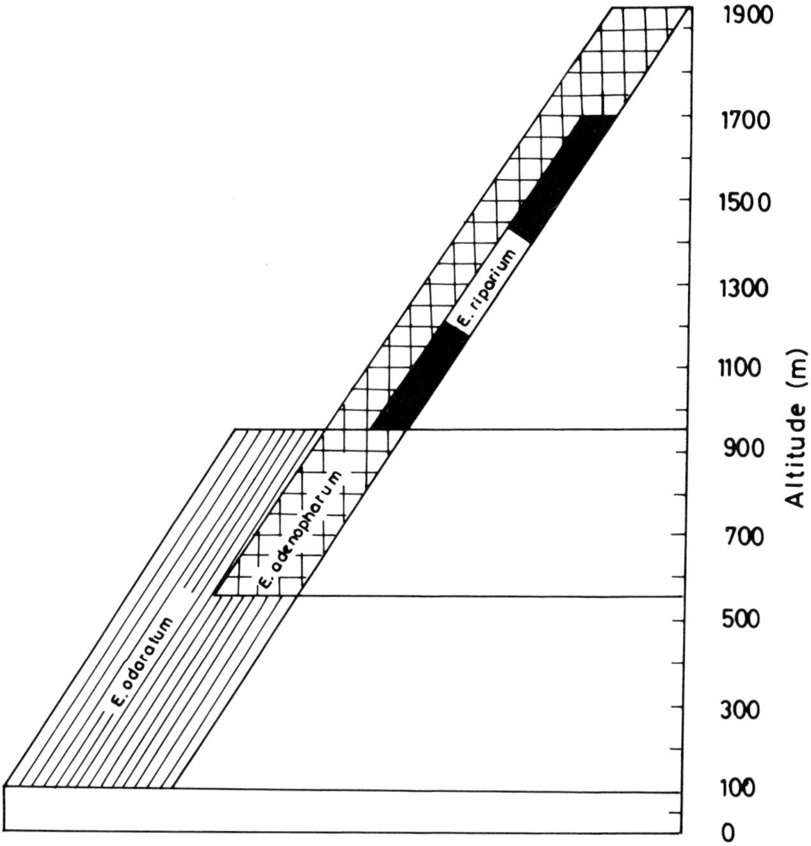

Figure 8.8 Altitudinal distribution of *Eupatorium* species in the Khasi Hills of Meghalaya (from Dev and Ramakrishnan, 1987a)

results with a RYT value of less than one, suggest that *Eupatorium odoratum* is more affected than might be expected from crowding (Kushwaha *et al.*, 1983c). Allelopathic inhibition of *Eupatorium odoratum* by *Imperata cylindrica* (Eussen and Wirjahardja, 1973) has been implicated.

Of the weeds found in the jhum fallows of north-eastern India, *Eupatorium* is an important genus with three species, namely, *E. odoratum* (up to an altitude of 950 m), *E. adenophorum* (between 55 and 1,960 m) and *E. riparium* (between 950 and 1,700), with overlapping zones of distribution (Fig 8.8). In a series of studies involving coexistence of these three species in the overlapping altitudinal zones (Dev and Ramakrishnan, 1987a,b), we were able to show that competitive interactions between species play an important role in the restriction of these species to their altitudinal limits. Two major conclusions arose from these studies : (1) where two species are competing in their marginal habitats, such as *Eupatorium odoratum* and *E. adenophorum* at Barapani in Meghalaya (950 m altitude), the one from the higher altitudinal range (*E. adenophorum* in this case) always had a competitive advantage over the other and (2) where intraspecific populations are involved, such as in the case of *E. adenophorum* from different altitudinal situations, the native population had an advantage over the introduced population. Apparently, a variety of competitive mechanisms play a role in the spatial and temporal distribution of species and populations as shown here and in earlier studies (Ramakrishnan, 1970).

C_3/C_4 strategy and coexistence

The manner in which the C_3 strategists represented by the dominant exotic weeds such as *Eupatorium* spp. and *Mikania micrantha* and C_4 strategists, represented by natives such as *Imperata cylindrica, Thysanolaena maxima* and *Saccharum* spp. are able to colonize a heterogeneous soil environment of early succession has been discussed in detail in Chapter 7 (Saxena and Ramakrishnan, 1983a, 1984c). The more nutrient-use efficient C_4 species are able to colonize nutrient-poor microsites whereas the less efficent C_3 species largely occur in nutrient-rich microsites. This is discussed in relation to the differential nutrient-use efficiency of the C_3 and C_4 species (cf Fig. 7.3). This is also reflected in the relationship between dry matter yield and nutrient uptake during growth of the two categories of species (Fig. 8.9). The slopes of the regression lines indicate that for a given dry matter yield *Eupatorium odoratum* and *Grewia elastica* (both C_3 species) used a greater quantity of nutrients than *Thysanolaena maxima* and *Imperata cylindrica* (both C_4 species). This is further reflected in the shift towards C_4 weeds during the second cropping in a year under the terrace agroecosystem at Nayabunglow in Meghalaya. Under the reduced soil fertility status after one crop harvest, C_4 weed species are more successful than C_3 plants (cf Chapter 5; Patnaik, 1988), because of the higher nutrient-use efficiency of the former.

However, two aspects related to biomass need to be emphasized in this discussion, which has not been considered elsewhere. Firstly, C_3 species such as *Eupatorium odoratum* have much higher relative growth and net assimilation rates than C_4 species such as *Thysanolaena maxima* and *Imperata cylindrica* (Table 8.1). The C_4 photosynthetic pathway need not always be superior to the C_3 strategy (Wilson, 1967). The higher relative growth rate and net assimilation rates in this C_3 species may be due to its greater leaf area ratio, which expresses the proportion of photosynthetic surface to respiratory mass (Saxena and Ramakrishnan, 1983a). For comparing productive potentials of C_3 and C_4 plants, the importance of

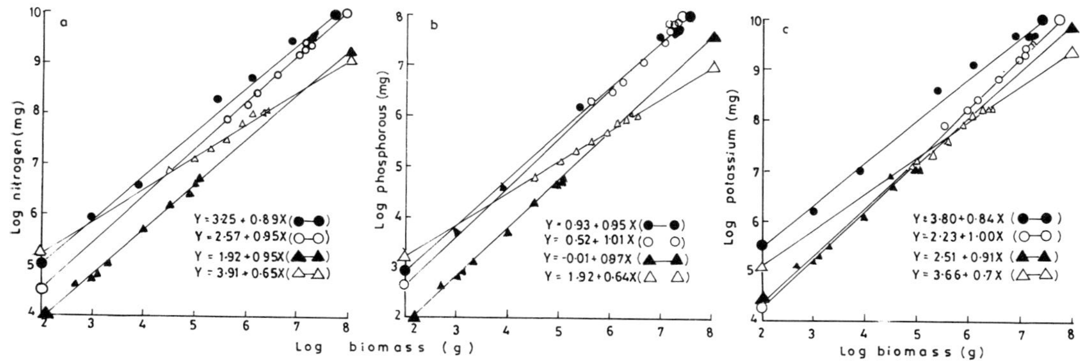

Figure 8.9 Relationship between dry matter yield (g) and nutrient uptake (m^{-2}) during the growth of *Eupatorium odoratum* (●), *Grewia elastica* (○), *Imperata cylindrica* (▲) and *Thysanolaena maxima* (△) after slash-and-burn of a 20-yr-old jhum fallow (from Saxena and Ramakrishnan, 1983a)

the extent of light interception surface has been emphasized by others too (Bull, 1971; Hofstra and Steinstra, 1977), rather than the photosynthetic rate per unit leaf area as stressed earlier (Black, 1971).

Secondly, an important feature of the study done on C_3/C_4 categories of plants in the jhum fallows at Burnihat in Meghalaya is the marked difference in the phenological patterns of the production by the two photosynthetic groups (Saxena and Ramakrishnan, 1984c). The biomass of C_4 species increased sharply following the burn and depended upon its intensity, in contrast to the C_3 plants (Fig. 8.9). Thus in the older fallows (10-and 20-year-old) that were burnt, there was a 10-fold increase compared to a 5-fold increase only in the younger fallows. Further, the pattern of biomass accumulation during the year was different in the two photosynthetic groups. While the C_3 group showed a substantial increase in September (180 days) to December (270 days), the growth of C_4 species was quite slow during this period. In other words, dry matter production by the C_4 species occurred more efficiently during the early part of the growing season (April–June) when solar radiation would be more intense accompanied by warmer temperatures. On the other hand, biomass accumulation by C_3 species was pronounced during the later part of the growing season (October–December) when solar radiation would be less intense coupled with cooler temperatures; a situation where the C_3 photosynthesis may be superior to the C_4 photosynthesis (Ehleringer and Bjorkman, 1977).

A few conclusions arise from this study with respect to species strategies for coexistence. The nutrient-use efficiency differential permits apparent coexistence of C_3/C_4 strategies in space, in a heterogeneous soil environment. Secondly, phenological differences in the productivity patterns of C_3/C_4 species permit temporal coexistence. Finally, the generally higher biomass of the C_3 species in the community (Fig. 8.10) compared to C_4 species in the jhum fallows suggests that the productivity here is not a function of C_4 photosynthesis, an observation made also by Caldwell *et al.*, (1977) while analysing desert shrubs communities. Thus, as discussed earlier, the productive potential of plants cannot be predicted solely on the basis of photosynthetic strategy (Saxena and Ramakrishnan, 1984c).

Allocation strategy in arrested seral grasslands at Cherrapunji

At Cherrapunji in Meghalaya, the degraded grasslands that respresent an arrested stage in succession (Ram and Ramakrishnan, 1988) and which are derived from the mixed broad-leaved climax forest (Khiewtam, 1986) are under extreme stress. This stress, as discussed earlier, is largely related to a highly nutrient-deficient soil developed under extreme climatic conditions created by exceptionally high rainfall (cf Chapter 7). The predominantly

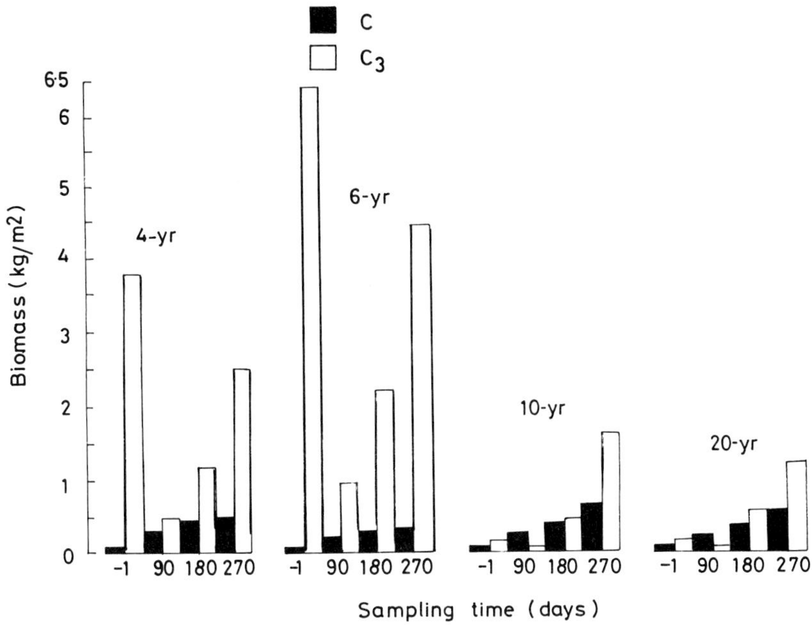

Figure 8.10 Changes in biomass of C_3 and C_4 species on plots under 4-, 6-, 10- and 20-year jhum cycle, a day before the burn after slashing (−1) and at different times after the burn (from Saxena and Ramakrishnan, 1984c)

graminaceous vegetation developed under these stress situations may be expected to have an allocation strategy geared for quick growth and reproduction (Ram and Ramakrishnan, 1988a).

In a comparative analysis of five species of the grassland types of Cherrapunji Meghalaya (1,300 m altitude) (cf Chapter 6), *Arundinella bengalensis*, *Carex cruciate* and *Chrysopogon gryllus* are found as part of the *Osbeckia* type of grassland on a nutrient-rich soil as well as on a nutrient-poor desertified soil (*Eragrostiella* type) of the grassland. *Ischaemum goeblii* is characteristic of frequently burnt *Ischaemum* grassland type, and *Eragrostiella leoptera* is found only in the nutrient-poor desertified soils of the *Eragrostella* grassland type. A comparative study of the allocation strategies of these five species was done on both nutrient-rich and nutrient-poor soils for the first three species and in their own respective nutrient-poor soil types for the latter two species (cf Table 6.11 for soil characteristics).

A useful strategy for survival of plants species growing in nutrient-poor soils is to have a greater allocation of biomass to the belowground organs as compared to the aboveground parts (Fig 8.11). This is a mechanism to maximize nutrient uptake through a large root surface (Chapin, 1980) rather than through a high nutrient absorption capacity (Nye and Tinker, 1977). Thus, *Eragrostiella leoptera* growing in a desertified site had maximal

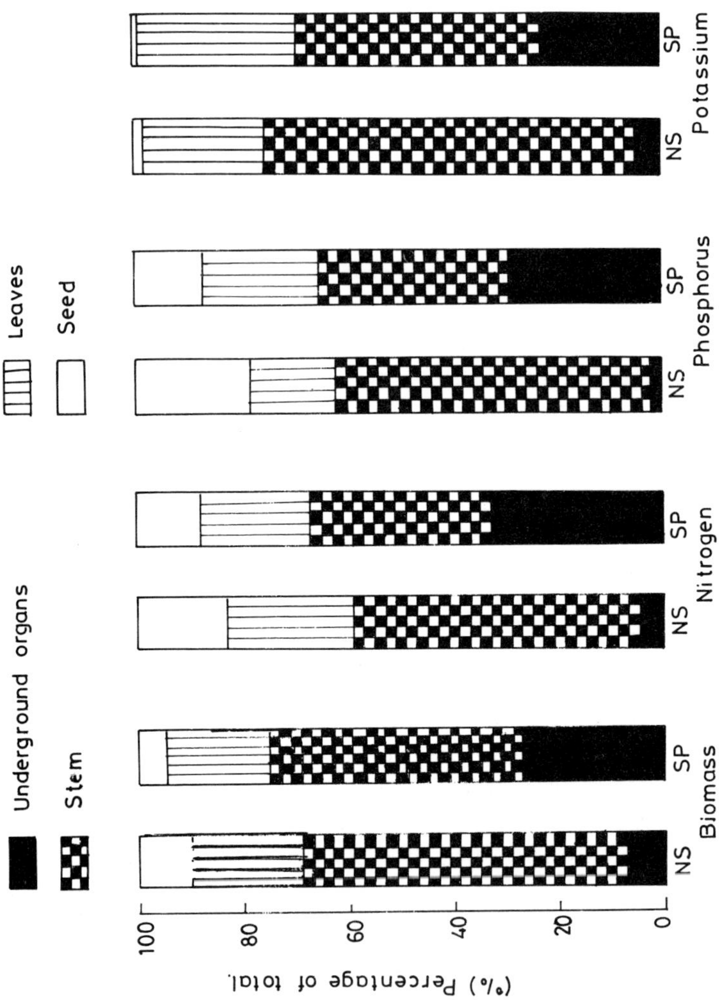

Figure 8.11 Allocation of biomass and nutrients (expressed as a proportion of total biomass or nutrients allocated to different plant components) in the early successional herbaceous species. The observations are the mean of 15 species of the non-sprouting (NS) and five species of the sprouting (SP) category during the peak growth period (from Saxena and Ramakrishnan, 1988)

Table 8.8 Mean values (± SE) of growth functions for the different species at Cherrapunji in Meghalaya (from Ram and Ramakrishnan, 1988a)

	Growth functions		
Species	*Relative growth rate* ($mg\,mg^{-1}\,day^{-1}$)	*Net assimilation rate* ($mg\,(cm^2)^{-1}\,day^{-1}$)	*Leaf area ratio* ($cm^2\,mg^{-1}$)
Arundinella	0.0135 ± 0.009	0.123 ± 0.0114	0.117 ± 0.0084
bengalensis	(0.0095 ± 0.0006)	(0.105 ± 0.0035)	(0.085 ± 0.0054)
Carex cruciata	0.0245 ± 0.0017	0.221 ± 0.0151	0.159 ± 0.0114
	(0.0147 ± 0.0011)	(0.135 ± 0.0115)	(0.251 ± 0.0181)
Chrysopogon	0.0106 ± 0.0008	0.121 ± 0.0115	0.082 ± 0.0075
gryllus	(0.0026 ± 0.0002)	(0.031 ± 0.0024)	(0.070 ± 0.0062)
Ischaemum goeblii	0.0116 ± 0.0012	0.125 ± 0.0101	0.013 ± 0.0011
Eragrostiella			
leoptera	0.0078 ± 0.0071	0.095 ± 0.0071	0.005 ± 0.0003

In parentheses are values for species in nutrient-poor soil

root biomass compared to all other species. Greater allocation of biomass to belowground organs is also a mechanism to store organic food and nutrients in the belowground rhizomatous tissue and thus contributes to more efficient vegetative reproduction using the food reserves. Greater allocation of resources to belowground organs therefore an adaptation for survival under stress environment, the *s*-strategy of Grime (1974). With larger allocation of the available resources to the belowground organs of regeneration, species such as *Carex cruciata* and *Arundinella bengalensis* have less allocation to sexual reproduction (Keeley and Keeley, 1977). In fact, for the three tussock perennial grasses, *Chrysopogon gryllus*, *Ischaemum goeblii* and *Eragrostiella leoptera* of stressed sites, with sprouting occurring from the base, the emphasis on sexual reproduction is minimal.

An important characteristic of value for survival in a nutrient-poor soil is a slow growth rate (Chapin, 1980). Thus, species like *Eragrostella leoptera* have very slow growth rates compared to the other species (Table 8.8). Further, those species occurring on nutrient-rich and nutrient-poor soils have drastically reduced relative growth rates and net assimilation rates in the latter soil type. This results in comparable tissue concentration of nutrients in both soil types; indeed in some cases the concentration is even higher under nutrient stress (Table 8.9). Thus, individuals of the species surviving under stress situations may possibly respond through luxury consumption; contrary to this more rapid growth of the species in fertile soils may have a dilution effect on the plant nutrient pool. Consequently, tissue concentrations of wild plants are less sensitive indicators of nutrient availability in the soil (Van den Driessche, 1974).

The importance of the C_4 photosynthetic pathway is evident in the vegetation organization of these degraded and successionally arrested grassland types. A majority of the species being C_4 grasses (Ramakrishnan

Table 8.9 Mean tissue concentration of nutrients (± SE) during the growth period of different species at Cherrapunji in Meghalaya (from Ram and Ramakrishnan, 1988a)

	Nutrient concentration (%)		
Species	*Nitrogen*	*Phosphorus*	*Potassium*
Arundinella	0.667 ± 0.045	0.098 ± 0.0075	0.733 ± 0.061
bengalensis	(0.983 ± 0.075)	(0.099 ± 0.0071)	(0.630 ± 0.050)
Carex cruciata	0.925 ± 0.065	0.102 ± 0.009	1.091 ± 0.091
	(0.801 ± 0.071)	(0.135 ± 0.0101)	(1.121 ± 0.101)
Chrysopogon	0.917 ± 0.081	0.168 ± 0.085	0.630 ± 0.051
gryllus	(1.283 ± 0.101)	(0.152 ± 0.089)	(0.82 ± 0.071)
Ischaemum goeblii	0.85 ± 0.071	0.114 ± 0.0095	0.59 ± 0.041
Eragrostiella			
leoptera	0.667 ± 0.051	0.106 ± 0.0080	0.53 ± 0.003

In parentheses are values for species in nutrient-poor soil

Table 8.10 Nutrient-use efficiency (mg dry matter produced per mg nutrient absorbed) of different species at Cherrapunji in Meghalaya (from Ram and Ramakrishnan, 1988a)

	Nutrient-use efficiency		
Species	*Nitrogen*	*Phosphorus*	*Potassium*
Arundinella	149.9 ± 12.5	1,020.4 ± 95.4	136.4 ± 12.5
bengalensis	(101.7 ± 9.7)	(1,010.1 ± 91.7)	(158.7 ± 14.5)
Carex	108.1 ± 9.1	980.4 ± 75.9	91.7 ± 8.5
cruciata	(124.8 ± 11.1)	(740.4 ± 61.4)	(89.2 ± 7.1)
Chrysopogon	109.1 ± 9.4	595.2 ± 41.5	158.7 ± 13.7
gryllus	(77.94 ± 5.4)	(657.9 ± 51.3)	(1,220.0 ± 10.5)
Ischaemum goeblii	117.6 ± 9.7	877.2 ± 73.4	169.5 ± 14.3
Eragrostiella			
leoptera	150.9 ± 12.7	943.40 ± 81.4	188.7 ± 17.1
LSD ($p = 0.01$)	3.99	31.1	4.86

In parentheses are values for species in nutrient-poor soil: LSD = least significant difference

and Ram, 1988; Ram and Ramakrishnan, 1988a) are adapted to nutrient-deficient soils because of their high nutrient-use efficiency compared to the C_3 species, an aspect we considered earlier in this Chapter and in Chapter 7. Thus, *Carex cruciata*, a C_3 species, showed generally lower nutrient-use efficiency values compared to the other C_4 grasses (Table 8.10). Indeed, the *Ischaemum* and *Eragrostiella* types of grasslands developed on more nutrient-depleted soils compared to *Osbeckia* and *Arundinella* types of grasslands, have a larger proportion of C_4 grasses than C_3 species. Thus, seral grassland communities at Cherrapunji in Meghalaya are adapted for survival under nutrient stress : (1) through their allocation strategy being geared for clonal reproduction, and (2) through a greater emphasis on C_4 species that are more nutrient-use efficient (Table 8.10).

Conclusions

The early successional herbaceous species coming up under the jhum system which involves perturbation due to slash-and-burn and mixed cropping on the site show a variety of strategies for quick exploitation of the rapidly depleting resources of a transient environment. Regeneration could be through sprouting species such as *Imperata cylindrica* and many others that rapidly colonize the site, or through a highly developed seed reproduction represented by *Eupatorium* spp. However, even predominantly sprouting species resort to some sexual reproduction that would ensure population heterogeneity and this often could be fire-adapted as in *Imperata cylindrica* and *Mikania micrantha.* The biomass and nutrient allocation strategies (Fig. 8.11) generalized for sprouting versus non-sprouting species only illustrates how limited resources are channelled for varied growth and reproductive functions, and also highlights how important it is to consider different currencies such as biomass and nutrients separately in such studies.

C_3/C_4 photosynthetic pathways of different species are important for spatial and temporal separation of ecological niches of species so that their apparent coexistence is ensured through mutual avoidance. The C_4 strategy with a high nutrient-use efficiency is important for effective growth and survival under nutrient-stressed situations.

In conclusion, it may be emphasized that the patterns and processes during succession could be better understood if the community/ecosystem level analyses are complemented with ecophysiological and strategy analyses at the individual and population levels. Unfortunately, ecosystem studies at individual, population, community and ecosystem levels have largely been done in isolation and they need to be integrated effectively to bring out the varied possibilities that exist for a better understanding of the vegetation development and its dynamics, and for effective ecosystem management, as discussed in this and the earlier chapters.

GENERAL CONSIDERATIONS

Under the jhum system, the early weedy colonizers of secondary succession have evolved a variety of strategies to survive through fire and other perturbation effects. These diverse strategies, to a certain extent, determine the structure of the early successional communities; this is in addition to a number of other factors such as fire intensity and frequency, perturbation during the cropping phase, and depletion/enhancement of propagules in the neighbourhood over a period of time.

The strategy analysis as discussed here has many management implications. Reproductive potential and strategies of species in terms of seed versus vegetative reproduction are important for adopting appropriate weed control measures, through physical operations or biological/chemical control. In the case of species that resort to seed reproduction under varied perturbation regimes, management of perturbation could be of some limited value. Changes in allocation strategy of some species during succession were shown through some case studies and are significant for biological suppression of weeds, due to reduced light availability during natural succession. Manipulation of species through agroforestry systems or through accelerated succession using appropriate tree cover is a possibility for weed management utilizing the information obtained through strategy analysis. Traditionally jhum farmers have used such techniques during agriculture by conserving tree species through tree stumps left in the jhum plots for quick forest regeneration.

In restoration ecology, strategy analysis of species as discussed here is very important in species selection. Thus, species that have a sprouting strategy for regeneration may have an advantage under nutrient stress conditions because of resource storage in the underground rhizomes that could be effectively capitalized upon. Further, a quick ground cover through vegetative reproduction could be critical for soil conservation in a transient environment. Differential nutrient uptake and nutrient-use strategies by different species such as those shown by C_3 and C_4 strategists may be important for species selection and the ability to cope with soil nutrient heterogeneity. Indeed in the jhum plots, the farmer combines these two strategies in the mixed cropping for nutrient conservation (cf Chapter 1). In restoring a degraded habitat, the early successional weedy species have a major role to play in conserving nutrients (cf Chapters 5 and 7), so much so that the jhum farmer retains about 20% weed biomass in his jhum plots during weeding operations (the non-weed concept discussed in Chapter 4). Therefore, quick colonization giving a good ground cover should be the immediate objective when recovering a steep slope, before consideration is given to the introduction of larger shrubs and trees.

CHAPTER 9

SHRUB AND TREE GROWTH STRATEGIES AND ARCHITECTURE

INTRODUCTION

Much of our knowledge of the growth characteristics of tree species is based upon temperate representatives (Cook, 1941; Kozlowski, 1964; Kramer and Kozlowski,1979). Recently, the canopy form of trees in relation to environmental conditions has received some attention (Brunig, 1976; Nelson *et al.*, 1981); several quantitative studies on branching pattern have been made (Whitney, 1976; Fisher, 1978; Steingraeber *et al.*, 1979; Pickett and Kempf, 1980; Borchert and Slade, 1981). However, temperate trees offer only a limited variety in growth pattern whilst the shapes and crown forms of tropical species are more variable. Although some studies on the growth pattern of tropical trees are available (Hallé and Oldeman, 1970), the need for a deeper understanding of growth patterns in tropical trees was emphasized by Tomlinson and Gill (1973). The works of Hallé and Oldeman (1970) and Hallé *et al.* (1978) deal with architectural models of trees in a qualitative way and have helped in classifying tree growth in an orderly fashion. Because these classical studies are based on qualitative morphometric analysis, they are not meant to explain adaptive strategy of trees and therefore to relate tree growth to the environment in which they are located. Indeed the classical work of Hallé *et al.* (1978) has, however, helped in creating a renewed interest on tropical tree architecture among ecologists, physiologists and morphologists interested in trees, for reasons of their own.

Studies on the growth pattern and architecture of shrubs are even more limited. The only studies that I am aware of, on growth and architecture of shrubs related to their environment from the temperate region, are the studies done on broad-leaved forests of north-west America (Pickett and Kempf, 1980; Kempf and Pickett, 1981), over a successional gradient.

Studies from the tropics are extremely limited. Apart from the architectural classification on tropical trees done by Hallé, *et al.* (1978), a few scattered morphometric studies have been done by others on trees (Holdsworth, 1963; Borchert, 1969; Gill and Tomlinson, 1971). Architecture is a dynamic concept. The ultimate architectural form of a tree or shrub is determined by the rate of growth of the leader axis versus lateral branches, the bud dynamics and the production of branches of different order, the birth and death rates of leaves borne on them, and the final display of branches and the leaves borne on the trunk. Such a dynamic concept would, therefore, be determined by micro-environmental changes that occur around it as the tree or shrub grows. The adapative strategy itself is an evolutionary process; therefore, it is species-specific and a consequence of natural selection, both in space and time. It follows from this that each species has an ecological range of its own which would permit it to adjust to a greater or lesser extent to the changing environmental conditions in the forest, over a short time as the tree/shrub grows, or over a longer time-scale over a successional gradient. Light obviously is an important factor as architectural form determines leaf display which would determine the ability of the species to optimize photosynthesis under varied light regimes of the forest environment. Since light availability changes during forest succession, architecture as related to successional status of the species becomes significant. Studies related to successional status of trees (Marks, 1975) and shrubs (Kempf and Pickett, 1981) have not received much attention to form generalizations; studies from the tropics are not available. It is in this context that we initiated studies on tropical tree architecture as related to successional status in fallows developed after jhum (Boojh and Ramakrishnan, 1982a, b,c,d, 1983a; Singh and Ramakrishnan, 1983; Ramakrishnan *et al.*, 1982; Das and Ramakrishnan, 1986; Ramakrishnan and Shukla, 1982; Shukla and Ramakrishnan, 1984a,b, 1986; Ramakrishnan, 1986a, 1987b). This was subsequently followed up by a series of studies on shrubs (Baruah and Ramakrishnan, 1988b, 1989a,b,c,d,). As bamboo is an important shrub of the early successional environment in the jhum fallows of north-east India, we have also considered the architectural plasticity of important bamboo species over a successional gradient (Rao and Ramakrishnan, 1988a).

GROWTH STRATEGY AND ARCHITECTURE OF TREES DURING SUCCESSION

This study was done at two altitudinal situations in Meghalaya – at Upper Shillong (1,900 m) and at Lailad (296 m). At Upper Shillong, two early successional 5-year-old individuals of *Alnus nepalensis* and *Schima khasiana*, and three late successionals, *Machilus kingii*, *Quercus dealbata* and *Quercus griffithii* were studied from a sub-tropical montane, humid, evergreen forest (Boojh and Ramakrishnan, 1982a). This forest had both sub-tropical elements like *Schima khasiana*, *Machillus kingii*, *Manglitia insignis*, *Eugenia terragona* and *Ficus nemoralis*, along with many temperate species such as *Quercus* spp., *Alnus nepalensis*, *Rhododendron arboreum* and *Betula alnoides* (Boojh and Ramakrishnan, 1982a). Of the five species identified, the two early successionals were studied under open situations only, as they do not grow in shade, the three late successional species were studied both from open and forest situations.

The lower elevation (296 m) study area was at Lailad in Meghalaya. The study site is part of a forest reserve which was clear-felled about 50 years ago (Singh and Ramakrishnan, 1982a). Outside the reserve are stands at different stages of early succession after jhum (cf Chapter 6). Five-year-old trees of two early successional (*Anthocephalus cadamba*) and *Duabanga sonneratioides*) and two late successional (*Dillenia pentagyna* and *Artocarpus chaplasha*) species were studied. Open-grown trees of all the four species and forest-grown individuals of *Anthocephalus cadamba* and *Artocarpus chaplasha* were considered (Shukla and Ramakrishnan, 1986). The following discussion on tree growth, architecture, leaf dynamics and allocation strategy is largely based upon these examples, although a more generalized study involving 22 tree species was also done at Lailad in Meghalaya (Ramakrishnan and Shukla, 1982).

Growth and architecture related to successional status

Architectural models

At lower elevations at Lailad in Meghalaya, the four species considered belong to four different models of Hallé *et al.* (1978). Just as there are variations from the Hallé's models in the species considered for the high elevation study and discussed below, here too the morphological characteristics do not always agree. Thus, rhythmicity is associated with such morphological features as:

(1) Axis articulation (pronounced segmentation);

(2) Intermittent branch production; and

(3) Regular change in leaf morphology.

The absence of these characteristics is considered to indicate continuous growth. *Anthocephalus cadamba* and *Duabanga sonneratioides* do not conform to this, as discussed below. Of the four species considered for the lower elevation study, *Anthocephalus cadamba* and *Duabanga sonneratioides* are light demanders, *Dillenia pentagyna* is shade tolerant after establishment, while *Artocarpus chaplasha* can regenerate in the shade.

The heterogeneous axes with both an orthotropic leader axis and plagiotropic branches, acrotonic branching (branches not at the base but distal on the axis) without modular construction (leader axis distinct from the branches), monopodial trunk with a continuous pattern of growth (leaves with similar morphology and with regular branch production), and plagiotropic, monopodial long-lived branches are features that assign *Anthocephalus cadamba* to Roux's model (Fig. 9.1). The leader produces decussately arranged leaf pairs and branches. Branch dimorphism with an orthotropic trunk and plagiotropic branches is evident. First-order branches arise sylleptically (by the continued growth of lateral buds laid during the current year's growth with a long first node (hypopodium) before bearing foliage leaves. All the other branch orders arise proleptically (by the activation and growth of lateral buds laid during the previous year's growth). The decussate leaf arrangement on plagiotropic units shows dorsiventrality by secondary orientation of blades in a horizontal plane, exhibiting apparent discordancy with that of the trunk where the decussately arranged leaves may be oriented in several planes.

Morphologically, *Duabanga sonneratioides* conforms to Massart's model and shows many basic characteristics in common with Roux's model. The main difference is that the leader axis shows rhythmic growth (leaves with changing morphology and branches organized in tiers) (Fig. 9.2). The radially symmetrical main axis produces spirally arranged leaves at short intervals and tiers of branches, each tier consisting of five branches. During the 1st year, only one orthotropic trunk bears plagiotropic branches. The first-order branches are extra-axillary and sylleptic and the higher order branches are axillary and proleptic. The leaves are at long intervals on the branches and are dorsiventral. The growth units (branch tiers) are neither demarcated by scale leaves nor is there a rest period between successive tiers, unlike that reported by Hallé and Oldeman (1970) for this species from Malayasia.

The leader axis of *Dillenia pentagyna* shows a rhythmic pattern of growth with a change in leaf morphology, and the branches are arranged in loose tiers. The homogeneous axes are all orthotropic with terminal inflorescence resulting in sympodial branches at the periphery of the crown. This conforms to Scarrone's model (Fig. 9.3). All branches are produced proleptically only after the fall of axillant leaves. The arrangement of leaves and branches is spiral. A 6-year-old tree produces about three to five first-order branches in a year. The branch growth is monopodial during the period of vegetative juvenility, but becomes sympodial after the onset of the terminal inflorescence. In older trees the lower branches may become pendulous.

The architectural model of *Artocarpus chaplasha* differs from that of *Dillenia pentagyna* in that the inflorescence is lateral and, therefore, the

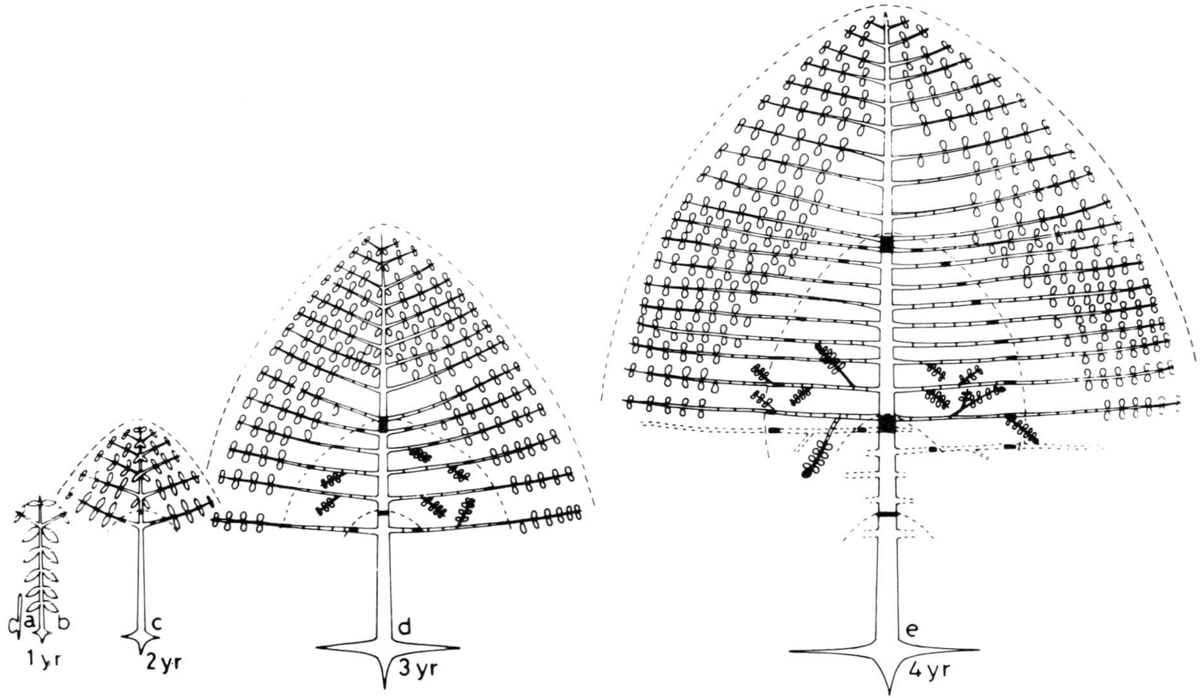

Figure 9.1 Architectural development with excurrent crown form (Roux's model) of *Anthocephalus cadamba* at Lailad in Meghalaya during the first 4 years of growth (each demarcated by broken line). Short internode (articulation) shows cessation of annual growth (from Shukla and Ramakrishnan, 1986)

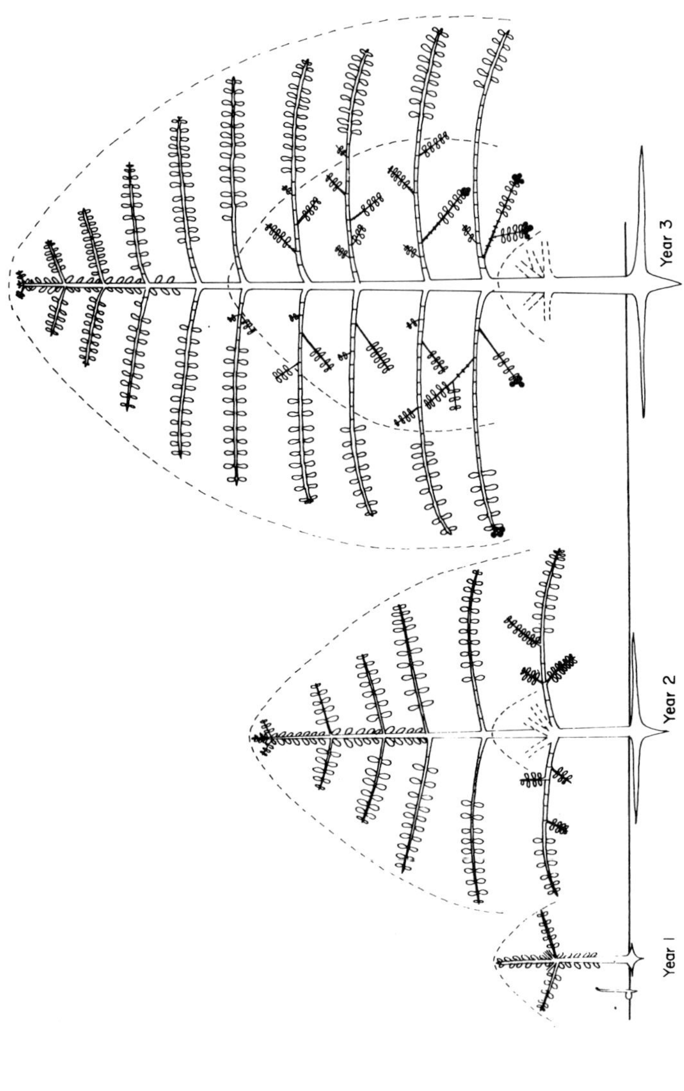

Figure 9.2 Architectural development with excurrent crown form (Massart's model) of *Duabanga sonneratioides* at Lailad in Meghalaya during the first 3 years of growth (each demarcated by broken line). During the 1st year only one tier of branches is produced. In subsequent years five tiers of branches are produced per year. Only two of the five branches in each tier are shown (from Shukla and Ramakrishnan, 1986)

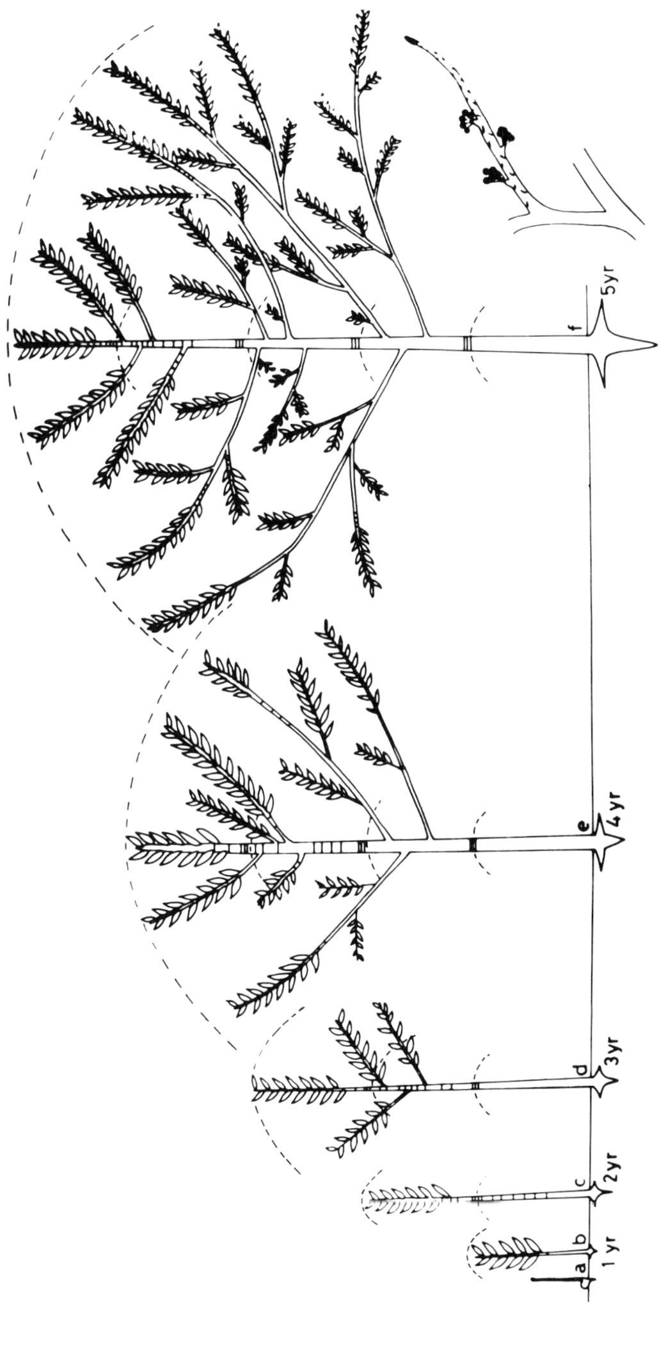

Figure 9.3 Architectural development with decurrent crown form (Scarrone's model) of *Dillenia pentagyna* at Lailad in Meghalaya during the first 5 years of growth (each demarcated by broken line). Aggregation of nodes indicate cessation of annual growth. Terminal flowers on a third-order branch of an old tree are shown separately in the final drawing (from Shukla and Ramakrishnan, 1986)

branches are monopodial. Hence this species conforms to Scarrone's model (Fig. 9.4). Axial conformity and the arrangement of the leaves and branches in this species is similar to *Dillenia pentagyna.* The trunk remains unbranched until the 3rd or 4th year. Unusually, a single tier of proleptic first-order branches is produced each year, but occasionally two tiers may be produced through syllepsis. A single tier has three to seven branches, depending upon the enviromental conditions and the age of the tree, and separated from the next tier by a gap of two to 11 internodes.

The architectural development pattern of *Alnus nepalensis* (Fig. 9.5) at higher elevations follows Attim's model (Hallé *et al.*, 1978). In this species, the continuous activity of the formative meristem results in a continuous branching pattern which becomes intermittent during the winter due to suppression. This helps in determining the age of the tree. The continuous activity of the meristem is expressed in the more or less constant bud composition and little variation in leaf shape and size along the axis. However, the growth rate is not uniform as is clear from the rhythmicity of internode length and leaf area pattern on the main axis (Fig. 9.6) as well as the temporary suspension of growth during the winter. Branches are produced through syllepsis (Hallé *et al.*, 1978). The branches are inherently orthotropic with spiral phyllotaxy and radial symmetry, but may subsequently become horizontal, with leaves arranged in one plane through twisting of the petioles (secondary orientation). The inherent orthotropy of such a branch is revealed in damaged trees where the leader is readily replaced by an adjacent lateral which conforms to the original architecture. Attim's architectural model followed by *Alnus nepalensis* is particularly geared to achieve fast and efficient growth through continued activity of the apical meristems.

Schima khasiana, Quercus dealbata, Quercus griffithii and *Machilus kingii* all follow Rauh's model of architecture (Fig. 9.5). The early successional *Schima khasiana* which follows this model showed more rhythms of internodal elongation and leaf production than the late successional species. This species, produces branches through syllepsis whereas the three late successional species (*Quercus* spp. and *Machilus kingii*) branch through prolepsis. It is suggested that the switch from a 'lower' state which determines prolepsis of a lateral meristem to a 'higher' state which determines syllepsis is conditioned by a 'threshold' which in turn is determined by the growth vigour of the parent shoot (Tomlinson and Gill, 1973). Such a sylleptic branch production in early successional species seems to be intended to ensure quick growth under fast diminishing light resources in a developing community. The growth of a hypopodium in sylleptic branches of early successional species is presumably a mechanism by which the leaves are put outside the shade created by the leaves of the parent shoot.

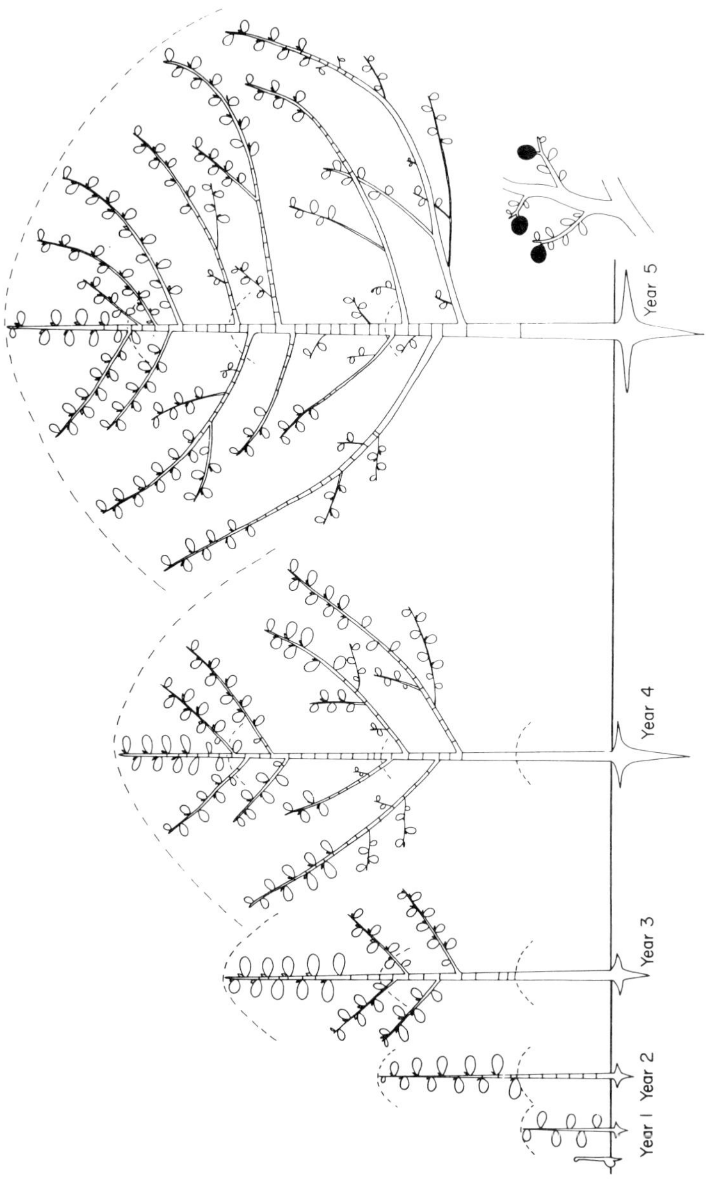

Figure 9.4 Architectural development with decurrent crown form (Rauh's model) of *Artocarpus chaplasha* at Lailad in Meghalaya during the first 5 years of growth (each demarcated by broken line). One tier of 3–5 alternate branches with a long gap between tiers demarcates annual growth. The lateral position of flowers on a third-order branch of an older tree is shown separately in the final drawing (from Shukla and Ramakrishnan, 1986)

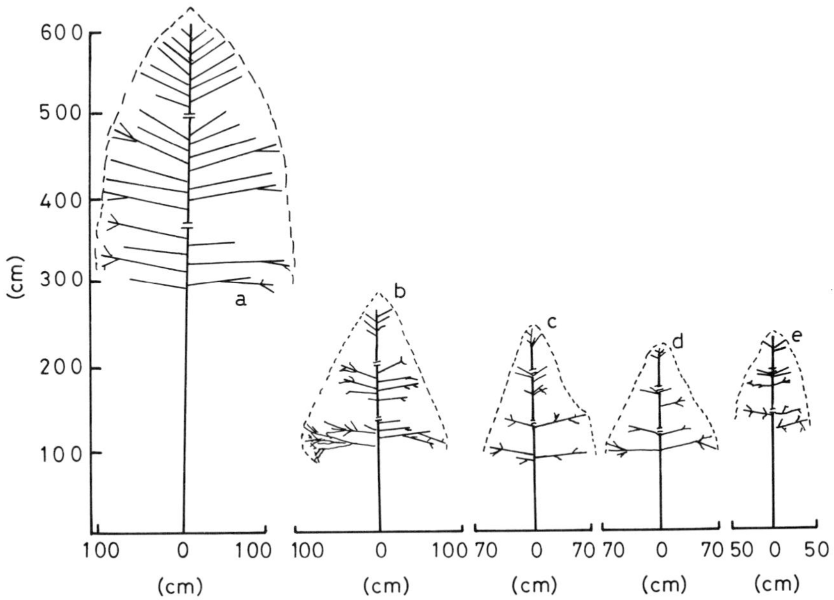

Figure 9.5 Architectural skeleton and branching pattern in open-grown trees of : (a), *Alnus nepalensis*; (b), *Schima khasiana*; (c), *Quercus dealbata*; (d), *Quercus griffithii*; and (e), *Machilus kingii*. Demarcations on the main axis represent the termination of annual growth delimited by the shortest internode of that growing season (from Boojh and Ramakrishnan, 1982a)

Reiteration of the architectural model (Hallé *et al.*, 1978) occurs after damage to the aerial parts in all the species. In early successional species, reiterated complexes generally show more vigorous growth with a prolonged growth period. In late successional species, such reiterated complexes exhibit an indeterminate growth, often with a bayonet-like joint. In *Alnus nepalensis*, reiteration always occurs through the substitution process, by a lateral branch adjacent to the damaged part. In other species, it may be achieved either through the activation of a dormant lateral meristem, substitution or coppicing. The reiteration which helps the tree to adjust its architecture to environmental modifications and disturbances may be advantageous for early successional species adjusting to changing light environments. Each wave of reiteration may enable them to activate dormant meristems and ensure vigorous growth. Likewise, in late successional species, this process may often be triggered in response to canopy openings. Thus, reiteration allows trees to exploit the changes in the environmental energy level. The trees utilize the larger quantity of incoming energy, which results from environmental changes or disturbances, by increasing the numbers or size of the producing organs. Branching and reiteration in the tree crown are examples of this process. Thus, branching within the model is a standard solution to the bioenergetic relationship,

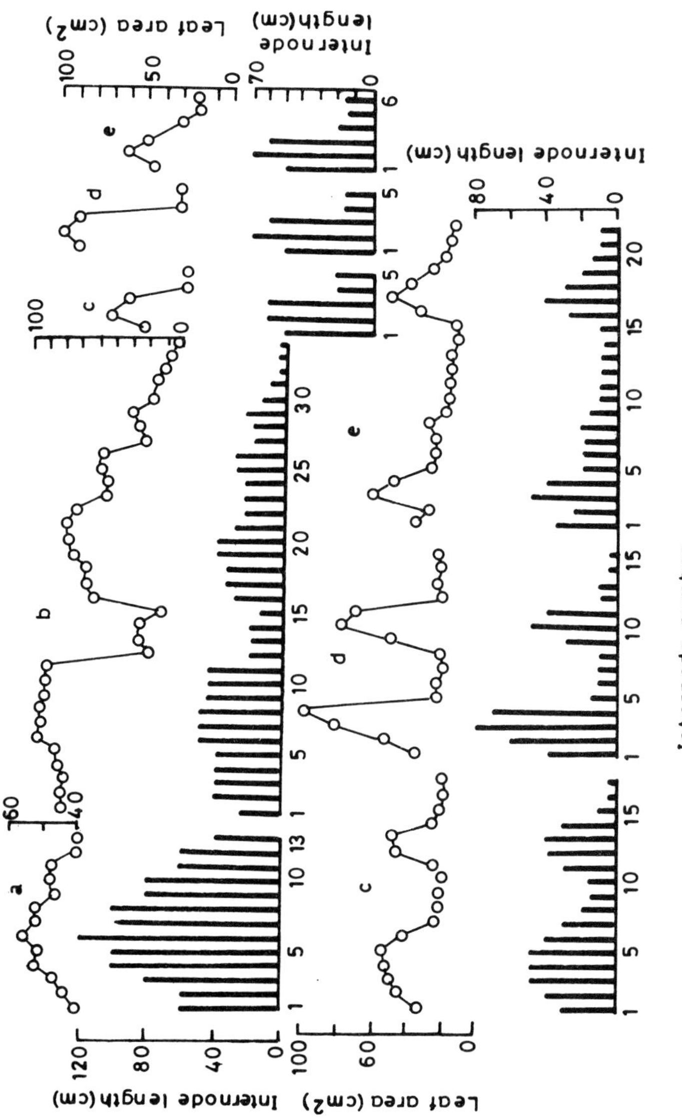

Figure 9.6 Annual growth rhythms in internodal elongation (bars) and leaf area changes (open circles) on the leader axis: of (a) *Alnus nepalensis*; (b) *Schima khasiana*; (c), *Quercus dealbata*; (d), *Quercus griffithii*; and (e) *Machilus kingii*. Inset figure is for forest-grown individuals (from Boojh and Ramakrishnan, 1982a)

whereas reiteration allows for ecological opportunism (Hallé *et al.*, 1978). Reiteration was found to be more frequent in late successional species (Shukla and Ramakrishnan, 1986).

A major conclusion arising from this analysis is that there is no relationship between the architectural models of Hallé *et al.* (1978) and the successional status of the tree species (Boojh and Ramakrishnan, 1982a; Shukla and Ramakrishnan, 1986). Our subsequent analysis would show that quantitative differences in tree architecture are more significant ecologically, than qualitative differences represented by these models; a dynamic approach to growth analysis is important. The phenomenon of continuous versus rhythmic growth is an example which illustrates this. The species which show a morphological rhythm such as *Duabanga sonneratioides* in reality have no seasonal bud dormancy in north-eastern India, unlike the same species from Malaysia studied by Hallé *et al.*, (1978). On the other hand, species considered to have continuous growth on the basis of morphology (e.g. *Anthocephalus cadamba*) in reality have a distinct period of seasonal bud dormancy.

Growth pattern

At a high altitude, the sub-temperate tree species show extension growth characteristics of the typical temperate broad-leaved tree species (Marks, 1975; Boojh and Ramakrishnan, 1982a). The shade intolerant early successional species, chiefly through an indeterminate pattern of growth and leaf production, achieve more extension growth than the late successional ones, which are determinate and have a brief growth period. Such a strategy helps the former to gain in height and exploit a high light regime in the forest. However, the two determinate flushes in the open-grown late successional trees and only one in the forest-grown trees are indicative of their ability to adjust their growth strategy, at least to some degree, depending upon light availability (Fig. 9.7). Thus, although the late successional species are adapted to grow under low light regimes where resources are often limiting, they also have the ability to adjust their growth pattern favourably under open situations such as openings in the forest canopy.

The date of bud break in temperate tree species is relatively constant, whereas that of growth cessation is usually more variable (Kramer, 1943). In the present study, the time of bud break showed more variation. For example, in *Anthocephalus cadamba*, it occurred between late January (winter) and late March (a warmer period), suggesting that temperature has less control on bud break in this species than endogenous factors which may be related to age. In contrast, *Duabanga sonneratioides* had continuous growth throughout the year.

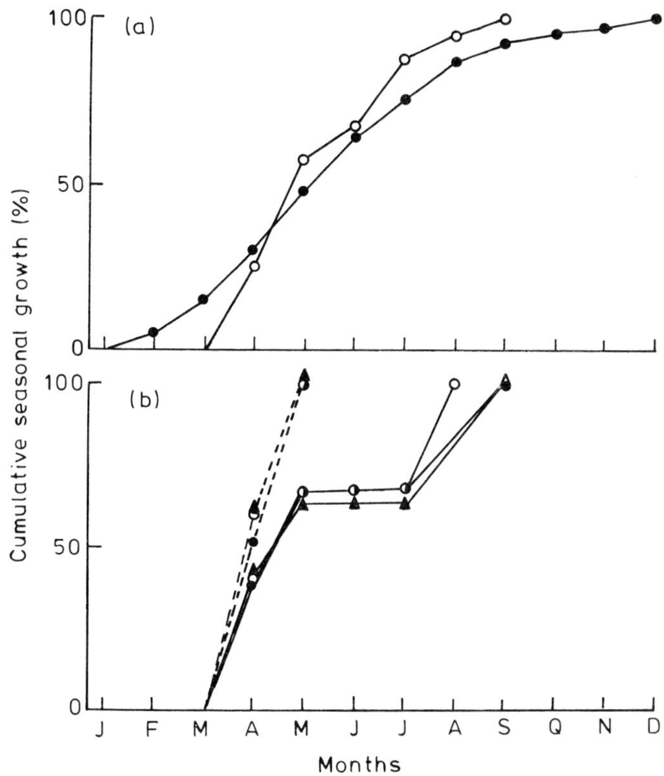

Figure 9.7 Cumulative percentage of extension growth completed during the growing season at various dates: (a) early successional species: *Alnus nepalensis* (●), and *Schima khasiana* (○); (b) late successional species : *Quercus dealbata* (●), *Quercus griffithii* (○), and *Machilus kingii* (▲); forest-grown trees (broken lines) (from Boojh and Ramakrishnan, 1982a)

In the three species where dormancy is evident, growth initiation occurred during the drier period (January–May) and this agrees with the observation on other tropical tree species (Longman and Jenik, 1974). The important growth characteristics of early and late successional sub-temperate tree species are summarized in Table 9.1. These observations also apply to tropical tree species of lower altitudes. The early successional species have a prolonged growth period, a shorter dormancy period and more extension and radial growth than the late successionals. The tree height/crown width ratio was higher in early successional species, while the ratio of the total height to height of lowest branch was higher for the late successional species.

Contraction of the growth period by the gradual bringing forward of dormancy and postponement of growth initiation with age is interesting (Fig. 9.8) and this agrees with the observations on other tropical tree species

Table 9.1 Growth characteristics (± SE) of early and late successional species namely *Alnus nepalensis* (An), *Schima khasiana* (Sk), *Quercus dealbata* (Qd), *Q. griffithii* (Qg) and *Machilus kingii* (Mk) at upper Shillong in Meghalaya (from Boojh and Ramakrishnan, 1982a)

	Early successional		*Late successional*					
	An	*Sk*	*Qd*		*Qg*		*Mk*	
Growth characteristics	*Open*	*Open*	*Open*	*Forest*	*Open*	*Forest*	*Open*	*Forest*
Dimensions of the experimental trees								
Height (m)	5.05 ± 0.68	2.03 ± 0.39	1.96 ± 0.75	1.54 ± 0.90	1.74 ± 0.35	1.25 ± 0.60	1.85 ± 0.25	1.37 ± 0.26
Diameter at breast height (cm)	3.55 ± 0.35	1.28 ± 0.34	1.14 ± 0.29	0.73 ± 0.29	1.05 ± 0.29	0.68 ± 0.30	1.14 ± 0.30	0.74 ± 0.30
Growth period (days)	334.00 ± 10.50	190.00 ± 9.18	73.00 ± 1.88	35.00 ± 0.69	63.00 ± 2.00	30.00 ± 1.35	72.00 ± 4.63	33.00 ± 1.00
Dormancy period (days)	31.00 ± 10.50	175.00 ± 9.18	292.00 ± 1.88	330.00 ± 0.69	302.00 ± 1.99	335.00 ± 1.35	293.00 ± 4.63	332.00 ± 1.00
Extension growth per year (cm)	104.00 ± 6.50	64.00 ± 3.09	49.00 ± 2.50	21.00 ± 1.90	46.00 ± 1.60	21.00 ± 1.60	44.00 ± 2.30	23.00 ± 2.35
Radial growth per year	0.95 ± 0.09	0.57 ± 0.15	0.55 ± 0.06	0.27 ± 0.03	0.42 ± 0.05	0.25 ± 0.05	0.49 ± 1.12	0.20 ± 0.02
Total height to crown width ratio	3.12 ± 0.22	2.75 ± 0.12	1.71 ± 0.04	1.56 ± 0.06	1.98 ± 0.20	1.62 ± 0.02	1.78 ± 0.09	1.62 ± 0.02
Ratio of total height to height at lowest branch	2.56 ± 0.16	2.63 ± 0.24	4.32 ± 0.74	5.66 ± 0.22	3.61 ± 0.28	5.25 ± 0.23	3.02 ± 0.30	4.86 ± 0.36

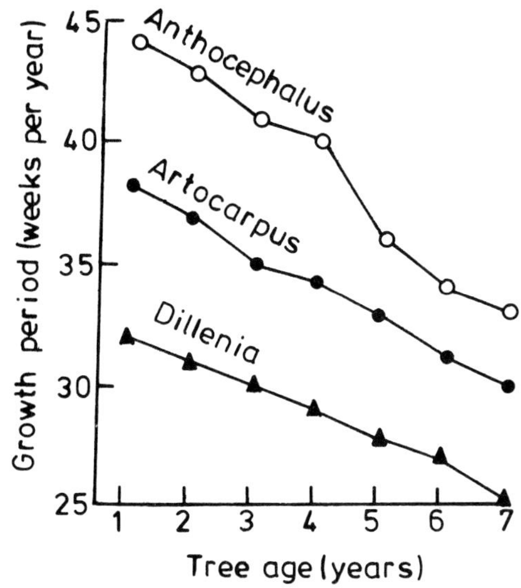

Figure 9.8 The length of the growing period in relation to trees age for : *Anthocephalus cadamba* (○); *Artocarpus chaplasha* (●) and *Dillenia pentagyna* (▲) at Lailad in Meghalaya (from Shukla and Ramakrishnan, 1986)

(Longman and Jenik, 1974). This obviously is related to intrinsic factors rather than external causes. The decrease in photosynthetic biomass in relation to non-photosynthetic biomass (Shukla and Ramakrishnan, 1984a) with age, resulting in reduced vigour (source: sink ratio) could be a possible explanation. This needs to be tested further.

The early successional tropical trees had extension growth over a longer period than the late successional species. This and the higher extension growth rate resulted in quicker upward movement of the canopy for the open-grown early successionls compared to late successionals. (Fig. 9.9). The forest-grown trees had slower growth rate and this was more pronounced in the early successional *Anthocephalus cadamba* than the late successionl *Artocarpus chaplasha.* The early successional species showed continuous radial growth throughout the year whereas in the late successional species and in forest-grown *Anthocephalus cadamba*, this ceased for about 2 months during the winter. These observations suggest that extension and radial growth need not always be related to one another (Hummel, 1946) and that they are probably controlled by environmental factors.

The four tropical species of the lower elevation showed rhythmic fluctuations in internodal length and leaf area within a growth period (Fig. 9.10). Obviously, this rhythmic fluctuation is intrinsic. More rhythmic fluctuations of this kind occurred in early successional species compared to late successional ones, and in open-grown compared to forest-grown

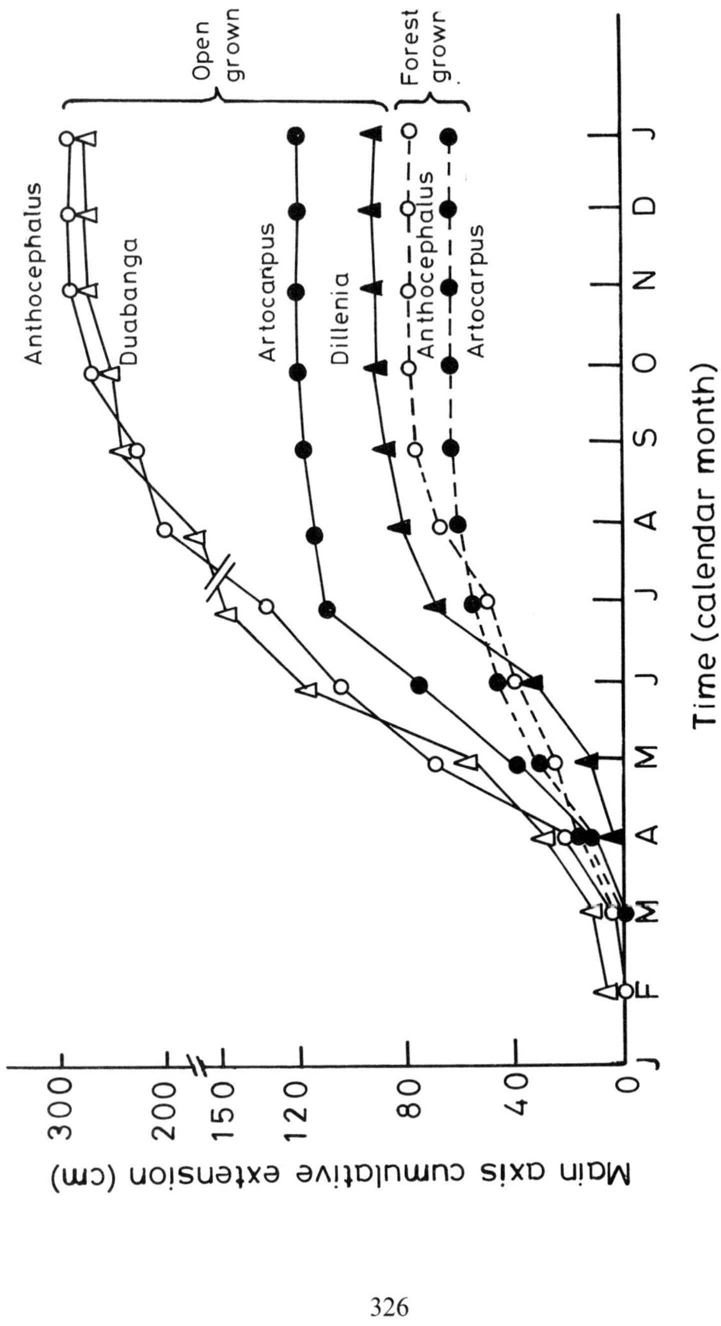

Figure 9.9 Cumulative extension growth of the main axis of open-grown (solid line) and forest-grown (broken line) trees of *Anthocephalus cadamba* (○); *Duabanga sonneratioides* (△); *Artocarpus chaplasha* (●); *Dillenia pentagyna* (▲) at Lailad in Meghalaya (from Shukla and Ramakrishnan, 1986)

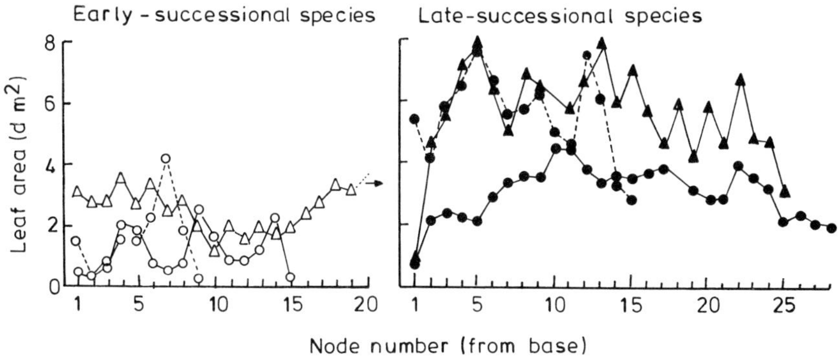

Figure 9.10 Rhythmical changes in leaf area on the main axis of open-grown (solid line) and forest-grown (broken line) trees of *Anthocephalus cadamba* (○); *Duabanga sonneratioides* (△) *Artocarpus chaplasha* (●); and *Dillenia pentagyna* (▲) at Lailad in Meghalaya. Only one tier of branches with a between-tier gap on the main axis is presented here for *Duabanga sonneratioides*. The arrow suggests that the rhythmicity continued for the other four tiers produced during the year in similar way (from Shukla and Ramakrishnan, 1986)

trees of the same species, suggesting that the rhythm frequency is dependent upon the rate and duration of growth of the individual concerned. Similar results were shown for the sub-temperate trees of the higher elevations.

Various terms such as intermittent (Koriba, 1958), episodic (Romberger, 1963), articulate (Tomlinson and Gill, 1973) and rhythmic (Hallé *et al.*, 1978) have been used for discontinuous growth of tropical trees and have caused considerable confusion. Basically, the four tropical species have indeterminate growth strategy (Shukla and Ramakrishnan, 1986), unlike the indeterminate/determinate strategy of the early/late successional temperate species. The indeterminate strategy of the four tropical trees are of two types: (1) evergrowing–with no dormant phase (*Duabanga sonneratioides*) and (2) periodic–with a dormant phase (*Anthocephalus cadamba, Dillenia pentagyna* and *Artocarpus chaplasha*). Leaf production (Shukla and Ramakrishnan, 1984a) and extension growth in all these species is continuous throughout the growth period although the rate may fluctuate. However, in a comprehensive analysis of the growth characteristics of 22 tree species at Lailad (296 m altitude) in Meghalaya (Ramakrishnan and Shukla, 1982), it was observed that many other late successional tropical trees may show determinate growth similar to the sub-temperate species of the higher altitude, with leaf production and extension growth restricted to predetermined leaves from an earlier dormant-bud phase. Such species may have one flush per year as in *Engelhardtia spicata*, *Mesua ferrea* and *Myristica linifolia*, two flushes as in *Actinodaphne angustifolia*, three flushes as in *Castanopsis indica*, or even four as in *Chikrasia tubularis*.

The differential shade tolerance of early versus late successional trees is reflected in the extension growth features of the different branch orders at

Table 9.2 Annual extension growth (cm) of the main axis and of first-and second-order branches in three positions in the canopy of open-grown trees of four species namely, *Anthocephalus cadamba* (Ac), *Dudbanga sonneratioides* (Ds), *Artocarpus chaplasha* (Ac) and *Dillenia pentagyna* (Dp) at Lailad in Meghalaya. The growth rates for forest-grown individuals were available for two species only (from Shukla and Ramakrishnan, 1986)

Order of shoot and canopy position	*Ac*	*Ds*	*Ac*	*Dp*
Zero (main axis)				
open	286 ± 36.4	298.2 ± 32.4	123.0 ± 28.1	102.2 ± 13.7
forest	79.5 ± 10.2	–	66.5 ± 8.6	–
First				
Top canopy				
open	262 ± 24.6	259.4 ± 32.8	104.2 ± 32.6	35.7 ± 14.0
forest	122.0 ± 16.0	–	74.0 ± 17.8	–
Middle canopy				
open	208.6 ± 28.4	174.1 ± 34.3	86.0 ± 29.0	31.7 ± 11.5
forest	86.0 ± 0.4	–	48.4 ± 6.4	–
Bottom canopy				
open	127.4 ± 28.8	44.4 ± 16.5	54.6 ± 18.6	22.2 ± 12.9
forest	42.4 ± 8.6	–	29.0 ± 17.2	–
Second				
Top canopy				
open	–	–	43.2 ± 20.2	6.3 ± 7.0
forest	–	–	27.5 ± 5.9	–
Middle canopy				
open	34.6 ± 13.2	53.8 ± 21.6	65.2 ± 20.8	6.9 ± 3.3
forest	–	–	47.5 ± 13.1	–
Bottom canopy				
open	45.2 ± 14.5	23.0 ± 9.7	23.6 ± 8.5	2.3 ± 1.4
forest	–	–	12.6 ± 8.5	–

different canopy positions (Table 9.2). What is given here for the tropical species is also true for the sub-temperate trees (Boojh and Ramakrishnan, 1982a). There is a general decrease of extension growth with increase in the order of branching and at lower canopy positions. However, in early successional species the decrease in extension growth with increase in branch order was more marked at lower canopy levels than that for late successional species.

Branching pattern

Bud dynamics, which takes into account the number and longevity of meristems, their position, time and potential of development leads directly to the dynamic concept of architecture and allows plant growth to be interpreted in terms of an organized flux of constructional units. Whilst Hallé *et al.* (1978) have used this for developing the tropical tree models, attempts to relate it to successional status are very few (Boojh and Ramakrishnan, 1982a). Figure 9.11 shows the dynamics of the bud populations for both early and late successional sub-temperate tree species of higher elevations. While in the early successional species proliferation from the axillary buds occurs synchronously with terminal bud growth, late successional species proliferate after a dormant period for the bud, and the activity depends upon the resources of the environment. Thus, the branches proliferate continuously in *Alnus nepalensis*, due to continuous meristem activity, while in *Schima khasiana* branching is rhythmic due to rhythmic activity of the terminal leader. This allows the two early successional species to deploy an extensive branching system through syllepsis by utilizing the current year's energy. The lesser number of reserve buds and their high mortality or inability to proliferate in the following year also suggests that all resources are directed exclusively towards the current year's growth and branch production. The late successional species in open situations are able to make some bud proliferation during the current year through lamma shoot formation in the second flush; shade-grown trees, however, produce branches in the next year only. This may be due to the limiting light conditions that exist in late successional environments. The pattern of branch production on the main axis and the first order branches is similar. However, because of syllepsis in the early successional species (in contrast to prolepsis in late successionals), the second order branching patterns are different.

The differential branch production pattern between early and late successional tropical species (Fig. 9.12) also suggests that like extension growth, branch production strategy is also geared to maximize growth in the former category when light is still adequate and not yet a limiting factor. Under open-grown conditions, the seasonal pattern of first- and second-order branch production of early successional species was spread over many months, whereas it was confined to a short period for late successional species. Apart from the lower level of branch production in forest-grown trees, in *Artocarpus chaplasha*, branch production was a month earlier in the forest-grown than in open-grown trees. This would provide a longer time for the leaves to photosynthesize during the growth season.

An early successional species such as *Pinus kesiya* at higher elevations at Shillong (1,250 m altitude) showed recurrent flushing, with three growth flushes occurring in a year. The three flushes together resulted in leader axis extension by about a metre in a 3-year-old pine sapling (Das and Ramakrishnan, 1986). A fresh set of needles (leaves) produced with each flush ensured active photosynthesis throughout the year. Such an exploitative growth strategy is characteristic of all light-demanding early successionals.

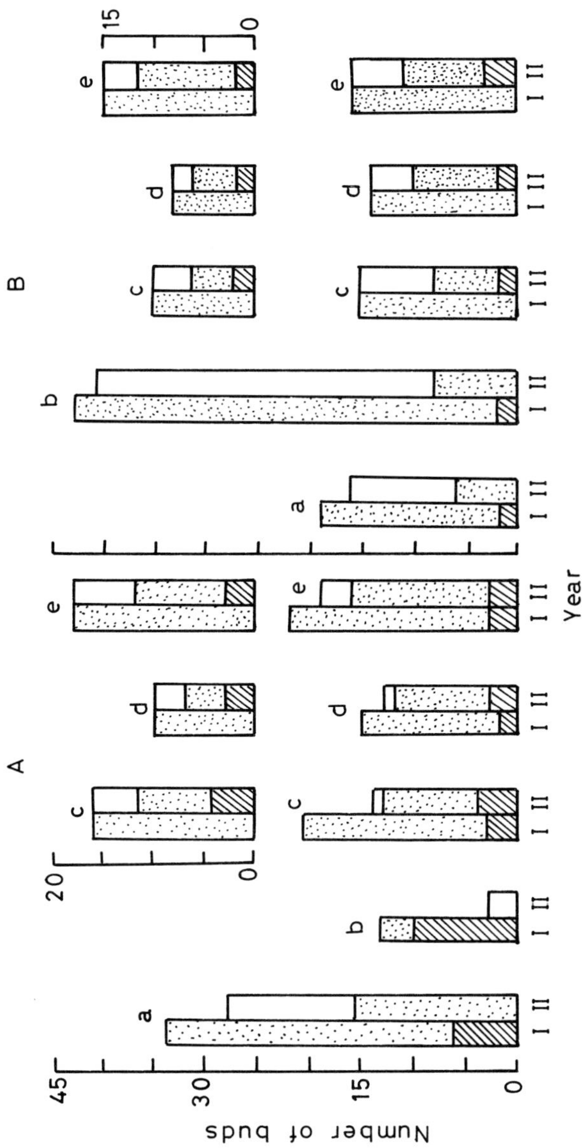

Figure 9.11 Dynamics of axillary bud populations on the leader axis (A) and the first-order shoot of the previous year (B) of: (a), *Schima khasiana*; (b) *Alnus nepalensis*; (c), *Quercus dealbata*; (d), *Quercus griffithii*; and (e), *Machilus kingii*. Inset bar diagrams are for forest-grown trees. Hatched bars, branches produced; stippled bars, dormant buds; open bars, dead buds (from Boojh and Ramakrishnan, 1982a)

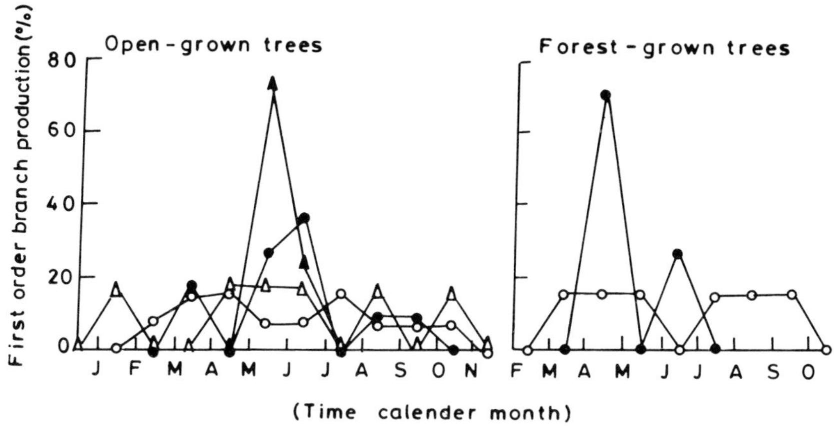

Figure 9.12 Monthly production (%) of first order branches of *Anthocephalus cadamba* (○); *Duabanga sonneratioides* (△); *Artocarpus chaplasha* (●); and *Dillenia pentagyna* (▲) growing in open- and forest-grown conditions at Lailad in Meghalaya (from Shukla and Ramakrishnan, 1986)

Table 9.3 Branch angles (from main axis from the apex) of first-order branches at two canopy positions of open-grown and forest-grown trees of four species at Lailad in Meghalaya, (from Shukla and Ramakrishnan, 1986)

	Open-grown trees		*Forest-grown trees*	
Canopy position	*Top*	*Bottom*	*Top*	*Bottom*
Early successionals				
Anthocephalus cadamba	68.2 ± 2.4	86.3 ± 1.8	72.4 ± 3.4	89.5 ± 1.4
Duabanga sonneratioides	72.2 ± 1.5	88.6 ± 2.4		
Late successionals				
Artocarpus chaplasha	55.0 ± 2.7	60.4 ± 2.6	54.2 ± 2.0	64.6 ± 2.7
*Dillenia pentagyna**	51.0 ± 1.8	63.8 ± 3.1	50.6 ± 1.7	62.2 ± 2.5

* Includes some trees of sprout origin

Branch organization and crown form

The overall geometry of a living tree, which depends partly upon branch orientation, is a manifestation of its adaptive strategy (Honda and Fisher, 1978). Branch angle in early successional species could differ depending upon the species (Boojh and Ramakrishnan, 1982a; Shukla and Ramakrishnan, 1986). However, certain generalizations could be made. In open-grown early successional species, first-order branch angle increased markedly at lower canopy positions compared with late successional species where the differences at different canopy levels are often not significantly different (Table 9.3). When open- and forest-grown trees are compared, such a difference between early and late successional species is again obvious. Similar results on the differential plasticity in branch orientation also

Table 9.4 Contribution of first-and second-order branches to the total length of the branch system of open-grown trees of four species at Lailad in Meghalaya (from Shukla and Ramakrishnan, 1986)

	First		*Second*		
Branch order	*Length* (cm)	*Proportion* (%)	*Length* (cm)	*Proportion* (%)	*Total length of all branch system* (cm)
Early successionals					
Anthocephalus cadamba	4,619	81	2,042	31	6,661
Duabanga sonneratioides	5,006	81	1,208	19	6,214
Late successionals					
Artocarpus chaplasha	1,441	57[a]	1,015	39[a]	2,625
Dillenia pentagyna	1,269	52	1,149	48	2,418

[a]4% were third-order branches

observed for sub-temperate species (Boojh and Ramakrishnan, 1982a) is of adaptive value. In a multilayered leaf organization, as in early successional species, an increase in the branch angle under low irradiance would help to open up the canopy for more efficient light capture. On the other hand, in late successional species where the leaves are peripherally placed, the branch angle may not make much difference to the arrangement of leaves and the efficiency of light capture.

The crown form of trees is partly determined by the differential elongation of different orders of branches. In early successional species, the excurrent (narrow) crown form is achieved partly through stronger correlative growth inhibition with apical control over the growth of the branches beneath (Brown *et al.*, 1967; Boojh and Ramakrishnan, 1982a). This crown form of early successional trees is also partly due to rapid extension of the first-order branches at the expense of the second-order branches, when this differential is compared with late successional species (Shukla and Ramakrishnan, 1986). Thus the total length of all the branch systems and the contribution by the first-order branches was more for early successional tropical tree species than for late successional ones (Table 9.4). Conversely, the contribution of the second-order branches was more for late successional species compared with early successional ones. Further, the length classes of the first order branches of early successional species had a wider range and a larger number of greater length classes than did the late successional species (Fig. 9.13). The production of proleptic first- and second-order branches in late successional species over a brief period of time, characterized by increased production and extension of second-order branches results in the development of a broad tree crown (decurrent crown) from where the leaves are peripherally placed. Thus they are able to increase their photosynthetic efficiency under the shade of the forest canopy (Shukla and Ramakrishnan, 1986).

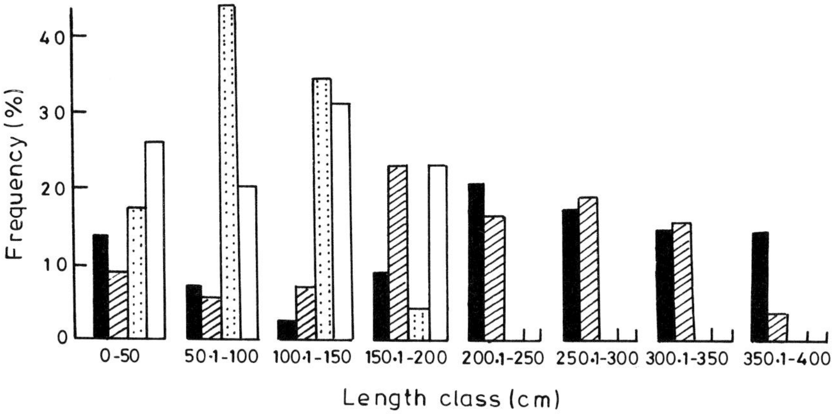

Figure 9.13 Percentage distribution of first-order branches of open-grown trees of different length classes of : *Duabanga sonneratioides* (dark column); *Anthocephalus cadamba* (hatched column); *Dillenia pentagyna* (stippled column); and *Artocarpus chaplasha* (open column) at Lailad in Meghalaya (from Shukla and Ramakrishnan, 1986)

The bifurcation or branching ratio (Rb) which was first used by geomorphologists to quantify drainage patterns of stream basins is based on Horton's (1945) suggestion that in a drainage network the number of segments (streams) of each order forms an inverse geometric series, which is constant throughout a river system. More recently, this concept has been used to characterize the branching networks of biological systems, such as trees (Holland, 1969; Pickett and Kempf, 1980; Boojh and Ramakrishnan, 1982c). Therefore, for the calculation of the bifurcation ratio (Rb) of trees which is, strictly, a quotient – Strahler's (1957) method was followed in which the ultimate branch was designated as first order. Where two branches of unequal orders meet, the resulting branch maintained the higher order. In fact, this kind of ordering of trees is the major weakness in using this (Rb) parameter, for, any ordering system which does not keep track of the actual physical dimensions of a tree and their morphological equivalences may have limited value in understanding mechanical design. Borchert and Slade (1981) have questioned the usefulness of this ratio in explaining adaptive strategies of trees. It completely ignores the information pertaining to architectural development of trees (Hallé *et al.*, 1978), according to which the main axis would represent the lowest order followed by increase in order number with each bifurcation in each branch system. All the same, bifurcation ratios have been used from time to time in evaluating tree-crown form (Oohata and Shidei, 1971; Whitney, 1976; Steingraeber *et al.*, 1979; Pickett and Kempf, 1980; Boojh and Ramakrishnan, 1982a,c; Shukla and Ramakrishnan, 1986). We ourselves, have obtained results of a conflicting nature. While our studies on early versus late successional trees at lower elevations growing under similar light environments suggest that the former category of species are less forked than the latter (Shukla and Ramakrishnan, 1986), the results for the high elevation species do not show such a difference (Boojh and Ramakrishnan, 1982a). We often do not find differences in the forking of open-versus forest-grown trees (Boojh and Ramakrishnan, 1982c). More critical studies are obviously needed to understand the ecological significance of this ratio.

Leaf dynamics related to successional status

Higher plants are built up of modules of structures or iterations of a basic constructional module in which the construction unit in the shoot is the leaf and its axillary bud (Harper and White, 1974; White, 1979). As a plant grows, new modules are added on as old ones die. The growth pattern of trees considered as a metapopulation of parts, therefore involves the quantification of the birth and death of these components.

The leaf, with its axillary bud, is the smallest module of organized structure in higher plants, and leaves have many of the properties associated with members of a population, e.g., they may increase in number exponentially in a rapidly growing plant, they have juvenile, mature and senescent phases, death rates and survivorship curves. Bazzaz and Harper (1977) in the case of *Linum usitatissimum* have shown that the use of demographic techniques to study populations of leaves on a plant is feasible. Such a study may lead to a more comprehensive understanding of plant growth and primary production processes in the canopy (Ramakrishnan, 1987b).

Although individual trees have been considered as aggregates of repeating and somewhat independent units in the pipe model theory of Shinozaki *et al.*, (1964), other authors have considered branches as individuals (Gill and Tomlinson, 1971) or dealt with leaf characteristics such as census (White, 1979), longevity (Gill and Tomlinson, 1971), age-structure (Kinerson *et al.*, 1974) or abscission rates (Kozlowski, 1973). However, comprehensive demographic studies of leaves in a canopy have received little attention (Boojh and Ramakrishnan, 1982b; Shukla and Ramakrishnan, 1984a).

Leaf area available for photosynthesis plays a crucial role in biomass production, the change in the total leaf area being primarily determined by the patterns of production and fall, and the longevity of leaves (Watson, 1956; Newhouse and Madgwick, 1968). Seasonal or other variations in leaf size will also affect the total photosynthetic area (Smith and Nobel, 1977). Several studies on leaf expansion of temperate angiosperm trees suggest variations based on the type of shoot and the environment in which they are exposed (Kozlowski, 1971). Much less is known about the leaf dynamics in tropical trees. While temperate angiosperm trees characteristically shed their leaves during autumn, tropical trees show a wide variety of patterns of leaf production, longevity and fall (Holttum, 1940; Alvim, 1964; Longman and Jenik, 1974). Because of the implication of light in leaf dynamics (Addicott and Lyon, 1973), strategies found in trees over a successional gradient might be expected to differ. It is in this context the leaf dynamics of the sub-temperate high altitude tree species at Upper Shillong in Meghalaya (Boojh and Ramakrishnan, 1982b) and low elevation tropical tree species at Lailad in Meghalaya (Shukla and Ramakrishnan, 1984a) were evaluated over a successional gradient.

Leaf production and fall

In the low elevation tropical trees, the monthly leaf production, as a percentage of the total number of leaves produced in a year was different on different axes and between early and late successional trees (Fig. 9.14). Early successionals had more steady leaf production throughout the year, whereas the late successionals had a major peak during April and June on the main axis and the first-order branches. Such a continuous and steady production of new leaves in early successional species, also noted for the high elevation sub-temperate trees of this category, may be significant in maximizing photosynthesis under the high irradiance of the early successional environment. While the late successional low elevation tropical species had more restricted leaf production during the year, the high elevation sub-temperate species of this category had even more restricted two short flushes of growth in the open, and just one single short flush of leaf growth in the shade (Boojh and Ramakrishnan, 1982b).

The apparent plastochron in days, which is a measure of the rate of leaf production (Gill and Tomlinson, 1971) shown for the leader axis (Table 9.5) suggest that it was least during July and August for eary successional tropical species. The shorter apparent plastochron in the early phase of the shorter growth period of late successional species results in a larger leaf production on the tree right at the beginning, and may be significant for optimizing photosynthesis in the low light regimes within the forest.

The leaf fall in the tropical trees occurred during the dry period of March–April (Fig. 9.15). The peaking was more pronounced in the late successional species than in the early successional ones, followed by a sharper decline in the former than in the latter.

Leaf demography

If one were to do a total demographic life table analysis as was done for the sub-temperate trees (Table 9.6), then the percentage survival of leaves was higher for the late successional species with a generally higher turnover time for the forest-grown late successionals compared to open-grown early successionals. Even amongst open-grown individuals of early and late

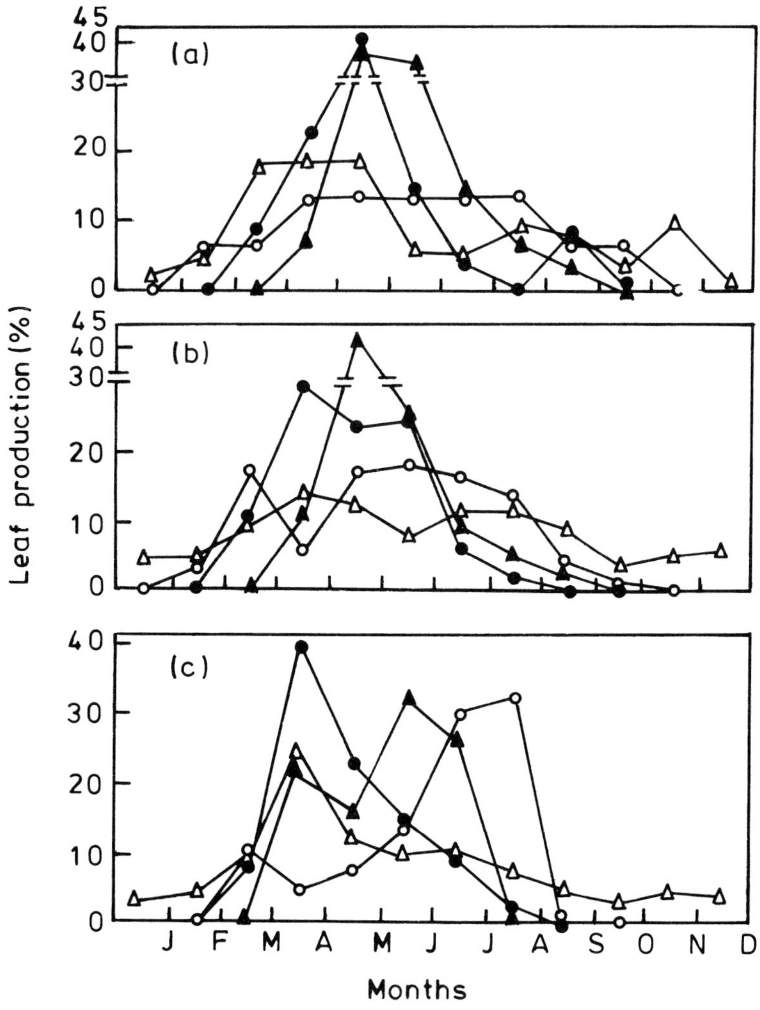

Figure 9.14 The pattern of leaf production in early successional (open symbols) and late successional (closed symbols) species over the growth period, expressed as percentages of total number of leaves produced in a year on : (a), main axis; (b), first-order; and (c), second-order branches of open-grown individuals. Early successional *Duabanga sonneratioides* (△) and *Anthocephalus cadamba* (○); and late successional *Dillenia pentagyna* (▲); and *Artocarpus chaplasha* (●) from Shukla and Ramakrishnan, 1984a)

successional species, as at the low elevation tropical trees, clear-cut differences in the leaf longevity are found, the former category of species having a shorter life span than the latter category (Fig. 9.16).

Fig 9.17 shows the net population size with time and the cumulative gains and losses that determine them, along with their seasonal patterns.

Table 9.5 Apparent plastochron (in days) (± SE) for the main axis leaves of early and late successional open-grown trees at Lailad in Meghalaya (from Shukla and Ramakrishnan, 1984a)

Month	Dubanga sonneratioides	Anthocephalus cadamba	Dillenia pentagyna	Artocarpus chaplasha
J	6.2 ± 1.2	–	–	–
F	3.5 ± 0.4	–	–	–
M	2.8 ± 0.2	23.0 ± 0.8	–	–
A	3.3 ± 0.4	16.0 ± 0.5	–	3.8 ± 0.4
M	2.3 ± 0.4	17.0 ± 0.6	4.4 ± 0.2	3.9 ± 0.4
J	1.9 ± 0.2	19.0 ± 0.8	6.0 ± 0.4	5.0 ± 0.4
J	2.8 ± 0.3	11.0 ± 0.8	5.2 ± 0.5	6.2 ± 0.9
A	2.2 ± 0.2	13.0 ± 0.5	4.4 ± 0.3	15.5 ± 0.7
S	3.3 ± 0.3	20.0 ± 0.9	15.0 ± 0.4	15.0 ± 1.1
O	3.4 ± 0.4	24.0 ± 0.8	31.0 ± 1.6	15.5 ± 1.6
N	6.0 ± 0.4	28.0 ± 1.2	–	–
D	7.8 ± 0.6	–	–	–

– denotes a month without leaf production

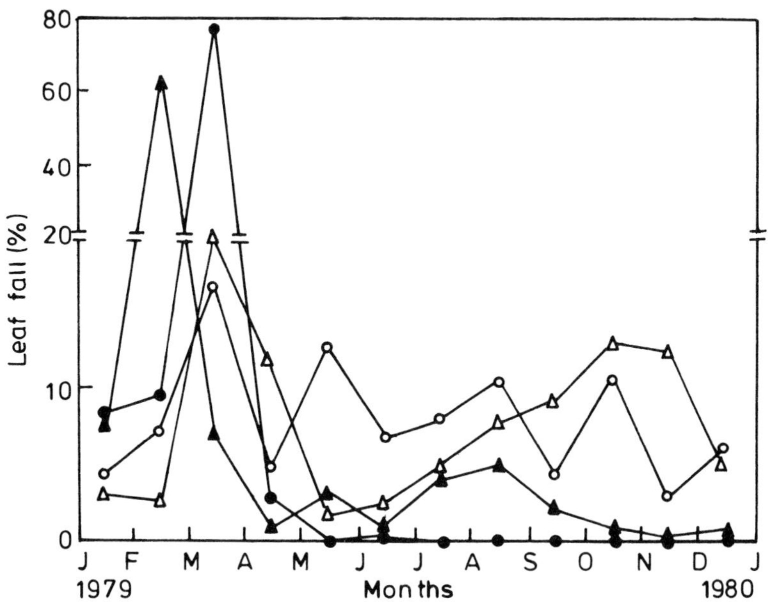

Figure 9.15 The pattern of leaf fall over the year expressed as the percentage of total leaves fallen in a year from the early successional *Duabanga sonneratioides* (△) and *Anthocephalus cadamba* (○); and late successional *Dillenia pentagyna* (▲); and *Artocarpus chaplasha* (●) (from Shukla and Ramakrishnan, 1984a)

The flux was more pronounced in the early successional species, as observed for the high elevation sub-temperate trees, than for the late successionals. This was also reflected in the survivorship curves (Boojh and Ramakrishnan,

Table 9.6 Population flux of leaves of early and late successional tree species namely *Alnus nepalensis* (An), *Schima khasiana* (Sk), *Quercus dealbata* (Qd), *Q. griffithii* (Qg) and *Machilus kingii* (Mk) at upper Shillong in Meghalaya (from Boojh and Ramakrishnan, 1982b)

	Early successional		*Late successional*					
	An	*Sk*	*Qd*		*Qg*		*Mk*	
	Open	*Open*	*Open*	*Shade*	*Open*	*Shade*	*Open*	*Shade*
(a) No. of leaves per tree April 1979	320	97	354	198	55	96	226	181
(b) No. of leaves per tree April 1980	331	129	356	228	70	108	286	201
(c) Net change (b – a)	±11	±32	±2	±30	±15	±12	±60	±20
(d) Rate of increase (b/a)	1.03	1.33	1.01	1.15	1.27	1.12	1.26	1.11
(e) No. of leaves produced between April 1979 and April 1980	647	1,167	225	114	169	78	206	92
(f) Total No.of leaves lost between April 1979 and April 1980	626	1,160	229	93	169	75	155	79
(g) Leaves present April 1979, alive by April 1980	9	0	204	166	0	21	71	102
(h) Percentage survival in (a), (× 100 g/a)	2.81	0	57.63	83.84	0	21.87	31.42	56.35
(i) Expected time for complete turnover (year) (1/100 – h × 100)	1.03	1.00	2.36	6.19	1.00	1.28	1.46	2.29
(j) Total leaves recorded during study	842	1,192	500	251	169	133	361	220
(k) Percentage annual mortality of leaves (f/j × 100)	74.35	97.31	45.80	37.05	100	56.39	42.94	35.91

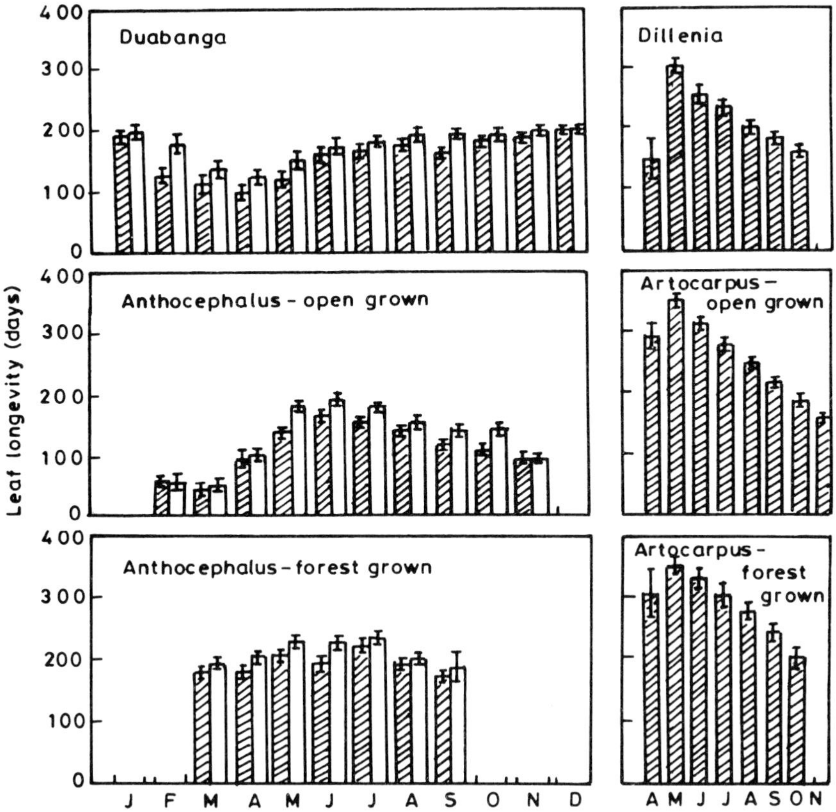

Figure 9.16 Changes in longevity of leaves according to the month of emergence on main axis (hatched bars) and first-order branches (open bars). Vertical lines indicate SE values (from Shukla and Ramakrishnan, 1984a)

1982b) of the Deevey Type I (Deevey, 1947), mortality being very slight during early life but increasing with age.

Based upon the leaf demographic analysis, the age structure of the living and dead leaves was calculated for early and late successional sub-temperate trees (Fig. 9.18). The early successional trees had a narrow range in age structure compared to the late successionals which had a wider age class present. Within the late successionals, open-grown *Quercus dealbata* and *Machilus kingii* had wider age classes present, but with a larger proportion originating prior to the study period. Similar results were obtained for the dead leaves (Boojh and Ramakrishnan, 1982b).

Leaves change their activity with age; in particular, their photosynthetic activity rises to a plateau and then declines. Hence the population of leaves that constitute a canopy should have a function determined by the age structure and age-specific photosynthetic activity of the leaves. Thus, apart

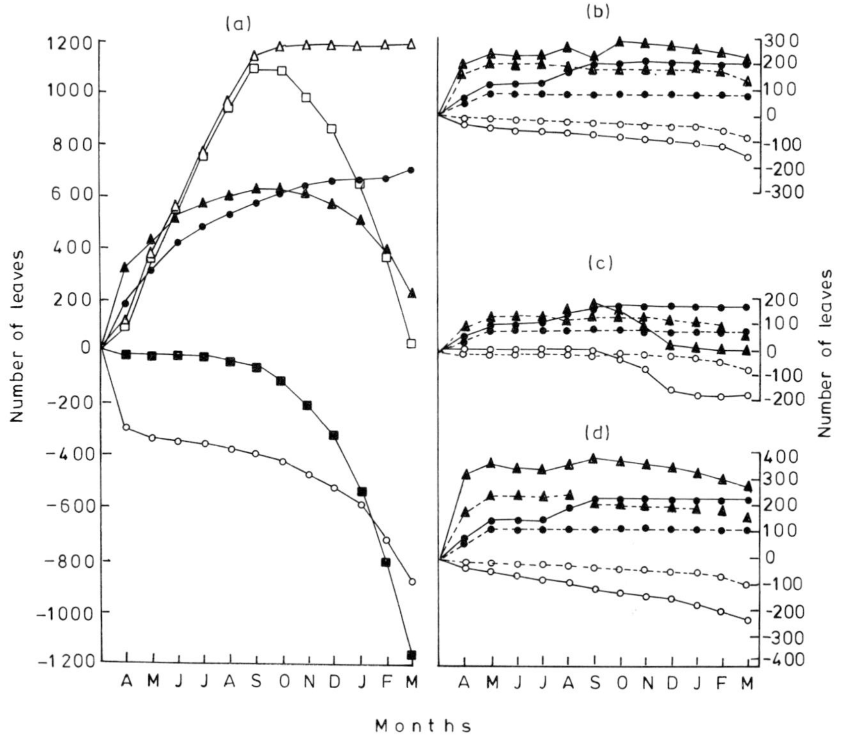

Figure 9.17 Flux of leaf modules in early and late successional tree species. Early successional (a): *Alnus nepalensis* (●, cumulative gains; ○, cumulative losses; ▲, net population size); and *Schima khasiana* (△, cumulative gains; ■ cumulative losses; □, net population size); and late successional : *Quercus dealbata* (b); *Quercus griffithii* (c); and *Machilus kingii* (d) (●, cumulative gains; ○, cumulative losses; ▲, net population size). Solid line, open-grown; broken line, forest-grown trees (from Boojh and Ramakrishnan, 1982b)

from having more leaves of younger age groups, early successional species have fewer age groups, as mutual shading of a large number of categories would reduce their performance. The late successional species, being shade adapted and having a monolayer arrangement of their leaves (Horn, 1971), pack leaves of various age groups in their crown, and this is more pronounced in the forest-grown trees of this category.

With greater fluxes of births and deaths, the net leaf population size of early successional tree species was greater during the favourable growing season (Table 9.6). Such a large needle population flux and size was also noted for the early successional pine (*Pinus kesiya*) at higher elevation of Meghalaya, the needles produced in three flushes in a year, having less than a year of life span (Das and Ramakrishnan, 1986). These species have thus developed a competitive attribute in the form of large leaf populations. This in turn depends upon an extended period of photosynthetic activity

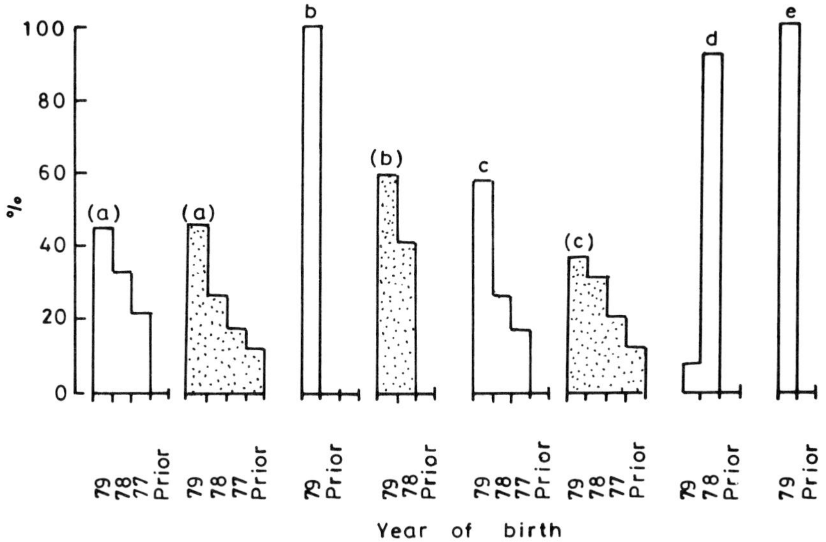

Figure 9.18 Age structure of living leaf population for late successional species : *Quercus dealbata* (a); *Quercus griffithii* (b); and *Machilus kingii* (c); and early successional : *Alnus nepalensis* (d); and *Schima khasiana* (e). Open column, open-grown; and stippled column, forest-grown trees (from Boojh and Ramakrishnan, 1982b)

under conditions conducive to high productivity. In late successional species, the lower flux of leaf modules, especially in forest situations, can be attributed to their rapid adjustment in growth in response to local depletion in resources arising during competition. Short-term experiments involving shading have also shown such adjustments (Grime and Jeffrey, 1965; Loach, 1970). An aspect that has received little attention in relation to shading effects is the problem of heterophylly in tropical trees and the demography of leaves of different types. A preliminary study on a late successional species, *Artocarpus chaplasha*, suggested that leaf-lobing is related to the degree of shading (Shukla and Ramakrishnan, 1981c). Lobing may increase the total leaf area exposed to light and light penetration through the canopy. Thus a larger number of deeply lobed leaves were found under shade than in the open, even on the same tree. More detailed demographic analysis of species with heterophyllous leaves is obviously needed.

Leafing characteristics

Leafiness considered on the basis of leaf production and leaffall follow certain general patterns for the early and late successional species. A climax sacred grove forest at Upper Shillong (1,900 m altitidude) in Meghalaya had an average leaf life expectancy of 20 months for dominant tree species, the life span extended up to 30 months in late successional *Quercus* spp. (Boojh and Ramakrishnan, 1982 f). This forest is predominantly evergreen in nature. We do often find that a large number of tropical early successional species are evergreen and many late successionals are deciduous (Shukla and Ramakrishnan, 1984a). However, many very late successional climax species of tropical and sub-temperate forests are also evergreen. Tropical trees such as *Michelia punduana*, *Garcinia anomala* and *Mesua ferrea* are evergeen climax species (Ramakrishnan, 1986a). But there is a major qualitative difference between the evergreenness of early successionals and climax species. The evergreen nature of early successional species is a consequence of reduced longevity and faster turnover rate of leaves whereas the evergreen habit of the late successionals is due to the extremely long leaf life span and a slower turnover rate.

Biomass allocation strategy related to successional status

In a general comparative growth analysis of 22 tropical tree species at Lailad (296 m altitude) in Meghalaya (Ramakrishnan and Shukla, 1982), 4–6-year-old trees were evaluated. Of these trees, six were early successional light demanders, seven were late successional shade tolerants and nine were mid-successional species of intermediate shade tolerance. Apart from the fact that the shoot growth (extention and girth) rate and duration decreased for the mid- and late successional category of species, root characteristics also differed. Early successionals had superficial spread, with roots occurring mostly in the upper 20 cm of the soil profile, and a less distinct tap root. The mid-successionals had a distinct tap root but with considerable lateral spread, while the late successional species showed a strong and deep tap root system with little lateral spread.

The shoot growth of early successionals was continuous, but the mid- and late successionals showed distinct flushing separated by a rest period, ranging from 1 month in *Chikrassia tubularis* to over 2 months in *Actinodaphne angustifolia*. The number of flushings also varied from one to four, as discussed earlier. The number of leaves per flush also ranged from four to five in *Amoora wallichii* to 14 in *Myristica linifolia*. This number was almost constant for each flush of a given species. The leaf life span also increased with increase in the successional status of the tree species, as discussed earlier.

An important aspect that is of value in predicting the success of the tree species in its successional environment is the relative apportionment of biomass (MacArthur and Wilson, 1967; Abrahamson and Gadgil, 1973). Whilst many of these studies are on herbaceous species, a few studies available on trees are for temperate species (Horn, 1971; Marks, 1975). Information on tropical trees is from the studies arising from the secondary

Table 9.7 Mean (± SE) belowground and aboveground productivity of plant^{-1} yr^{-1} of tree species divided into three distinct successional status at Lailad in Meghalaya (from Ramakrishnan and Shukla, 1982)

Species	*Belowground productivity*	*Aboveground productivity*
Early successional		
Sapium baccatum	82.0 ± 2.31	518.8 ± 38.46
Sterculia villosa	74.4 ± 3.16	430.8 ± 12.46
Zanthozylum rhetsa	71.8 ± 2.08	398.4 ± 16.48
Hibiscus macrophylla	77.2 ± 3.24	539.2 ± 24.56
Litsea sebifera	86.4 ± 5.78	392.3 ± 16.45
Vitex oltissima	78.5 ± 5.24	304.0 ± 24.68
Mid-successional		
Amoora wallichii	83.1 ± 4.21	384.9 ± 13.24
Dysoxylum procerum	69.6 ± 5.24	265.7 ± 8.26
Chikrassia tubularis	88.3 ± 8.24	447.1 ± 28.26
Castanopsis indica	74.2 ± 3.04	398.6 ± 29.08
Englehardtia spicata	71.5 ± 3.28	308.0 ± 31.45
Elaeocarpus robustus	82.2 ± 9.25	324.0 ± 21.60
Myristica linifolia	89.5 ± 8.48	464.2 ± 36.84
Actinodaphne angustifolia	74.5 ± 6.24	402.4 ± 24.20
Cinnamomum spp.	83.2 ± 8.04	308.5 ± 42.14
Late successional		
Saprosma ternatum	86.4 ± 3.01	244.4 ± 22.46
Michelia punduana	71.2 ± 5.62	234.2 ± 16.45
Goniothalamus simonsii	66.5 ± 5.05	204.2 ± 30.18
Garcinia anomala	96.5 ± 4.62	320.0 ± 24.20
Millusa roxburghiana	62.4 ± 4.45	184.0 ± 16.48
Travesea palmata	38.08 ± 3.68	84.8 ± 8.54
Mesua ferrea	56.8 ± 2.04	138.2 ± 12.04

successional jhum fallows in north-eastern India (Ramakrishnan and Shukla, 1982; Shukla and Ramakrishnan, 1984b).

On the basis of a comparative analysis of the 22 trees species studied at Lailad in Meghalaya, considering 1- to 5-year-old saplings, (Ramakrishnan and Shukla, 1982), we were able to show that the belowground and aboveground productivity of the late successional trees was generally lower than that of early and mid-successional species (Table 9.7), an observation also confirmed through another study (Shukla and Ramakrishnan, 1984b). However, the belowground productivity when compared to the aboveground productivity (belowground/aboveground ratio) was relatively higher for late successional species when compared to early successional ones. This is also reflected in the biomass allocation strategy differences based on successional status, and confirmed through another detailed comparative analysis of two early and late successional species (Fig. 9.19). The percentage

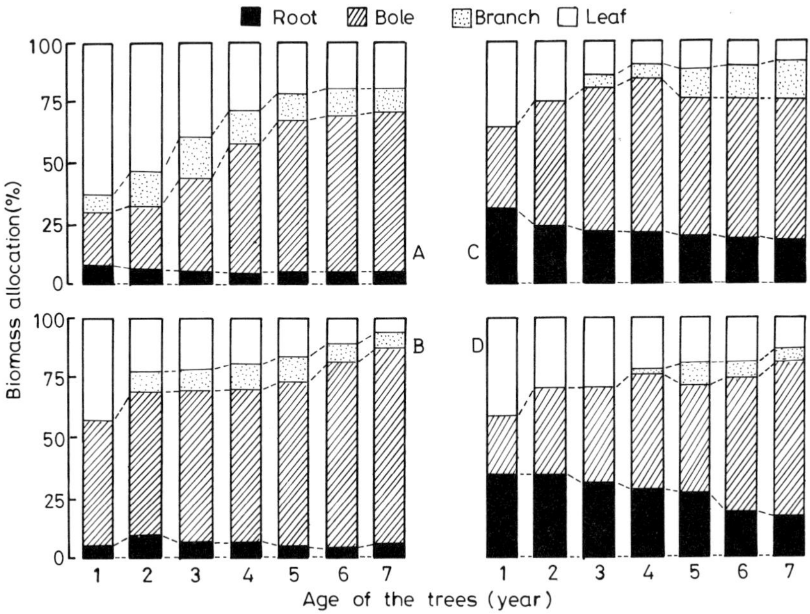

Figure 9.19 Allocation pattern of biomass to four different components, viz., root, bole, branch and leaf, of early successional (A, *Duabanga sonneratioides*; B, *Anthocephalus cadamba*) and late successional (C, *Dillenia pentagyna*; D, *Artocarpus chaplasha*) species of different age groups (Shukla and Ramakrishnan, 1984b)

biomass of the total allocated to the roots of early successionals was lower (3–8%) than that to late successionals (12–21%), and this allocation declined with age of the tree for late successionals only. Allocation to the branches was generally greater and started earlier from the 1st or 2nd year of growth in the case of early successionals and started off later in the case of the late successionals. Allocation to the bole component was also higher for the early successionals than for the late successionals. Consequently, allocation to the shoot by the early successional species was greater than that made by the late successional species. The allocation to the leaf component declined and that to the bole increased with age in all the species; this was, however, more pronounced for early successionals. Greater allocation of biomass to the bole by early successionals helps them to raise the canopy for maximum light exploitation. Consequently, the shoot/root ratio of the early successional species was consistently higher than that for late successional species (Table 9.8). The two categories of species showed some degree of stability in the shoot/root ratio (Wareing, 1970) in spite of fluctuations with age. However, the markedly higher values for this ratio

Table 9.8 Changes in mean shoot/root biomass ratio with increase in age for early and late successional tree species at Lailad in Meghalaya (from Shukla and Ramakrishnan, 1984b)

	Early successional		*Late successional*	
Age (yr)	Duabanga sonneratioides	Anthocephalus cadamba	Dillenia pentagyna	Artocarpus chaplasha
1	12.35	20.41	2.17	1.92
2	15.15	10.10	3.20	1.95
3	16.40	14.08	3.70	2.33
4	22.22	13.51	3.70	2.56
5	18.18	18.86	4.00	2.70
6	17.85	22.22	4.55	4.55
7	16.95	16.13	4.76	5.00

Table 9.9 The ratio of leaf area or leaf weight per unit branch weight (± SE) (after Shukla and Ramakrishnan, 1984a)

Species	*Leaf area/ branch weight*	*Leaf weight/ branch weight*
Early successionals		
Duabanga sonneratioides	272.08 ± 36.20	2.71 ± 0.38
Anthocephalus cadamba	242.93 ± 36.00	2.30 ± 0.31
Late successionals		
Dillenia pentagyna	152.30 ± 23.40	1.30 ± 0.19
Artocarpus chaplasha	118.60 ± 21.60	1.22 ± 0.08

in early successionals suggest that they tend to maximize allocation to the shoot system in order to produce a canopy as rapidly as possible and to exploit the high light energy environment of early successional communities. If one considers the leaf as the source and the branch as the immediate sink (Montenegro *et al.*, 1979), their biomass ratio (leaf/branch) for early successional tropical trees is different from the late successionals (Table 9.9). The markedly higher ratio for early successional species is again a strategy for highest photosynthetic rate capitalizing upon the high light regime in the forest.

A higher biomass and productivity for early successionals is to be expected because of faster extension growth rate, larger leaf area and faster leaf turnover rates, as discussed earlier. However, greater productivity in younger trees of early successionals followed by a decrease beyond 5 years, but continued exponential growth in biomass in the case of late successionals is in accordance with the results from Zaire where pioneers have declining growth rates after 5–8 years (Leburn and Gilbert, 1954).

The early successional species had a higher proportion of the biomass of their root system (80% of the total root biomass in 7-year-old early successionals) in surface layers of the soil profile up to a depth of 20 cm

(Fig. 9.20). Although this may offer poor physical support to the shoot system, it probably provides enough absorptive surface to exploit the short-term increase in nutrient and water availability associated with early successional disturbed sites (cf Chapter 5 and 7). On the other hand, more uniform distribution of the root system through the soil profile in late successional species and the higher proportion of the total biomass at depths greater than 20 cm may reflect their ability to draw nutrients from deeper layers of the soil over a longer period of time.

The overall growth form related to changes in leaf area index (LAI) clean bole length, crown depth and maximum crown width (Fig. 9.21) may all affect the light interception strategy of a tree. Leaf are index is an important measure of the active leaf area of a given species. The high LAI of the late successional species compared to early successionals in younger age classes up to 3 years is due to rosette clustering of leaves on the unbranched sapling of the former, in contrast to the early branching initiated in the latter group of species. Differences in LAI in older age classes may be explained by the greater extension growth of lateral branches and consequently more sparse distribution of leaves over a large ground area in early successionals compared to the late successionals. This results in lower LAI for early successional species even though they have higher total leaf area compared to late successional species of the same age.

The greater crown width of the early successional species is due to more rapid extension growth of the plagiotropic branches. This, with the lesser degree of forking in early successional species, may lead to more open canopies which contrast with the late successional species where the degree of forking of the branches is greater (Shukla and Ramakrishnan, 1986) and the clean bole length, crown depth and maximum crown width are less.

Conclusions

The patterns of growth and architectural development of early and late successional species are well suited to their strategy of niche occupancy in the forest ecosystem. The shade intolerant early successional species chiefly through sylleptic branch production and rapid leaf production achieve more extension growth than the late successional ones which show proleptic branch production and growth over a brief period. Thus, the early successional trees are able to exploit a high light regime in the forest, making quick growth before light becomes limiting. However, the differences observed between open-grown and forest-grown late successional species are suggestive of their ability to adjust their growth strategy, at least to some degree, depending upon light availability; this category of species although adapted to grow under low light regimes, has the ability to adjust

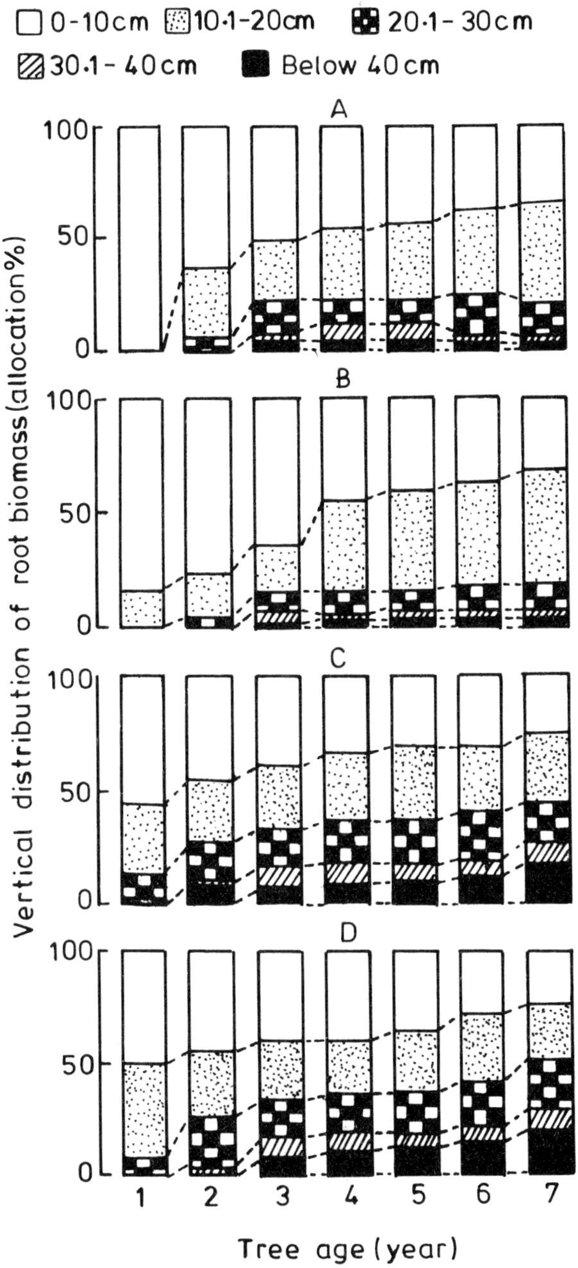

Figure 9.20 Allocation pattern of root biomass down soil profiles to below 40 cm in early successional *Duabanga sonneratioides* (A); and *Anthocephalus cadamba* (B); and late successional *Dillenia pentagyna*, (C); and *Artocarpus chaplasha*, (D) at Lailad in Meghalaya (from Shukla and Ramakrishnan, 1984b)

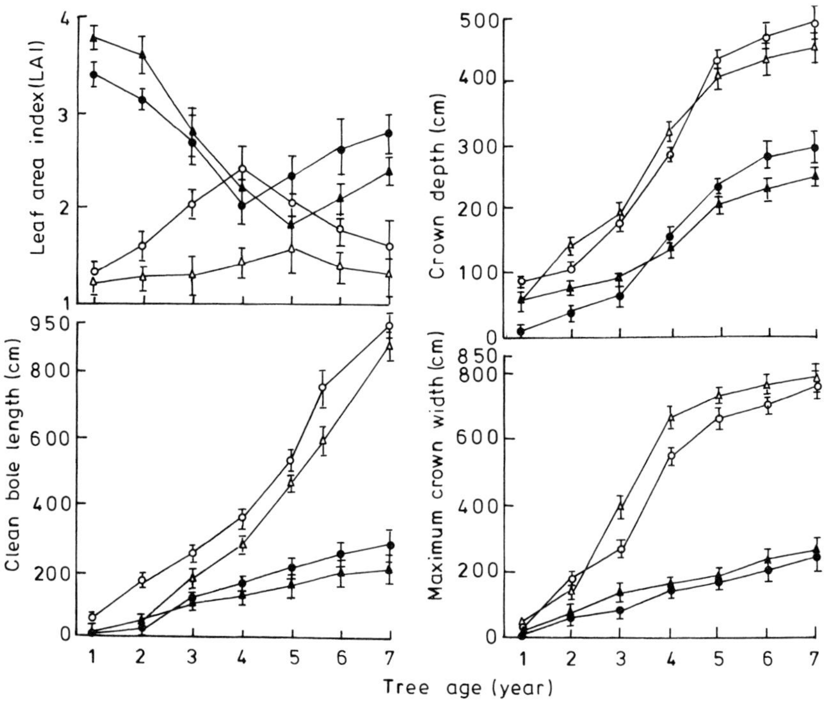

Figure 9.21 Changes in leaf are index (LAI) (a); clean bole length (b); crown depth (c); and maximum crown width (d) of early successional (open symbols) and late successional (closed symbols) species with tree age △, *Duabanga sonneratioides*; ○ , *Anthocephalus cadamba*; ▲, *Dillenia pentagyna*; ●, *Artocarpus chaplasha* (from Shukla and Ramakrishnan, 1984b)

its growth pattern favourably under open situations such as openings in the forest canopy.

Self-pruning of the first-order branches is better developed in early successional than late successional species, perhaps due to shading, resulting in a faster upward growth shift of the canopy. This, together with the greater extension of the first-order branches of the early successionals, when compared with the second-order branches results in a an excurrent (narrow) crown form. By comparison, the proleptic branching pattern of late successionals over a brief period of time, characterized by increased production and extension of second-order branches results in the development of a broader tree-crown (decurrent crown form) where the leaves are more peripherally placed. Thus, they are able to increase their photosynthetic potential under the shade of the forest canopy.

With faster production and fall of leaves of early successional species, the leaf turnover rate is higher than that of the late successional species. Thus, the rapid growth of early successional trees seems to be more related to their ability for unrestricted leaf production than to efficient energy conversion per unit leaf area.

The biomass allocation strategy differences between early and late successional trees are also critical for the success of these two categories of species in their respective environments. Early successionals with greater biomass allocation to the aboveground parts, greater apportioning of biomass to the bole component which have faster self-pruning lower branches (first-order) on them, have an exploitative strategy for optimizing photosynthesis under high light environments, before this factor becomes limiting. This exploitative strategy of early successionals is also reflected in the allocation of biomass to the root system compared to the shoot component, in a manner that the superficially-placed roots are able to exploit a temporarily nutrient-enriched surface soil layer after slash-and-burn under jhum (cf Chapter 5). The late successionals with slow growth, peripherally placed leaves on the skeletal frame of the tree and slow turnover rates for leaves, are geared to maximize production under a low light regime of a late successional environment where conservation of resources is critical. Such a 'conservative strategy' of late successionals is also reflected in the greater allocation to the deep penetrating root system compared to the shoot, so that these species are able to obtain nutrients from deeper soil layers through the better developed tap root system.

The diverse strategies of early versus late successional trees are compatible for exploiting the diverse light regimes available at different canopy positions in a mixed plantation forestry. In addition, native early successional tree species could form the basis for agroforestry systems. At this point it is pertinent to note sharp differences in the growth characteristics of two closely related early successional species of *Schima* – *Schima wallichii* and *Schima khasiana*, the former having a faster growth rate than the latter (Boojh and Ramakrishnan, 1983a). Indeed, within the same species *Schima wallichii*, we could identify two altitudinal populations (Boojh and Ramakrishnan, 1982d), a low elevation ecotype (100 m altitude in Meghalaya) which had double the growth rate of the high altitude population (1,600 m altitude), suggesting thereby that such differences between species and within species could be exploited for agroforestry and plantation forestry systems. In doing so, it is important to consider the total biology of the species starting with seed germination and establishment characteristics, (Boojh and Ramakrishnan, 1981a,b, 1982e; Shukla and Ramakrishnan, 1981a,b, 1982a) aspects that have received little attention.

Species compatability is ultimately reflected in community organization and function. This is evident from the two studies on high elevation sub-temperate climax vegetation at Shillong (1,900 m altitude) (Boojh and Ramakrishnan, 1981, 1982) and a low elevation tropical forest at Lailad

(296 m altitude) (Shukla and Ramakrishnan, 1982) in Meghalaya. Correlation of phenological characteristics with niche separation in a community may best be illustrated through leaf production and fall patterns. Thus, the leaves in the understorey are retained over a longer time period compared to overstorey trees, because of more homogeneous environmental conditions with less risk in the understorey environment. For similar reasons, many overstorey species shed their leaves earlier in the season compared to understorey species. In the tropical forest at the lower elevation, a large proportion of overstorey species exhibit dry season leaffall–dry season flushing while the understorey species generally have dry season leaffall–wet season flushing. The continuous leaffall flushing strategy of leaf replacement occurred only for a few early successional gap species in the forest, such as *Anthocephalus cadamba* and *Duabanga sonneratioides*. The former two strategies that predominate here have been suggested to be adaptive to small seasonality in moisture regime (Jackson, 1978). Indeed such a temporal strategy differential is even reflected in flowering and fruiting behaviour. Thus, in the tropical climax forest, peaking in fruiting that occurred, had predominantly species producing fleshy fruits with large seeds that are animal dispersed. The dry season fruiters largely produced small winged seeds whose dissemination is favoured by wind during the dry summer. It may be concluded that the species organization of the community with an assemblage of different patterns of phenophases fit well for a balanced ecosystem function. Indeed, such a diversity could best be ensured in man-made forest ecosystems through use of mutually compatible tree species mixtures, as suggested through our tree growth strategy and architecture analysis.

GROWTH STRATEGY AND ARCHITECTURE OF SHRUBS DURING SUCCESSION

In our studies on the shrub stratum of the secondary successional jhum fallows at Lailad in Meghalaya (296 m altitude), we considered two early successional species (*Melastoma malabathricum* and *Mussaenda frondosa*) and two late successional species (*Litsaea khasiana* and *Oxyspora vagans*). The general conclusions arising from these studies on shrub growth strategy, architecture and leaf dynamics (Baruah and Ramakrishnan, 1989a,c) are essentially similar to the studies on trees discussed earlier. Therefore, we shall consider for this discussion, salient differences only.

Growth and architecture

Of the four species considered here, the three species other than *Litsaea khasiana* belong to Koriba's model, although with important departures in flower organization from the original one of Hallé *et al.*, (1978). While resembling Rauh's model in many features, *Litsaea khasiana* is significantly different from it because of its sympodial growth which it shares with Koriba's model. Further, this species does not conform to Massart's model referred to by Hallé *et al.*, (1978) with respect to another closely related *Litsaea sebifera*. These results confirm that the models of Hallé *et al.*, (1978) could at best be considered only as a useful set of reference points with many intergrading forms. With morphometric features similar to that of trees, the early successional shrubs follow an exploitative strategy to optimize production under a higher light regime through quick growth; the late successional shrubs have a conservative strategy for survival under shade. However, with light availability being more restricted for the early successionals at the shrub stratum of the forest and with deeper shade at this level in the late successional forests, when compared with the tree stratum, the differences between these two categories of shrubs are more pronounced than for trees.

Leaf dynamics

The leaf dynamics of the two shrubs categories are broadly similar to that of the trees. However, the leafing behaviour of these tropical shrubs, (Baruah and Ramakrishnan, 1989c,e) differ from that for tropical trees discussed earlier (Shukla and Ramakrishnan, 1984a). The four species studied in detail here, as well as phenological observations on over 70 shrub species in successional forests of 5–60 years of age (Baruah and Ramakrishnan, 1989e), suggest that early successionals here are predominantly deciduous or leaf-exchanging types, and that the late successionals are largely evergreen. This contrasts with trees changing from evergreen of the fast leaf turnover type, to deciduous, and again to evergreen of the slow leaf turnover type during succession, discussed earlier. Such a contrast between trees and shrubs may be related to micro-environmental differences for light at upper and lower canopy positions in the forest. These results also signify that evergreen versus deciduous habit in the tropics is often the result of a complex of interacting factors.

Allocation strategy

The biomass production and allocation strategies of early versus late successional shrubs generally followed the same pattern as that for trees (Baruah and Ramakrishnan, 1989d). The major point of departure was that related to productivity. Both early and late successional shrubs showed increase in productivity with age of the individual, which was faster in the beginning but continued at a slower rate after about 3 years, in contrast

Table 9.10 Mean monthly (± SE) nutrient uptake efficiency (mg nutrient absorbed per g root dry weight) of early and late successional shrubs in open and shade environments at Lailad in Meghalaya. Values in parentheses are use efficiency values (from Baruah and Ramakrishnan, 1989b)

		Nutrient uptake efficiency		
Species	*Treatment*	*Nitrogen*	*Phosphorus*	*Potassium*
Early successional				
Mallotus indica	Open	137.2 ± 1.3 (67 ± 1)	7.1 ± 0.04 (1,062 ± 15)	122.0 ± 1.0 (82 ± 1)
	Shade	80.9 ± 0.6 (80 ± 1)	4.0 ± 0.02 (1,268 ± 10)	65.6 ± 1.3 (105 ± 1)
Clerodendron infortunatum	Open	94.8 ± 1.0 (82 ± 1)	5.0 ± 0.06 (1,320 ± 14)	91.3 ± 1.4 (97 ± 3)
	Shade	58.5 ± 1.6 (99 ± 2)	3.2 ± 0.01 (1,547 ± 18)	45.0 ± 1.1 (131 ± 2)
Late successional				
Litsaea khasiana	Open	67.3 ± 1.4 (102 ± 2)	4.1 ± 0.02 (1,464 ± 14)	59.4 ± 0.8 (125 ± 3)
	Shade	39.6 ± 0.3 (140 ± 2)	2.2 ± 0.01 (1,909 ± 29)	46.6 ± 1.6 (183 ± 3)
Combretum flagrocarpum	Open	74.8 ± 1.0 (116 ± 2)	4.6 ± 0.03 (1,657 ± 12)	66.6 ± 1.6 (136 ± 3)
	Shade	42.6 ± 0.6 (162 ± 3)	2.4 ± 0.01 (2,273 ± 38)	35.0 ± 0.6 (210 ± 2)

to a differential behaviour observed between early versus late successional trees (Shukla and Ramakrishnan, 1984b). This was related to the generally reduced light availability with a reduced differential in the shrub stratum for both early and late successional shrubs, compared to a larger differential for the tree stratum (Baruah and Ramakrishnan, 1989d).

A comparative study between two early successional (*Mallotus indica* and *Clerodendron infortunatum*) and two late successional (*Litsaea khasiana* and *Combretum flagrocarpum*) shrubs under two light regimes (Baruah and Ramakrishnan, 1989b) showed that the former category of light demanding species is more sensitive to shading than the latter category (Loach, 1970). However, increased biomass production in the open by the shade-tolerant shrubs is perhaps suggestive of their ability to survive under shade rather than to maximize light interception.

The higher nutrient uptake efficiency and the lower nutrient-use efficiency of early successionals and the reverse situation for the late successionals (Table 9.10) suggests a close adaptation to the soil environment in which they occur; the former category occupies a temporarily nutrient-enriched soil and the latter occurs in a nutrient-depleted soil under more competitive conditions. Nutrient-use efficiency of all species improved under shade compared to the open situation, but this improvement was more pronounced for the shade-tolerant late-successional species.

Finally, shrub species such as *Clerodendron infortunatum* occurring in 2- to 25-year-old jhum fallows and *Ficus fulva* occurring in 5- to 60-year-old jhum fallows showed much plasticity in biomass and nutrient allocation strategies (Baruah and Ramakrishnan, 1988b). The vigour of the two species declined with fallow age, perhaps related to reduced light availability and greater competition in older fallows. Higher allocation of resources to the root component in older fallows than in younger ones would help in regeneration of the species after another slash-and-burn cycle. On the other hand, more allocation of resources to the leaf component in younger fallows would be helpful to capitalize upon the high light regime before this resource becomes limiting. The low nutrient uptake efficiency in older fallows is, to a certain extent, compensated by higher nutrient-use efficiency. These results on allocation strategies are similar to those obtained for a herbaceous perennial, *Eupatorium odoratum* (Saxena and Ramakrishnan, 1984a; cf Chapter 9).

Conclusions

The shrub species occurring over a successional gradient in the jhum fallows showed close adaptation to the niche they occupy in the gradient. Starting with germination and seedling establishment stages (Baruah and Ramakrishnan, 1988a), the adaptive characteristics were reflected throughout the life history of the species. Phenotypic plasticity over a successional gradient observed for individual species (Baruah and Ramakrishnan, 1988b) is also reflected in the phenologies of the shrub stratum at the community level (Baruah and Ramakrishnan, 1989e).

A comparative study of 5-, 25- and 60-year-old jhum fallow communities showed three distinct leafing patterns. The predominance of deciduousness in the early successional environment agrees with the presumably more xeric environment prevailing there, when compared to the closed 60-year-old fallow where a majority of the species are largely evergreen (Baruah and Ramakrishnan, 1989e). In agreement with the expected relationship of deciduousness and xeric conditions, maximum leaffall coincided with the dry summer (Frankie *et al.*, 1974). The distinct leafless phase in the deciduous shrub elements of early successional fallows is compensated, to a certain extent, by a larger production of leaves, which peaked early in the growing season so that high productivity is ensured; this contrasts with the late successional evergreen elements. However, in a closed 60-year-old fallow where soil nutrient levels are low (Singh and Ramakrishnan, 1982a) and competition for nutrients is likely to be intense, the evergreen habit of the shrubs may represent an adaptation for retention and internal cycling of nutrients (Woolhouse, 1967). The important point that emerges from this community level analysis is the phenological separation of growth types over a successional gradient.

ARCHITECTURAL PLASTICITY AND GROWTH STRATEGIES OF BAMBOOS IN SUCCESSIONAL ENVIRONMENTS

Bamboos play an important role during early succession in the jhum fallows of north-eastern India, both in determining community structure (cf Chapter 6) and community function such as nutrient cycling (cf Chapter 7). Of all the bamboo species found in north-eastern India, *Dendrocalamus hamiltonii* is one of the predominant ones coming up in large numbers during secondary succession in 10–25-year-old fallows, declining sharply in older fallows. In contrast, *Neohouzeoua dulloa* is more plastic with vigorous growth in the early jhum fallows of up to about 30 years, but occurring in jhum fallows up to 60 years because of its tolerance for some shade. An appreciation of the growth strategy of bamboos is important in evaluation of their success in early successional communities; the two above-mentioned bamboo species with their differential plastic responses would provide the basis for such an understanding. This understanding could form a basis for bamboo utilization. It may be noted here that bamboos constitute a largely neglected group of species, although some qualitive analysis on architectural form is done by Hallé *et al.* (1978). As far as we know, no information is available on their architectural plasticity over a successional gradient except for our own studies at Lailad (296 m altitude) in Meghalaya (Rao and Ramakrishnan, 1988a); even their role in ecosystem properties has yet to receive adequate attention outside the north-eastern India (Toky and Ramakrishnan, 1982b, 1983a,b; Rao and Ramakrishnan, 1987a,b, 1988a, b, 1989; Ramakrishnan, 1989).

Growth and architecture

Neohouzeoua dulloa and *Dendrocalamus hamiltonii* have extensive sympodially-branched underground rhizomes. Each branch 9–14 cm long in the former and 12–24 cm in the latter has a narrow neck which broadens into a thick base (Fig. 9.22). Rhizomatous branches are produced from two of the lateral buds whereas the apical bud would give rise to the aboveground shoot system. With thinner rhizome and aerial branch systems in *Neohouzeoua dulloa*, the clumps are more loosely organized in this species than in *Dendrocalamus hamiltonii*. Hallé *et al.* (1978) established McClure's model to accommodate the basic architecture of bamboos. The essential feature of this model is that all aerial parts are vegetatively determinant in their growth, like an elaborate compound leaf. Although there is much confusion

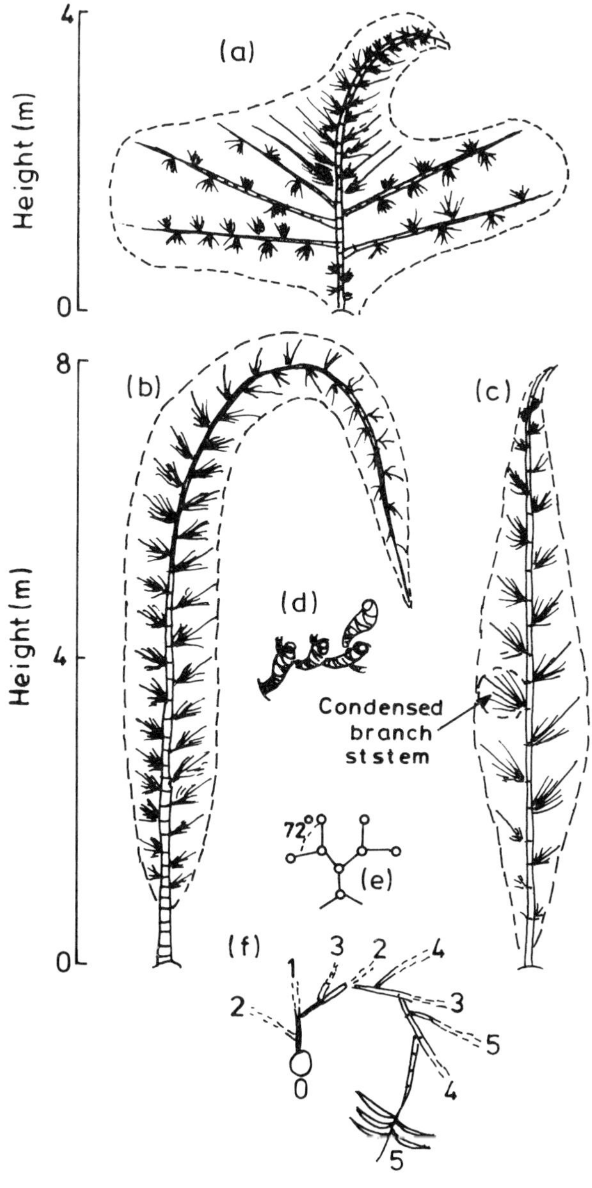

Figure 9.22 Crown architecture of *Dendrocalamus hamiltonii* in a 5-year-old jhum fallow (a); in a 15-year-old fallow (b); and of *Neohouzeoua dulloa* in 15-year-old fallow (c); along with a diagrammatic representation (e), of the branch orders in a branch complement; zero to fifth-order branches are shown along with leafy shoots (f) (from Rao and Ramakrishnan, 1988a)

regarding branch ordering in bamboos, which may include the underground rhizome, for the present analysis we have considered only the aboveground branches, the main axis being designated as the zero-order branch.

The canopy form of *Dendrocalamus hamiltonii* showed greater plasticity depending upon the fallow age. In a 5-year-old fallow, the culm assumes a conical form which is related to more space and light availability. In older fallows where space and light are limiting, the culm had a cylindrical form. Partly because of the shade intolerance in this species and partly because of the intermast phase of about 30 years (Janzen, 1976; Varmah and Bahadur, 1980), this species is eliminated during succession after 25–30 years. *Neohouzeoua dulloa*, on the other hand, with a reported intermast phase of 15–17 years (Janzen, 1976) regenerates through seeds under shade surviving even in older fallows, although with reduced vigour.

The orthotropic zero-order branches of the two species complete their elongation within 90–120 days after initiation through syllepsis in *Neohouzeoua dulloa*, but through prolepsis in *Dendrocalamus hamiltonii*. The sylleptic bud dormancy in the former species was minimal (6% of the total buds) in a 15-year-old fallow, whereas the proleptic buds of the latter had declining bud dormancy with fallow age up to 25 years. Reiteration of the zero-order branch may occur only in *Dendrocalamus hamiltonii*. All the branches on the zero-order are produced and elongation completed, simultaneously. The branch systems on the zero-order are generally of a condensed type, looking as if arising in a cluster.

Leaves on both species are borne on all branches towards their tip, except the zero-order branch. These branches are shed every year and new leafy shoots arise in their place simulataneously in *Neohouzeoua dulloa* (evergreen leaf-exchanging type) but after a dormancy period of up to 60 days in *Dendrocalamus hamiltonii* (deciduous type). Peak growth was attained by the former species in a 15-year-old jhum fallow, but in a 25-year-old fallow by the latter species (Table 9.11).

Leaf dynamics

The early successional shrubs and trees have been shown to have unrestricted leaf production occurring throughout the growing season, as discussed earlier. This, rather than the efficiency of energy conversion is considered important for faster growth rate of this category of species and for exploiting a high light environment. The two bamboo species follow two distinct strategies in this regard (Rao and Ramakrishnan, 1988b). *Neohouzeua dulloa* has unrestricted leaf production throughout the year although the total leaf area attained is lower than that by *Dendrocalamus hamiltonii*. The latter species has a restricted leaf production period of 6 months in a year, but

Table 9.11 Leaf population flux on current year's (CY) and previous year's (PY) ramets of *Nehouzena dulloa* in 5- to 60-year-old successional fallows. (from Rao and Ramakrishnan, 1988b)

	Fallow age (yr)							
	5		*10*		*25*		*60*	
Parameters	*PY*	*CY*	*PY*	*CY*	*PY*	*CY*	*PY*	*CY*
(a) Number of leaves ramet^{-1} at the beginning of study	90	0	169	0	65	0	33	0
(b) Number of leaves produced ramet^{-1} during study period	411	191	1,853	456	531	228	170	90
(c) Number of leaves lost ramet^{-1} during study period	482	127	1,943	281	562	144	189	62
(d) Number of leaves ramet^{-1} at the end of study period	19	64	200	175	82	84	26	28
(e) Total number of leaves ramet^{-1} recorded during study period	501	191	2,022	456	596	228	203	90
(f) Net change (d – a)	–71	64	31	175	17	84	–7	28
(g) Rate of change (d/a)	0.21		1.18		1.26		0.77	
(h) Number of leaves survived in (a) at the end of study period	0		0		0		0	
(i) Expected time for complete turnover (yr)(1/100 – hx100)	1		1		1		1	
(j) Percentage annual mortality (c/e × 100)	96.2	66.5	96.0	61.6	94.3	63.2	93.1	68.9

with a much faster production rate.

Leaf population flux, with maximal production and losses occurred in a 15-year-old jhum fallow for both the species. Whereas all the leaves on *Dendrocalamus hamiltonii* are lost during the year, the leaf population of up to 7% from the previous year's culms and up to 40% from the current year's culms of *Neohouzeua dulloa* are carried over to the following year.

Neohouzeua dulloa was more plastic with limited adverse effects on leaf population dynamics even in a 60-year-old jhum fallow, unlike *Dendrocalamus hamiltonii* which is restricted only to much younger fallows of up to 25–30 years. Whilst evergreen habit with fast leaf turnover rate is advantageous for quick growth of a species such as *Neohouzeua dulloa*, the deciduous habit of *Dendrocalamus hamiltonii* is compensated by faster leaf production and larger leaf area attained before the entire leaf population is shed during the dry season.

Allocation strategy

The biomass and nutrient allocation strategy for these two rhizomatous species generally followed that observed for many rhizomatous herbs such as *Imperata cylindrica* (Saxena and Ramakrishnan, 1983a). With rapid translocation resources occurring from belowground to aboveground components in the younger fallows, the aboveground/ belowground ratio, with respect to biomass and nutrients such as nitrogen and phosphorus, was higher in younger fallows, but declined sharply in older fallows because of rapid retranslocation. Such a strategy would help the survival of these rhizomatous species through another jhum cycle enabling quick growth. The clonally reproducing bamboos represent an extreme stage among the vegetatively reproducing categories of species, with seed reproduction occurring only once during the life cycle.

Conclusions

Bamboos as early successional species have a variety of strategies enabling early colonization after jhum : these light damanding species have quick growth, the initial advantage being through the utilization of reserves stored in the belowground rhizomes, and unlike other species, bamboos have their cell division and extension growth separated into two distinct events, the latter being exclusively through cell elongation of preformed tissue. Consequently, extension growth is remarkably faster, building a canopy that could exploit a high light regime. Internodal elongation starts from the base upwards, with peak elongation occurring during the 5th week as also reported by others (McClure, 1966); also the vegetative reproduction strategy through underground rhizome is an adaptation to disturbances such as fire events, the seed reproduction being highly restricted because of a single flowering event during the life cycle, high predation, rapid loss of seed viability and weaker seedlings derived from very small seeds.

GENERAL CONSIDERATIONS

A major outcome of this quantitative morphometric analysis of tree and bamboo architecture was that we could relate architecture as an adaptive strategy evolved due to natural selection under a given light environment and under a given nutrient flux in the soil. Considering that crown architecture is for optimizing photosynthesis by appropriate leaf display, light which becomes limiting over a successional gradient is obviously one of the key environmental factors. Similarly root architecture relates to soil nutrient flux in the soil. Thus, this study resolves issues related to strategies of species during forest succession and thus contributes to the understanding of the process of succession itself. This approach explains, to some extent, why certain species hold to the ground for a period of time during succession and others replace them subsequently (cf also Chapter 8), a question of continuing interest to ecologists since the concept of succession became the central theme in ecology starting with the classical work of Clements (1916). Relating tree-crown architecture to photosynthetic efficiency and biomass production is also important for a better appreciation of this important physiological function at an organismic level (Ramakrishnan, 1987b), and its evolution both in time and space.

In north-eastern India, as elsewhere in the world, jhum traditionally is an agroforestry system. Here, the farmer after slash-and-burn of the forest carries out cropping, and leaves many of the tree stumps *in situ* for quick regeneration of the forest fallow. These species obviously are early successional light demanders with a narrow (excurrent) crown form and with all the desirable attributes of a fast growing exploitative species discussed in this section. With a rapid shortening of the jhum cycle to about 4–5 years (cf Chapters 1, 6) tree species are unable to regenerate in the jhum fallow plots, due to weed take-over (cf Chapter 4). It is in this context that introduction of appropriate shrub and tree species during the early fallow phase becomes significant. Such an introduction of larger shrub/tree species would have the advantage of weed suppression through shading effect by the introduced species (cf Chapters 4, 6) and building upon soil fertility over a brief period of 4–5 years of fallow phase (cf Chapters 5, 7), so that jhum as a land use could be both ecologically and economically viable, particularly under the shortened jhum cycle as is prevalent now. Bamboo as an agroforestry species could form a peripheral fence around jhum plots to control damage to crops by wild animals and to prevent blow-off of ash liberated during slash-and-burn, a loss that could be as much as about 50% of the total ash produced (cf Chapter 4). Under longer jhum cycles in the past, a good forest cover all around acted as a wind-break to prevent ash loss through wind-blow. With forest rapidly being lost and converted into weedy growth, this function of a wind-break has to be done by an introduced agroforestry species; bamboo is appropriate because it is traditionally used by many tribal communities. A variety of early successional shrubs and trees could be used both during cropping and fallow phases to quickly recover soil fertility over a short fallow phase. Legumes and non-leguminous nitrogen fixers have a particularly crucial role in this agroforestry system. We have already seen the role of an early successional species such as the Nepalese alder (*Alnus nepalensis*) in soil nutrient recovery over a short 5-year period through nitrogen build-up (cf Chapter 5, 7). Indeed, such an early successional species could catch the imagination of the people for use as an agroforestry input into jhum because this species is traditionally conserved by the tribals in their jhum plots. Therefore, they could easily identify themselves with the alder technology. The obvious conclusion is that the social dimension in technology acceptability could often be critical. One may come up with a variety of species that may be biologically useful for agroforestry but all that is offered as technologically sound need not necessarily be socially acceptable.

In a larger context, the right choice of tree species for sedentary agroforestry systems is important. The early successional tree species have a crown architecture that permits maximum light penetration to the ground layer so that the growth and productivity of the crop species would not be limited by light. Also, a fast turnover of nutrients through fast leaf turnover would ensure a constant and steady supply of nutrients in the surface layers of the soil to be used by the crop species. If the root architecture of the tree species used is such that the roots are more uniformly distributed at different soil depths, it would make the tree-crop mixture more compatible. Although early successional species generally have a superficially-placed root system, variations do exist; some mid-successional trees are better suited. Identification of suitable root architecture from this continuum should not be difficult.

Social forestry in a developing country like India should aim at identification of fast growing native species for meeting the fodder, fuelwood and timber needs of the rural population. Use of early successional species and some of the mid-successionals should be appropriate. Apart from fast growth rate, rapid turnover rates for nutrients through leaf turnover on the tree would be an important criterion for restoring nutrient-deficient wastelands. The use of shrubs for wasteland reclamation has not received as much attention as it deserves. Use of appropriate architectural forms with the right growth strategy is important. The earlier discussion sets the basis for such an identification.

Mixed plantation programmes have distinct ecological advantages over monoculture forestry and could be economically as productive or even better than the latter, if compatible species are used. Light-demanding and fast-growing early successional species could form compatible mixtures

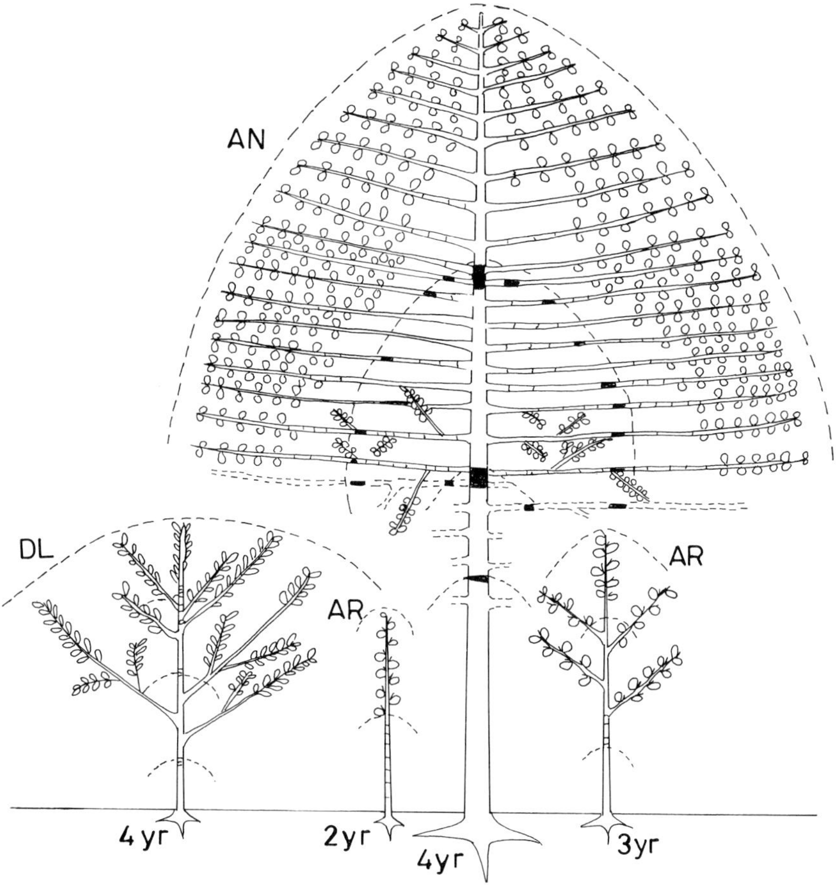

Figure 9.23 Model of mixed tree plantation involving early successional *Anthocephalus cadamba* (AN) and late successional *Dillenia pentagyna* (DL) and *Artocarpus chaplasha* (AR) (from Ramakrishnan, 1986a)

with shade-tolerant mid- or late successional species for exploiting light availability at different canopy levels and thus optimizing production per unit area (Fig. 9.23). Such a mixture would also allow optimum use of nutrients from the soil profile, the early successionals exploiting surface soil layers and the late successionals reaching down to deeper layers. It should even be possible to have a 'condensed succession' for revegetation of damaged ecosystems by an appropriate mix of different categories of trees and shrubs that are mutually compatible. The application of the concept of tree architecture and growth strategy analysis of shrubs and trees and their future elaboration for forestry management is immense.

SECTION 3

Management Implications

CHAPTER 10

JHUM AND RAIN FOREST ECOSYSTEM REDEVELOPMENT

INTRODUCTION

Agriculture, forestry and fisheries have been traditional activities in the Asian tropics; the first two are particularly emphasized by the tribals of north-eastern India. Fisheries in the north-east form either part of the restricted valley land wet rice cultivation systems or fish is collected from the fast-flowing rivers. In the recent past, increased population size has brought considerable pressure on the land and water resources. Forested ecosystems, in particular, have been one of the major casualities, because of rapid conversion of forests to meet the food, fodder and timber needs of the rapidly increasing human population, and industrial activities such as mining, and energy generation through large hydroelectric projects. Much of the conversion of forested areas is still due to extraction of timber for export outside the region. Large tracts of land have been desertified as a consequence. The degradation in north-eastern India has led to extensive formation of bamboo forests or invasion by exotic or native weeds, leading to arrested successional stages or bald landscape (Ramakrishnan, 1985b,d,e 1988c; Ramakrishnan and Ram, 1988; Ramakrishnan and Vitousek, 1989).

Problems of redeveloping disturbed land areas have been analysed from ecological, socio-economic and socio-cultural angles, indicating as to how this could be done with maximum benefits accruing to the people of the region. The concepts and principles developed here, and based on the north-eastern India case study, would largely be applicable to the forest ecosystem management in the humid tropics elsewhere.

FOREST–AGROECOSYSTEM INTERPHASE

The rain forest vegetation types are characterized by perturbation of different scales, both spatial and temporal (Whitmore, 1975). Intense human pressure on the rain forests, both by the traditional man for jhum

(Ramakrishnan, 1985b,c,) and the industrial man (Myers, 1980) has often resulted in over-exploited degraded ecosystems. Jhum, as in north-eastern India, has always been in harmony with the environment, when it was practised in its more traditional form with a long fallow period of about 20 years or more. In such a system, man is an integral component of the ecosystem function (Ramakrishnan, 1985c). With limited perturbations, a high degree of biological diversity could be maintained in these forests.

As discussed earlier (cf Chapter 1), the jhum farmer of the north-eastern India, as elsewhere, has always managed his agroecosystem for centuries with optimum yield on a long-term basis, rather than trying to maximize production on short-term considerations. However, the drastic shortening of the jhum cycle to about 4–5 years, and in extreme cases with a 0-year cycle where the slash arising from the herbaceous growth between two cropping seasons (December–May) is used along with lopped lower branches of sparsely distributed trees (cf the sedentary fallow systems in Chapter 2), has resulted in distortions, both ecological and economic. Although population pressure is a major factor for the shortened jhum cycle, desertification and the consequent shrinkage in available land area for agriculture has occurred largely due to intense logging activities of the industrial man during the last few decades (Ramakrishnan, 1985a), as has also happened in Indonesia (Kartawinata and Vayda, 1984), and elsewhere (Myers, 1980).

In the earlier Chapters we have made a detailed scientific analysis of the jhum system of agriculture, other land use systems, village organization and function, secondary successional patterns and processes, the problem of desertification and a variety of other ecological and social issues of north-eastern India. In this analysis, the objective is to link rural ecosystem redevelopment with redevelopment and management issues of the rain forest ecosystem of the region.

Land use redevelopment as an alternative to jhum

A major objection to traditional jhum is that it causes extensive deforestation, and that with increase in population under conditions of reduced land availability, the land is not given a sufficient fallow period for soil fertility to recover (cf Chapter 5; Ramakrishnan and Toky, 1981; Mishra and Ramakrishnan, 1983d, 1984). With degradation of the environment, the system breaks down and the farmer often moves away to another site.

Much has been written on the possible alternatives to regulate, modify or eliminate jhum from north-eastern India. One of the earliest attempts to regulate shifting agriculture in Burma was the development of the 'taungya' system by a forester, D. Brandis in 1956, where the forest farmer was required to raise desired plantations identified by the Forest Department in plots after the cropping (Puri, 1960). While this system is still prevalent in some areas of the Asian tropics (FAO, 1978), the taungya system often breaks down when the population pressures increase rapidly and land degradation occurs. Consequently, it has had less general acceptance.

The measures so far adopted to counter jhum are largely related to development of settled terrace cultivation in north-eastern India, (cf Chapter 2) as have also been tried out in Bangladesh (FAO, 1978). Resettlement outside forested areas in Korea or reorientation through plantation/horticultural crops as successfully tried out on a limited scale in India (cf Chapter, 2; Patnaik and Ramakrishnan, 1989; Kumar and Ramakrishnan, 1989b), are other options. A possibility for the north-eastern India, although of limited value because of the constraints imposed by the availability of land, is to intensify valley land agriculture wherever possible through more modern management techniques.

Some of these strategies try to directly attack the problem at its root (as for example, terracing the slope – a replacement to jhum) or try to circumvent the problem through parallely developed economic activities (example, plantation/horticulture).

Valley land agroecosystem

Realizing that a 10-year jhum cycle is the cut-off point for economic efficiency, energy efficiency (cf Chapter 1) soil fertility maintenance (cf Chapter 5) and weed suppression (cf Chapter 4) in the jhum plots, one way of taking the pressure off the land to ensure a longer jhum cycle than the prevalent 4- to 5-year cycle is to strengthen the valley land agriculture. With a rolling hill topography, the land area available in north-eastern India could be sustained ecologically and economically because of the constant nutrient wash-out into the valleys from the hill slopes. There is much variation available in this land use technology within the region. Some tribes are more efficient than others (cf Chapter 2; Toky and Ramakrishnan, 1981a; Mishra and Ramakrishnan, 1981; Patnaik and Ramakrishnan, 1989; Maikhuri and Ramakrishnan, 1990a,b; Kumar and Ramakrishnan, 1990).

The Apatani way of wet rice cultivation in Arunachal Pradesh, where they effectively recycle domestic waste, has an elaborate water management system. They combine long-duration rice varieties with fish culture, and raise millet (*Eleusine coracana*) on the raised partition bunds of rice plots. This is one of the most efficient production systems that I have come across. Indeed the ecological efficiency of this system is extremely high with an energy output/input ratio of about 80 (Kumar and Ramakrishnan, 1990). Like Apatanis, the immigrant Nepalis who are also traditionally sedentary rice farmers have a well organized and productive valley land system of agriculture.

Transfer of traditional technology from one community to another can increase the efficiency of this system. Additional input of technology from outside could enable one to obtain two or even three crops through intensive management, where traditionally now only one crop is raised annually.

Cash crop economy

A shift to horticulture/plantation crops along with forestry which would ensure permanent plant cover may, perhaps, be the ideal alternative to jhum in the hilly terrain. The north-eastern region in India offers much scope for a variety of tropical and temperate fruits. Traditionally, banana and pineapple are already being cultivated in the region. Orange plantations are to be found in the Cherrapunji area. A variety of temperate fruits are also grown at higher altitudes. Tea and coffee have been tried on an experimental scale as in the Khasi hills of Meghalaya. Cashew nut has been tried on a limited scale as found in the Garo hills of Meghalaya. In Tripura, the Forest Development Corporation has successfully established rubber plantations. It is important, however, to organize the cash crop economy as a small-scale family enterprise and on a co-operative basis, as the tribals have two contrasting traits in them – they are fiercely independent by nature and yet have a strong co-operative spirit, as seen from their jhum activities (cf Chapter 1) and their village ecosystem function (cf Chapter 3).

Elsewhere, the studies from Sarawak, Malaysia suggest that cash cropping could be intensified, eventually replacing shifting agriculture there. The Ibans in Sarawak grow enough rice for their needs with cash crops. Pepper farming has been taken up by them with little resistance. The Kenyah of the Buram area of Sarawak have taken to rubber plantations (Chin, 1982). Perhaps, in the north-eastern India and elsewhere, long-term strategies for land use redevelopment may enable cash crop economy to be developed with the peoples' participation.

Terrace agroecoystem

Terracing has been suggested from time to time as a land use system to replace jhum in north-eastern India. This has also been tried out in the adjoining Bangladesh (FAO, 1978). In north-eastern India, it has failed to take-off in a major way because of the following:

(1) The need for heavy input of costly inorganic fertilizers to maintain reasonable returns to the farmer;

(2) The poor fertilizer-use efficiency due to intense leaching of nutrients from the soil, even if run-off losses are checked;

(3) Exaggerated weed potential under continuous cropping of the same site;

(4) The gradual decline in soil physical properties over a period of 6–8 years;

(5) The recurrent problem and cost involved in maintaining terraces under high rainfall; and

(6) The often short life of the terrace itself (6–8 years) due to site desertification.

However, there are situations such as that at Nayabunglow where because of a well-developed soil profile and the possibility of a heavy input of cow dung by the cattle farming immigrant Nepali community, terraces could be sustained over longer time periods. Ever then, the cost involved is high.

The Indian Council of Agricultural Research (ICAR) complex located in north-eastern India has developed a model, in order to replace jhum in north-eastern India (Borthakur *et al.*, 1978). Their model suggests use of a three-tier land use for the hill slope. The upper third of the slope is designated for forestry, the middle portion for horticulture/plantation crops, and the lower third for terraced agriculture. This system demands, according to them, inorganic fertilizer of the order of 60 kg of N, 30 kg of P_2O_5 and 30 kg of K ha^{-1} $year^{-1}$; this is a major deterrent against it. The system also conflicts with the social structure of the tribal communities adversely affecting the independence of the family unit, as this system cannot be maintained by a single family. In view of this and in view of the fact that this alternate system totally ignores traditional technologies of the tribals (Ramakrishnan, 1984b), the ICAR model has not been able to make any significant impact so far, and has not been considered seriously.

Redevelopment of jhum

A basic feature of the jhum system in north-eastern India, in common with similar systems elsewhere in the humid tropics, is mixed cropping involving over 35 species in the extreme (Nye and Greenland, 1960; Watters, 1971; Ruthenberg, 1976; Ramakrishnan, 1984a, 1990). On the whole these systems are efficient from the point of crop organization for photosynthesis, and for nutrient uptake and use efficiencies. Indeed, the crop organization on a hill slope by the jhum farmer may have similarities with early successional natural herbaceous communities, in terms of species/population organization (Ramakrishnan, 1988a). Mixed cropping also helps in effective weed management, pest control, conservation of nutrients in the system through recycling of crop residues and checking losses through water. Indeed, these agroecosystems with jhum cycles of 10 years or longer are stable systems, and even where the cycle lengths are shorter, the farmer has adapted his cropping pattern to maximize production (cf Chapter 1; Ramakrishnan, 1984a). A point that should be emphasized is that jhum is not a single system but it varies markedly depending upon ecological, socio-economic and socio-cultural factors. With a large number of highly insulated tribes of diverse linguistic and socio-cultural backgrounds living in this large geographical region, it is not surprising that a variety of jhum systems with differences in finer patterns exist in north-eastern India. Some of these systems yield many times more than others. Mere mainpulation of the crop mixture alone could improve productivity. Thus, a low elevation jhum system with a 20-year cycle gave about 5,000 Rupees ha^{-1} yr^{-1} (Toky and Ramakrishnan, 1981a) compared with a modified version at a higher elevation that gave three times more return (Mishra and Ramakrishnan, 1981; cf also Chapter 1).

With many advanced features based on strong science (Ramakrishnan, 1984a), the jhum system is efficient under the given ecological context, provided the cycle length is longer than 10 years. In fact, modern agricultural technology has not been able to develop a more efficient way of maintaining soil fertility in the humid tropics than that possible through natural processes during the secondary successional fallow phase after jhum. With rapid shortening of the jhum cycle during the recent past that has resulted in an average cycle length of about 5 years, this land use system has become untenable because adequate soil fertility cannot be recovered over this shortened fallow phase (cf Chapter 5, 7), and weed potential of the site is aggravated with arrested succession at the weed stage (cf Chater 4, 6). The chief objective in redevelopment is to remove these distortions under the short jhum cycle so that the system can be stabilized and economic yield enhanced under the short cycle.

The soil fertility at a given site is generally lower under a short jhum cycle than under longer ones. Soon after jhum, soil fertility is already very low due to, slash-and-burn, losses that occur due to ash blown from the site, the loss through water due to run-off and leaching, and crop and weed removal (cf Chapter 5). During the early secondary successional phase of up to 10 years or so, further depletion in soil nutrients occurs due to rapid transfer from the soil to the developing biomass. It is only after 10 years that the soil nutrient recovery process starts, when litterfall occurs (cf Chapter 5,7; Ramakrishnan and Toky, 1981; Mishra and Ramakrishnan, 1983d). In one of the studies on nitrogen budget analysis under jhum (Mishra and Ramakrishnan, 1984), we have shown that the system loses up to about 600 kg of nitrogen during one cropping season. With a minimum of 10–15 years of natural fallow phase required to recover this entire quantity, under a 5-year jhum cycle the system is only able to recover through natural processes not more half of what is lost. In other words, under a 5-year cycle, the agroecosystem starts with a net deficit of 300 kg ha^{-1} of nitrogen after every jhum cycle. Eventually, site desertification occurs.

Realizing that a 10-year jhum cycle often cannot be sustained in the region because of the present level of pressure on land by the people, redevelopment of this system has to be done to make it viable at the 5-year cycle level. Jhum, which basically is an agroforestry system, could be redeveloped by additional agroforestry inputs alone. Based on a detailed evaluation of a large number of native tree species, we were able to identify many fast-growing tree species of value for incorporation into this agroforestry system (cf Chapter 9; Ramakrishnan, 1986a, 1987a; Ramakrishnan *et al.*, 1982). Apart from fast growth, a narrow crown form (excurrent crown form) with adequate light penetration to ground level, the ability of the bole to slough off lateral branches so that the clean bole length of the tree could be increased, and the ability to rapidly recycle nutrients through a rapid turn-over of the leaf population of the tree, are traits that allow complementality of the trees with crops below. Such an integration of woody species along with crops would help in conservation of resources such as soil and water, as trees enhance physico-chemical properties of the soil and provide protection against erosion and leaching losses. An obvious role of trees lies in efficient recycling of mineral nutrients through tree litterfall. Trees, once introduced into the cropping phase of the jhum system, could continue through the fallow phase so that the timber could be harvested on a short rotation basis. Of the many species identified, the Nepalese alder (*Alnus nepalensis*) has found ready acceptance with the farmers, because traditionally the jhum farmer conserves this tree species whenever it occurs in his jhum plot, during the slash-and-burn operations. The message that comes through very clearly is that although many technologies may be viable otherwise, unless the technology is appropriately integrated in the social and the cultural context, it may not find acceptance and may not be able to become established. In Nagaland, redevelopment of jhum through the use of Nepalese alder is now adopted extensively (Plate 10.1) because of governmental effort (Gokhale *et al.*, 1985). The species coppices profusely. It is a fast growing early successional tree with a loosely organized leaf population, with a narrow (excurrent) crown form that allows adequate light penetration to the ground level permitting crop growth. The leaf turnover rate is also fast (Boojh and Ramakrishnan, 1982b). Therefore, the nutrient turnover rate between the tree and the soil surface layers is fast. The superficial root system characteristic of early successionl species (Ramakrishnan and Shukla, 1982; Shukla and Ramakrishnan, 1984b) nodulates profusely and fixes nitrogen.

Plate 10.1 Nepalese alder (*Alnus nepalensis*) extensively being used in Nagaland as an agroforestry input to stabilize jhum

The dry fallen leaves that are curled up on the ground trap and retain moisture from dew and rain. This multipurpose tree species can be used as timber for furniture too. Tender leaves may be used as cattle feed.

The Nepalese alder fixes nitrogen in the soil at the rate of up to about $150\,kg\,ha^{-1}\,yr^{-1}$ so that all the $600\,kg\,ha^{-1}$ of nitrogen lost during a cropping season could be recovered during a 5-year-long fallow phase; other nutrient elements are also recovered. Therefore, a 5-year jhum cycle could be sustained. Over a period of 4–5 years, the tree grows to a height of more than 6 m with a diameter of 15 cm. Regular pollarding, every 4–6 years provides more firewood than many other fast-growing species in the region. Yet the tree could live for over 100–200 years, as recorded in the region. Many tribal farmers of Nagaland in the Angami, Chakhesang, Chang, Yin Chunger and Konyak areas now extensively use alder as part of the jhum system. In such an agroforestry system, even cash crops such as cardamum and black pepper could be integrated.

Agroforestry redevelopment would have the advantage of reducing the weed potential of the cropping system, since under short 5-year jhum cycles, weed potential of the site becomes aggravated leading to arrested weed succession during the fallow phase (Chapter 4, 6; Saxena and Ramakrishnan, 1984b). During prolonged secondary succession under longer jhum cycles, weed suppression occurs due to the shading effect by the shrubs and tree species that close up the forest canopy. Introduced trees under a redeveloped jhum would, to some exent, control weed growth through shading.

In the past, when good natural forest cover existed around jhum plots, this acted as a good wind-break. Such a wind-break would prevent loss of nutrients through wind-blown ash deposited on the soil surface soon after the slash-and-burn operation and just before the monsoon rainfall sets in (cf Chapter 5; Ramakrishnan and Toky, 1981). Bamboo or other fast growing trees used as a wind-break around the jhum plots help in checking losses through ash blow-off and at the same time helps in conserving nutrient losses through water.

Natural forest succession after jhum could provide a model for the construction of sustained yield agroecosystems (Ramakrishnan, 1988a). Such a system for the San Carlos de Rio Negro region of southern Venezuela has been described (Uhl and Murphy, 1981). During the 1st year of succession, forbs, grasses and some woody species are present. To parallel this, the crop plot could be planted with forbs (beans and pineapple), grasses (corn and sugarcane), woody species (yucca, cashew and papaya) or even larger trees (brazilnut and jackfruit). Over a period of time, fast growing woody species would form one or two additional strata above the ground layer as in early successional forest. This approach to tropical agriculture is suggested to avoid site degradation by maintaining continuous crop cover and at the same time providing yields over extended periods of time.

DESERTIFICATION AND FOREST ECOSYSTEM REDEVELOPMENT

Forest ecosystem degradation in north-eastern India has resulted in arrested succession of weeds over large areas under the influence of large-scale deforestation due to timber extraction in the past and the subsequent recent impact of shortened jhum cycle of 4–5 years imposed again and again on the same site (cf Chapter 4). Under a longer jhum cycle of 10–20 years constantly imposed on the same site, large tracts of land have been taken over by bamboo forests (cf Chapter 6). In extreme cases, total desertification has occurred over large tracts of land as in the west Khasi hills of Meghalaya including the Cherrapunji region, which has resulted in bald landscape with arrested and degraded grasslands (cf Chapter 6,7).

Totally desertified sites needs to be treated intially with herbaceous species and some shrubs, before tree species could be introduced. The early successional herbaceous species with a variety of growth strategies either related to reproductive capabilities (vegetative versus seed reproduction or *r*-versus *k*-strategy) or nutrient uptake and use capabilities (C_3 versus C_4 photosynthetic strategy of species with differential nutrient-use efficiencies), could be the basis for determining the choice of the appropriate species and their organization on a slope (cf Chapter 8).

The concept of architectural attributes of shrubs and trees is useful, when these components of the forest ecosystem are introduced during redevelopment. Early successional species have the following:

(1) Faster extension growth of the leader axis;

(2) Faster branch production;

(3) Greater allocation of resources for shoot growth compared to the root system;

(4) More allocation of biomass to the first-order branches;

(5) Greater plasticity in branch orientation;

(6) A large leaf population that is loosely organized on the skeletel framework of the tree; and

(7) A larger population, of younger leaves on the tree at a given time.

Late successionals, on the other hand, have contrasting attributes, where natural selection has been for shade tolerance.

If the root architectural changes are considered over a successional gradient, a few contrasting features between early and late successional species become obvious. The early successionals allocate more biomass to the shoot system compared to the root system so that they have a high shoot/root ratio; in contrast, late successionals have a low ratio. Further, early successionals have more root biomass located in the surface layers of the soil, whereas the late successionals have roots that are more uniformly distributed into deeper layers; the mid-successional species fall in between (cf Chapter 9).

All these contrasting features of early versus late successionals are geared to perimt early successional species to capitalize upon a high light regime of an open early successional environent making effective use of nutrient-enriched surface soil layers (exploitative strategy). On the contrary, the late successional species have a 'conservative strategy' to make adequate growth in an environment where resource availability is limited.

During redevelopment of a degraded forest ecosystem, introduction of shrubs and trees is important. If these species are natives rather than exotics, one is reasonably certain about the long-term ecological implications (Ramakrishnan, 1985f). I emphasize this here because, the temptation to use exotics such as *Eucalyptus* spp. or *Lucaena leucocephala* for afforesting degraded lands could be great, because they are often simpler to do. Often the temptation to use exotic species is because one is not aware of the potential of native tree species growing in the backyard. Mixed plantations, having many distinct ecological advantages over monoculture forestry (Singh and Ramakrishnan, 1981), could also be economically productive or even better than the latter, if compatible species are used. Fast-growing, light-demanding early successional species could form compatible mixtures with shade-tolerant mid-or late successional species for exploiting light availability at different canopy levels and thus optimizing production per unit area (cf Chapter 9). Such a mixture would also allow optimum use of nutrients from the soil profile, the early successionals exploiting surface soil layers, and the late successionals reaching deeper down. It should even be possible to have a 'condensed succession' for revegetation of damaged sites by appropriate mixing of the different categories of species and by appropriately adjusting species introduction into the mixture. More work in this emerging area of study would be most rewarding.

We have already considered agroforestry redevelopment in detail in the foregoing pages. Social forestry programmes in a developing country should aim at identifying fast growing native species for plantations for meeting fodder, fuelwood and timber needs of the rural population. Use of early successional species that have a fast growth rate with rapid turn-over of leaves (and consequently of nutrients through litterfall) are appropriate because resource mobilization and utilization would be optimal. Social forestry systems developed as a village woodlot, either on private land or on common land, along with agroforestry would take the pressure off the natural forests and contribute towards conservation. In the ultimate analysis restoration ecology, if it is to be effective, has to take into account not only ecological, but also social, economic and cultural aspects.

GENERAL CONSIDERATIONS AND CONCLUSIONS : PHILOSOPHY OF REDEVELOPMENT

Land use redevelopment should be closely linked to the development of other sectors of the village ecosystem. Redeveloped animal husbandry systems should be effectively integrated into redeveloped land uses. Introduction of technology to meet rural needs, such as improved cooking stoves for energy conservation, biogas, mini-hydel projects and rural industries for processing or semi-processing of village produce are some of the possibilities that are relevant for north-eastern India. With rainfall largely concentrated during 4–5 months of the monsoon season, rain water harvesting in cheaply constructed polythene sheet-lined water tanks costing about 10000 Rupees could channelize roof-top water from the houses as well as the run-off water from the hill slopes (Plate 10.2). The stored water in small tanks of about 80 m^3 could be used to meet a variety of domestic needs as well for irrigation of 2–3 families during the dry season (P.S. Ramakrishnan, K.G. Saxena, B.P. Kotyari, K.S. Rao and T. Kumar,

Plate 10.2 Rain water harvested and stored in cheap polythene-lined tanks could prove to be cost-effective for agroecosystem and forest ecosystem redevelopment through drip irrigation during the dry season. Covered tanks could even meet drinking water needs

unpublished). Indeed such cheap technology has been found to have remotely linked connections with ecosystem redevelopment. A redeveloped village ecosystem, otherwise constrained due to water shortage has indirect implications in regenerating natural vegetation on degraded hill slopes. A well-developed agroforestry system would take the pressure off the biomass of the regenerating hill slopes. Additional moisture provided on the hill slope through water harvest and drip irrigation, particularly during the dry season would accelerate natural/man-induced revegetation of the hill slope. This could often prove to be more cost-effective than other more elaborate revegetation technologies. Therefore, an integrated view is critical when considering any aspect of ecosystem redevelopment in tribal areas.

With a variety of ecological situations, socio-economic conditions and socio-cultural variations in the humid tropics, developmental strategies necessarily have to be location-specific. If they are based upon traditional technologies of the region, they would find ready acceptance by the people concerned. Only then would the peoples' participation be ensured, as it would be based on a value system that the people can appreciate and identify with.

The world's population is increasing at the alarming rate of 80 million people per year, and demographic pressures are more pronounced in the

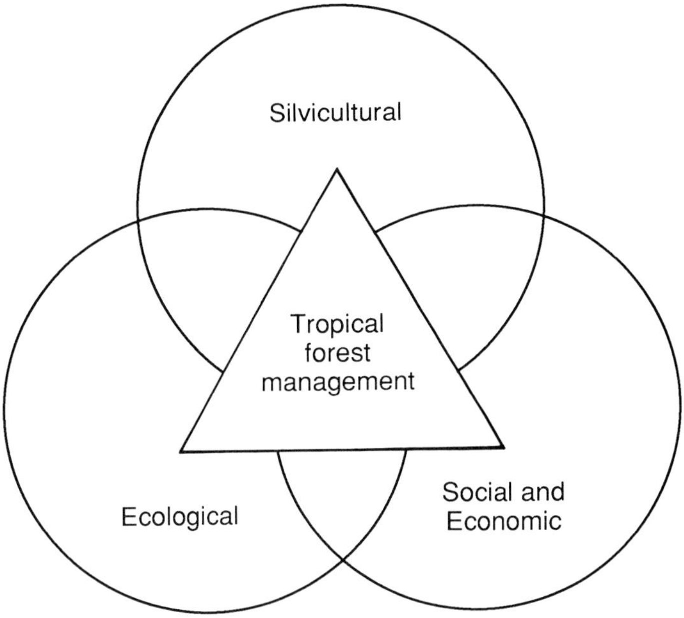

Figure 10.1 Interdisciplinary interactions called for in tropical forest management and conservation

developing world. Consequently, deforestation in the tropics is essentially a human problem, and a narrow-based forestry initiative alone cannot tackle it. A more broadly based interdisciplinary approach will facilitate the sharing of knowledge arising out of the three major components of the management strategy (Fig. 10.1). Linking up forestry with sustainable development of rural communities by making adequate provision for food fodder, fuelwood and timber to meet daily needs is critical. In this effort and in providing cash income for a better quality of life, multipurpose species can play a useful role. These species could also markedly reduce pressure on natural forests.

CHAPTER 11

CONSERVATION LINKED WITH SUSTAINABLE DEVELOPMENT

INTRODUCTION

The estimate on the species number on Earth varies from 5 to 30 million; no one doubts that the numbers are far higher than was thought until recently (World Resources Institute, 1988). With an estimated 15–33% loss of the total wildlife by 2,000 AD, the extinction rate is becoming critical for human survival (Lovejoy, 1986). It is in this context that conservation science assumes considerable significance. Apart from a number of biological problems such as community equilibrium and the fragility of ecosystems, there are many social challenges that impinge upon peoples' perception of the environment and development. However, it is only in recent times that efforts have been directed to build up the necessary, and obvious, contacts between ecological and social sciences.

In Chapter 10, we have considered in a comprehensive manner the question of ecosystem redevelopment in north-eastern India. Such an approach to restoration ecology in the humid tropics is important for an overall conservation strategy design. However, many questions pertaining to conservation need to be considered. What is it that we wish to conserve? Is it possible or even necessary to conserve all the species that are likely to be lost and about which we may or may not even be aware of ? How do we ensure that our conservation strategies are effective? What is sustainable development in the context of biological conservation? In this chapter, we shall consider some of these issues and attempt to answer them using the experiences from the north-eastern Indian case study considered in detail in the earlier chapters.

AGROECOSYSTEM CONCEPT AND CONSERVATION

The recent emergence of the agroecosystem concept provides a useful framework not only to design a relatively stable and productive system, as discussed in earlier Chapters, but also to design conservation strategies for plant species of food value. A human-manipulated ecosystem such as the agroecosystem in a traditional society is well harmonised with its ecological, socio-economic and cultural components and therefore reflects the coevolutionary aspects of culture and environment (Gliessman, 1989b). Therefore, an understanding of the ecological and social linkages related to agroecosystem function would form the basis, not only for their management but for conservation of biological diversity.

Traditional agroecosystems and biological diversity

Jhum agroecosystem and conservation of biological diversity

The jhum farmer in north-eastern India, as elsewhere throughout the humid tropics, cultivates many local varieties of traditional and non-traditional crop species. As discussed in Chapter 1, the number of species in the crop mixture may vary from 4 or 5, to over 35 (Ramakrishnan, 1985c). The stability and sustainability of this and other traditional agroecosystems are based on crop diversity (Altierie, 1983; Ramakrishnan, 1984a, 1988a). The crop mixture and the varieties used are location-specific native strains, differing in composition depending upon ecological, socio-economic and socio-cultural factors.

An important consideration with regard to the mixed cropping system is the optimal use of nutrient resources from the soil by appropriate changes in the crop mixture depending upon the length of the jhum cycle and the consequent high/low soil fertility levels. Thus, the jhum farmer emphasizes nutrient-use efficient tuber and rhizomatous species under short jhum cycles in contrast with the emphasis placed on less efficient cereals under long jhum cycles (Ramakrishnan, 1984a). At higher elevations in north-eastern India, high acidity under pine forests, lower temperature conditions restricting quick vegetation growth, and slower nutrient release related to low temperature and pine litter, all contribute to lower soil nutrient status. This would explain the farmer's emphasis on more nutrient-use efficient tuber crops.

The organization of diverse crop species on a hill slope of 2–3 ha itself often follows a pattern. This is unlike a monocrop of a high yielding variety where modern man has tried to exploit the ruderal nature of crop species by genetic manipulation of rice or wheat in order to maximize seed production at the expense of vegetative growth over the shortest possible time. Under jhum, the *r*-strategists with emphasis on seeds, such as cereals and legumes are combined with *k*-strategists with emphasis on vegetative growth, such as leafy vegetables. The aim of the farmer is to maximize production from the site through a mix of these two strategies. Most mixed cropping systems with cultivars of dissimilar reproductive strategies imitate early successional forest fallows following jhum.

During the cropping seasons under jhum, the soil on the steep hill slopes is highly heterogeneous. The availability of nitrogen is uncertain, since this is a highly labile element. The microdistribution of C_3 and C_4 plants on a slope enables them to exploit nutrients from such a heterogeneous soil. The C_4 species with a high nutrient-use efficiency could occur in nutrient-poor microsites while C_3 species with low nutrient-use efficiency are suited to nutrient-rich microsites, as shown for early successional herbaceous jhum fallows (Saxena and Ramakrishnan, 1983a, 1984c). Such a C_3/C_4 strategy helps in coexistence of species through mutual avoidance. The positioning of the C_3 and C_4 crop species under jhum in north-eastern India imitates the local natural herbaceous communities with the more nutrient-use efficient species being located in the nutrient-poor upper parts of the slope and less efficient species situated at the base of the slope (Ramakrishnan, 1988a).

Sedentary agroecosystem and conservation

In many areas in the humid tropics where jhum is the chief land use, attempts have been made to develop alternate land use strategies. This has been necessitated, partly due to the rapid decline in the jhum cycle. Sedentary terrace agriculture on hill slopes has been one of these land use strategies. Such alternative systems are often energy intensive and ecologically inefficient. They are conceived as intensive production systems with emphasis on high yielding crop varieties. These have often been rejected by the traditional farmers. Their tendency to hold on to traditional land use practices has to some extent contributed towards conservation of traditional crop varieties. Rice farmers of valley lands in Thailand often combine traditional varieties of rice during the monsoon season with modern semi-dwarf varieties during the dry season, optimizing production and at the same time maintaining stability of the system (Grigg, 1974).

A variety of valley land agroecosystem types are available in north-eastern India. These agroecosystems, apart from showing wide variations in economic and ecological efficiencies which are related to ecological conditions and socio-economic and cultural differences between the communities (cf Chapter 2), harbour many traditional rice varieties. Thus, for example, the Apatanis of Arunachal Pradesh who practise one of the most efficient wet rice cultivation methods make effective use of their irrigated land by planting early and late ripening varieties. The early variety is sown further away from the village where disturbance by animals and poorer irrigation facilities could be major constraints (Kumar and Ramakrishnan, 1990). Relatively poorer nutrient status of plots farther away from the village is another consideration in deciding upon the early variety. The late variety, grown closer to the village where waste recycling from the domestic sector ensures nutrient-enriched soil, also allows pisciculture integrated with the late ripening rice variety.

The renewed interest of recent times in traditional agroecosystems has started contributing towards their adaptation to modern times. The concept of a home garden which is well developed in the humid tropics offers possibilities of maximizing production per unit area through tight vertical packaging of species as done by many tribes in north-eastern India (Gangwar and Ramakrishnan, 1989a; Maikhuri and Ramakrishnan, 1990a). With a mixture of annuals and perennials forming a multistoreyed structure, they resemble a natural forest. An extreme example of this is the te'lom plots of about 6 ha maintained by the Huastee Indian farmers of north-eastern Mexico. With over 300 species in a te'lom plot, Huastee system is not only an alternative to land management in the humid tropics but provides protection for wild genetic conservation and contributes towards a combination of commercial and subsistence agriculture (Alcorn, 1984).

Lesser-known plants of food value

Of all the 3,000 plant species used as food at some time during human civilization, about 150 species are cultivated, of which less than 20 provide over 90% of the food needs (National Research Council, 1982). In fact, just about three species (wheat, rice and maize) meet over half of the human energy needs. Reliance on such a small number of plants carries great risks, for monocultures are extremely vulnerable to catastrophic failures brought about by diseases or climatic stresses.

Over large areas of the humid tropics, genetic uniformity of a few varieties is displacing many local varieties traditionally cultivated by the farmers (Wilkes, 1977). As the product of technology displaces the source upon which the technology itself is based, namely, the genetic diversity, short-term solutions often become long-term irreversible liabilities. In fact, 75% of the world's lesser-known crops are grown and consumed in tropical countries (National Research Council, 1975). With depletion of genetic diversity of crop plants (Sasson, 1987), there is considerable interest in lesser-known plants of food value for meeting the needs of an increasing human population particularly in the developing world (National Research Council, 1975; Raven, 1981). There is a need, not only to conserve them, but also to improve upon the economic yield of the more promising ones. Such a diversification of the agroecosystem base would go a long way towards meeting the nutritional needs of the people.

Some important under-exploited food plants are the grain Chenopods (National Research Council, 1975; Risi and Galwey, 1984; Partap and

Kapoor,1985), with a high protein value. These plants are being cultivated extensively in the Andes or in the Himalayas. Greater emphasis on cultivation, improvement and consumption by the local communities would countribute to enhancement of the quality of life of the people concerned. With considerable similarity in the preparation carried out for many Chenopods by the Himalayan communities and their Andean counterparts (Risi and Galwey, 1984; Partap and Kapoor, 1985), possibilities exist for preparation of marketable products. Needless to say that there is much scope for improvement of the grain Chenopods.

The tribals of north-eastern India, endowed with a rich natural resource base and with a diverse socio-economic and socio-cultural background, depend upon a variety of under-exploited plants and animals. Four tribal communities of Arunachal Pradesh (the Nishis, the Hill Miris, the Sulungs and the Apatanis) alone use over 70 such plant and 12 animal species from the wild (Gangwar and Ramakrishnan, 1990). The plant species often come in handy during the lean season of the year when traditional food items are in short supply. With a large number of tribes in a small geographic area, each being highly insulated due to language barriers and socio-cultural differences, at the same time occupying diverse ecological situations, north-eastern India is a veritable ethno-biological treasure.

Many lesser known plants of food value are cultivated under jhum (Ramakrishnan, 1983), sedentary agriculture and fallow systems (Gangwar and Ramakrishnan, 1987a; cf Chapter 2). Many tribes in Arunachal Pradesh, for example, cultivate *Dioscorea* spp. for tubers, *Amaranthus viridis* for leaves, *Chenopodium ambrosoides*, *Panicum miliaceum*, *Setaria italica* and *Coix lacryma-jobi* for seeds, apart from many others (Gangwar and Ramakrishnan, 1990).

Flemingia vestita and *Perilla ocimoides* are two important lesser-known crops of food value under sedentary fallow systems of agriculture of the Khasis of Meghalaya (Gangwar and Ramakrishnan, 1987a). Apart from the value of its tuber as a protein source, this legume considerably improves the nitrogen status of the soil through nodular fixation (Gangwar and Ramakrishnan, 1989b; cf Chapter 2). With up to 250 kg ha^{-1} yr^{-1} of nitrogen fixed, such multipurpose lesser-known crops are important both for sustainable agriculture and for human nutrition.

Redeveloped agroecosystems and village ecosystems linked with conservation

As discussed in the foregoing chapter, this multidisciplinary study has led to a number of practical guidelines for redeveloping jhum in north-eastern India, under very short jhum cycles. These guidelines noted below, and discussed elsewhere in much detail, would also contribute to effective resource management and conservation in north-eastern India. Such an approach to agroecosystem redevelopment would effectively linkup animal husbandry and domestic sectors with conservation stategies so that local peoples' needs are adequately taken care of. These guidelines, in summary are :

– Vary species composition in the crop mixture under jhum so that economic returns increase and ecological efficiency improved (e.g. emphasis on potato at high elevations, along with other species, compared to rice at lower elevations has resulted in a many-fold increase in economic yield, in spite of low fertility of the more acidic soils at higher elevations, and at the same time ensured crop diversity through mixed cropping). With wide variations in the crop mixtures and in cropping patterns (cf Chapter 1), the possibilities for agroecosystem redevelopment through transfer of indigenous jhum technology from one tribe to others exist.

– Redesign agroforestry systems incorporating ecological insights on tree architecture (cf Chapter 9). The canopy form of tree species should be compatible with crop species at the ground level, by permitting sufficient light penetration and through fast recycling of nutrients. Species such as the Nepalese alder (*Alnus nepalensis*) is a multipurpose agroforestry species of economic value to the tribals, building up nitrogen status of the soil through fixation, synchronizing elegantly with a 5-year jhum cycle system because of adequate growth it makes during this period, and yet coppicing readily (cf Chapter 1,5).

– Use bamboo and other fast growing native tree species as wind-breaks to check wind and water borne losses of ash and nutrients (Chapter 1, 5), so that biological diversity of the agroecosystem types as a whole is enhanced.

– Redevelop valley land wet rice cultivation and improve other land use systems such as home gardens by use of appropriate native crop species, transfer of indigenous technology from one area to another, and redesigned systems developed on the basis of traditional knowledge.

– Strengthen traditional animal husbandry practices such as swine husbandry and poultry (cf Chapter 3) so that more effective recycling of waste between this sub-system and the agriculture sub-system would result in improved agroecosystem types of greater conservation value.

– Introduce appropriate rural technology such as biogas units, energy efficient stoves, rain water harvesting tanks, and mini-hydel systems for strengthening the village ecosystem function so that natural resource conservation could be augumented through their impacts. To cite one example, we have put up cheap rain water harvesting tanks across the

Himalayas (unpublished data). This single activity has direct value not only for a redeveloped agroecosystem but has also contributed towards restoration of a degraded hill slope. This tinkering with one component of the village ecosystem could have tremendous impact on its other components and could enhance biological diversity through vegetation regeneration on hill slopes. This is so because, under the monsoonic climate, water often is a limiting factor in the Himalayas for a large part of the year.

– Encourge artisanal skills and products based on natural resources accessible to rural communities – products based on items such as bamboo and other woods, leather, etc. In the Kapkot region of the Kumaon hills in the central Himalayas, strengthening the locally available bamboo resource base through biological and technological inputs has already contributed remarkably to restoring degraded hill slopes and thus has helped in conserving locally perceived natural resource bases and biological diversity (P.S. Ramakrishnan, K.G. Saxena, K.S. Rao and B.P. Kothyari, unpublished data).

This is not in any way to suggest that agroecosystems, traditional or redeveloped are the only systems to be encouraged. In situations such as that prevalent in the north-eastern region, for the present at least, such an approach is desirable for conservation of biological diversity, particularly in the absence of alternate viable technologies to meet at least short-term developmental objectives. In any case, for agroecosystem redevelopment in developing countries, an absolute priority should be given to soil conservation so as to give benefit to small farmers (Sasson, 1990).

RAIN FOREST – A CRITICAL ECOSYSTEM

In recent times, ecological, evolutionary biologists and conservationists have focussed increasing attention on tropical rain forests. Although they cover only about 7% of the Earth's land area, they harbour more than half the species of the world's biota. With rapid conversion of these forests, they will mostly disappear by the turn of the century, if the present trends continue. This would result in the extinction of a large number of species, many of them not even catalogued or understood by the tropical biologists. Also the tropical rain forests are among the most fragile of all habitats. Often conversion results in wet deserts (Ramkrishnan, 1985d,e; Ramakrishnan and Ram, 1988), because of highly leached infertile soils subjected to wash-out by heavy rain.

Much of the area under rain forests is red and yellow earths which are acidic and nutrient-deficient. With high concentration of iron and aluminium, phosphorus availability is restricted because of the formation of insoluble compounds. Potassium is highly labile and therefore is leached out quickly (cf Chapter 5; Toky and Ramakrishnan, 1981b; Mishra and Ramakrishnan, 1983c; Ram and Ramakrishnan, 1988b). With a weak soil base, a dense forest cover is supported because of a thick fine-root mat located on top of the mineral soil which enables tight recycling of nutrients released by the quickly decomposing leaf litter, before the element could seep into the mineral soil (cf Chapter 7). With a fine-root biomass of up to $15\,t\,ha^{-1}\,yr^{-1}$, in the relatively undisturbed sacred-grove forest of north-eastern India (Khiewtam, 1986), and also found elsewhere in similar ecosystems (Ramakrishnan, 1984d), any large-scale disturbance in this system (in terms of intensity, spread and frequency) would make it unstable and recovery difficult.

In north-eastern India, large-scale disturbance of the rain forest ecosystem has resulted in varied levels of degraded arrested bamboo forests, with weed take-over or a totally bald landscape. During the last few decades, large-scale timber extraction for industrial purposes has cleared vast areas of land for invasion by exotic weeds (Ramakrishnan, 1985d, 1991; Ramakrishnan and Vitousek, 1989). The economic consequences of this for the community should be the responsibility of the timber extractors and the burden of proof should rest on them (Tisdell, 1990). Thus exotic weeds such as *Eupatorium* spp., and *Mikania micrantha* have taken over vast tracts of cleared land along with native weeds such as *Imperata cylindrica* and *Thysanolaena maxima.* Once this large-scale invasion has occurred, the jhum farmer is even more limited by the land area available for his jhum system of agriculture, as he prefers to avoid sites of high weed density. Because of this and increased population pressure, jhum cycle has dropped drastically in length from a more favourable 20 years or more, to an extremely short 5 years or even less. Having no other option, in the absence of an alternate agricultural technology that is viable from an ecological and social angle, the jhum farmer perforce had to resort to very short jhum cycles although the system operates below subsistence level and has caused further environmental degradation. Large-scale timber extraction and very short jhum cycles of 4–5 years have resulted in an arrested succession of weeds in north-eastern India. (cf Chapter 8; Toky and Ramakrishnan, 1982c; Saxena and Ramakrishnan, 1984b).

Cherrapunji: an example of extreme desertification

Where climatic/edaphic factors create stressful conditions, further deterioration could occurr leading to extreme desertification of the site. Indeed vast tracts of land in the Khasi hills of Meghalaya are represented by a totally bald landscape. Thus, for example, in Cherrapunji in Meghalaya, desertification has been more rapid and sudden. The human factor is accentuated by the peculiar geology with nutrient-deficient soil over limestone rock formations, with a network of underground tunnels with stalactites and stalagmites. The harsh climate with much of the annual average rainfall of 1,150 cm (with a very high 2,250 cm in an exceptional year as in 1974) coming down in about 4–5 months during the monsoon compounds the problem (Ramakrishnan, 1989e). For these reasons, the fragility of the Cherrapunji forest ecosystem is highly exaggerated (Khiewatam, 1986; Ramakrishnan and Ram, 1988).

The sacred grove of Mawsmai that which is protected for reasons of traditional religious beliefs is a sad reminder of what Cherrapunji looked like in the past. Although the forest is stunted because it is supported by an unbalanced soil derived from limestone, it does form a dense multilayered canopy protecting the soil from the ravages of the extreme climate. Had it not been for the traditional belief of the Khasis that the sacred grove is the abode of their Gods and the spirits of their dead ancestors, and therefore nothing should be removed, not even dead twigs of the forests, even the sacred grove of Mawsmai would have disappeared long ago. Unfortunately, much of these groves, which were part of every village in the Khasi hills, have already disappeared, with the advent of a modern value system! With that, we have almost lost a unique way of preserving this national heritage of rain forest ecosystems.

The Cherrapunji ecosystem that now stands desertified due to deforestation inflicted sometime in the distant past (Ramakrishnan, 1985c) now refuses to recover its original state, as represented by the relict sacred grove. The fact that jhum around Cherrapunji is banned by the village council is suggestive of the part played by this land use in creating the present landscape. The sharp boundary between the sacred grove and the bald landscape indicates that the system will never recover through natural processes of revegetation. Artificially, the cost could be enormous.

Linked with this drastic loss in biological diversity is the human suffering which now is immense. Water is a scarce commmodity during the dry months in this high rainfall spot of the world, a contender for being the wettest spot on Earth along with the near-by Mawsengram in the Khasi hills of Meghalaya and the far-away spot in the Hawaiian islands of the Pacific. All the water flows away into the plains down below, as here there is no vegetation to hold the water and the soil, and there is no soil to recover the forest. It is a vicious circle ! What one sees now is stunted grassy growth protected within rock crevices like pock-marks on a rocky landscape. For fuelwood, the tribal villager has to trek long distances. The ruins of villages all around remind one of the yester years and migrations of the past. The tribal, who is traditionally bound to the the land and forests has been forced to seek other avenues for survival.

Arrested bamboo forests and biological diversity

Even where jhum cycle could be maintained at a more reasonable 10–20 years, because of large-scale pressures the secondary forests are largely bamboo forests often maintained in an arrested stage partly because of constant imposition of jhum at this cycle length. Although economically important, bamboo thickets often tend to be biologically least diverse. They tend to form pure stands, particularly if perturbations occur at regular intervals.

An important contributory factor to stabilization of bamboo forests as arrested stages in succession is also the large-scale extraction of firewood by the tribal people (Ramakrishnan, 1985d). These extractions are small events leading to small-size gap formations but over large areas. Such a disturbance regime results in invasion of the openings by bamboo. Indeed much of the firewood consumed in this region comes from secondary forests, with a small component coming from primary rain forests. In India, as in many other parts of the world, it has been found that from the peripheral zones of the forest boundaries, about 70% of the firewood requirements are met; beyond 10 km the entraction of firewood steadily decreases until at about 15 km it is almost absent (Myers, 1980). However, the increased population pressure and rapid depletion of primary forest cover in the region is rapidly changing this situation (Ramakrishnan, 1985d). In the absence of hardwood species in this region, even bamboo is used as a firewood species, as well as being exported to the Assam plains and elsewhere in the country (Rao and Ramakrishnan, 1987b). Small-scale clearings of this nature that are not totally devastated help in sustaining bamboo in an arrested successional stage with obvious adverse consequences to biological diversity in the region.

Rain forest restoration and conservation

The concept of succession offers the basis for a meaningful restoration programme for conserving biological diversity. Early successional species with their diverse adaptive strategies offer an important basis for initiating restoration of bald and desertified landscapes (cf Chapter 9, 10). The herbaceous species are either *r*-strategists and select for greater reproductive capacity in more open habitats, or *k*-strategists and select for greater ability to compete for resources, although at the cost of lower reproductive potential in closed habitats. Further, while some species emphasize upon sexual reproduction, others have extensive insulated underground rhizomes. C_3 versus C_4 strategy of herbaceous species offers the kind of variability to the early successional herbaceous community to enable adaption to nutritionally heterogeneous soil environments. (cf Chapter 7, 8, 10).

If one were to build biological diversity upon an arrested successional community of herbaceous vegetation or bamboo forest, apart from bamboo species, one could consider a variety of shrub species with diverse growth strategies, related to light availability and nutrient status of the soil (cf Chapters 9, 10). This could then be elaborated upon using the differential architectural features of early versus late successional tree species, for light capture and nutrient uptake and use. The resultant mixed plantation forestry systems and indeed other forestry systems based on a 'condensed succession' could help in ensuring some degree of biological diversity.

THE HUMAN FACTOR IN CONSERVATION

Rural development

In north-eastern India, where a large number tribal communities of varied linguistic, social, economic and cultural backgrounds live in a geographical region, with communities often changing over very short distances, the human factor in conserving biological diversity is critical. This is apart from considerations related to ecological heterogeneity. We have seen, for example, how one has to deal with a variety of jhum systems (cf Chapter 1), a variety of valley land agroecosystems, home gardens and other land use systems (cf Chapter 2). In these and other land use linked activities such as animal husbandry and domestic sub-systems of the village ecosystem (cf Chapter 3), one comes across a rich biological diversity. This biological diversity should be conserved side by side with conserving biological diversity in the natural rain forest ecosystem. Indeed, as much as one needs to restore/redevelop a degraded natural forest ecosystem through successional concepts, one also needs to redevelop the agriculture, animal husbandry and the domestic sub-systems of the village ecosystem in order to make conservation more meaningful for the local community. In such a strategy, traditional knowledge would form an important basis for conservation. We have considered some of these strategies for a redeveloped agroecosystem, earlier in this chapter and in Chapter 10. In such a redesign of village-based activities, social considerations are critical. Thus, even though we came up with a number of native species of agroforestry value based upon our tree growth and architecture analysis, only one species, namely, the Nepalese alder (*Alnus nepalensis*) could obtain greater acceptability amongst the tribals for redeveloping jhum (cf Chapter 1, 5, 7, 9) because this species was traditionally conserved by them wherever they occurred in their jhum plots. Therefore, the people concerned could relate this species with a value system with which they were already familiar. Thus, when the Nepalese alder was used for strengthening the jhum as an agroforestry system, it rapidly became popular, and now in a state such as

Nagaland where it was used extensively by the governmental agencies (Gokhale *et al.*, 1985) it is wide-spread. A link-up between social and natural sciences is not only important to identify appropriate technologies for rural development, but is also important for identifying and appreciating remote linkages between introduced technology and its impact on other components of the rural ecosystem, e.g., the rain water harvesting technology introduced in the Himalayas (discussed earlier) and its impact on restoration of a hill slope.

Biosphere reserves

When the Man and Biosphere (MAB) programme of UNESCO was launched by UNESCO in 1971, it was also envisaged to have biosphere reserves as representative ecological areas, as part of an international network. The chief objective was to demonstrate the value of conservation and its relationship to ecodevelopment with local participation. This approach to conservation sets apart the concept of biosphere reserves from all other conservation strategies. Realizing that we would have to depend more and more upon such biosphere reserves for conserving much of the biological diversity, we need to recognize effective management strategies not only for the core and the buffer zone, but also for the interaction zone located outside the buffer zone which gradually merges into the typical rural environment (Ramakrishnan, 1987c). In developing countries, the buffer and the interaction zones would obviously be under intense biotic stress from the humans living there; even the core zone would not be free from it. Under these conditions, an integrated approach to management linking up environmental awareness with the food, fodder and fuelwood needs of the rural communities, and a diversified set of economic activities that have an input of appropriate rural technologies is important. Building upon traditional knowledge would facilitate peoples' participation and ensure that development is sustainable.

The sacred groves of the Meghalaya

The importance of sacred groves such as the Mawsmai forest in the largely desertified Cherrapunji in Meghalaya, discussed above, hardly needs to be emphasized. The Khasis, as well as the Jaintias of Meghalaya have many well-developed groves. These are often located on hill slopes. The third tribe of Meghalaya, namely, the Garos do not have this concept well developed, although they usually worship large monoliths placed along the periphery of dense forests. However, the large Balphakram forest reserve extending over a few hundred km^2 and with extremely high species diversity is suggested to have religious value, as the departed souls of the local Garos are said to reside there (Kumar and Rao, 1985).

The Khasis and the Jaintias of Meghalaya who traditionally are animists like the Garos believe that their gods representing natural forces responsible for creation of living beings and others like sun, wind and water live in these groves. They also believe that the good spirits of the ancestors also live in them and protect the village community. Therefore, removing any material from the groves is a taboo. The Jaintia tribe of Meghalaya have many sacred groves of which a few are up to about 2 km^2. In the Cherrapunji area alone, we have recorded (Khiewtam and Ramakrishnan, 1989) 21 such groves, some very strictly protected and others with varied levels of disturbance. The largest recorded grove, the Kryngah grove at Umwai village is about 12 km^2 although the average size of the rest of the 20 groves is about 185 ha.

Detailed accounts of selected groves are available only from Meyghalaya where floristic accounts by Bor (1942) and Hazra (1975), and detailed ecosystem analysis have been carried out (Boojh and Ramakrishnan, 1981c, 1982f; Khiewtam, 1986; Khiewtam and Ramakrishnan, 1989). The rest of the region has not been adequately looked at from social and ecological aspects, and the groves have never been catalogued. This needs to be done. Meghalaya with high species diversity and a high level of endemism has few parks, sanctuaries and reserve forests. Consequently the sacred groves are important for conservation of biological diversity. The ecosystem organization of the relatively undisturbed areas and those outside with varied disturbance regimes provide the necessary basis for a meaningful ecosystem redevelopment programme relevant to conservation (Ramakrishnan, 1988c).

Since the management of these groves is under the control of the village community and the overall supervision of the village headman, the sacred grove concept is a good example of the sociological basis for nature conservation. As long as traditional value systems are supported rather than tampered with, the sacred grove concept provides a good basis for environmental education and for reviving awareness (Ramakrishnan, 1986b) amongst the people in the region.

GENERAL CONSIDERATIONS AND CONCLUSIONS

The concept of biological conservation is closely linked with sustainable development. The launching of the world conservation strategy in 1980 was a significant turning point involving an integration of nature conservation with sustainable development of both human and natural resources. This implies that ecological integrity is to be maintained not only to meet human needs but also to achieve equity with social justice and provision for maintenance of cultural diversity (Jacobs *et al.*, 1987). Local peoples' participation in conservation is crucial and can only be ensured by making sure that they have access to the natural resources developed on a sustainable basis.

Sustaining cultural diversity is an important mechanism for promoting conservation. The wide variations found in the land use, animal husbandry and domestic sub-systems among the tribals of north eastern India based partly upon socio-cultural differences contribute to biological diversity in the region. The variety of crop germplasm of the conventional and non-conventional crop species thus conserved is remarkable indeed. Redeveloped agroecosystem types could also incorporate this biological diversity in them. Religion and culture are important for promoting conservation in traditional societies. The concept of the sacred groves, traditionally found in each tribal village of Meghalaya (Boojh and Ramakrishnan, 1983b; Khiewtam and Ramakrishnan, 1989) is a strong basis for promoting conservation issues amongst many tribal societies. With traditional religious beliefs suggesting that the spirit of the ancestors and the gods live in these groves, the tribals are deeply committed to conserving these forest patches which may sometimes be a few km^2. There is an urgent need to revive these traditional beliefs since much of it has been lost in the name of modernity.

Realizing that conservation of all species known and/or as yet not catalogued is an impossible task, priorities have to be determined. The criticality from a conservation viewpoint may be defined both in terms of ecosystem fragility and biological diversity. In such a definition, diversity sustained through human activities, such as agriculture is also significant. The concept of the biosphere reserve provides the necessary framework for conservation, since the approach is based upon the people as the focal point, through their sustainable development.

Which species are to be conserved would also depend upon the variety of ways in which the species could be used by the society. Thus a species such as *Flemingia vestita* cultivated by the tribals in north-eastern India (Gangwar and Ramakrishnan, 1989b) is not only a lesser-known crop of food value used by the tribals during the lean season, but is also an important component of the jhum system and the fallow system of agriculture, useful for building up the nitrogen status of the soil. Many such multipurpose species (National Research Council, 1979; Burley and Stewart, 1985) that are under-exploited and that abound in the tropics (National Research Council, 1975) offer untapped potential for future use by mankind.

Fragility of an ecosystem should be an important criterion for conservation. We have seen how a tropical rain forest is extremely fragile leading to desertification of the Cherrapunji region, an area receiving very high

Table 11.1 The guiding principles of ecological sustainable resource management (adapted from Hare *et al.*, 1990)

Intergenerational equity: providing for today while retaining resources and options for tomorrow.
Conservation of culture and biological diversity and ecological integrity.
Constant natural capital and 'sustainable income'.
Anticipatory and precautionary policy approach to resource use, erring on the side of caution.
Resource use in a manner that contributes to equity and social justice while avoiding social disruptions.
Limits on natural resource use within the capacity of the environment to supply renewable resources and assimilate wastes.
Qualitative rather than quantitative development of human well-being.
Pricing of environmental values and natural resources to cover full environmental and social costs.
Global rather than regional or national perspectives of environmental issues.
Efficiency of resource use by all societies.
Strong community participation in policy and practice in the process of transition to an ecologically sustainable society.

rainfall. Site desertification in a lesser form is also evident through weed invasion leading to arrested successional weed communities. There could be other levels of expression such as the arrested bamboo forests of the north-eastern region or the arrested pine forests of the western Himalayas. These are all expressions of fragility and such ecosystem types should be conserved on a priority basis, before damage is done.

In north-eastern India, agriculture as a land use is closely linked with other sub-systems of the village ecosystem function, and with forestry. Therefore, a holistic approach has a distinct advantage over all others for *in situ* conservation. It takes into account peoples' perception of ecology and conservation at a human level and at the same time enables conservation of natural plant and animal populations at the sub-specific level (Ramakrishnan, 1981). This latter aspect of biological conservation is particularly important because ecotype populations often could be as different between themselves, as two distinct species of contrasting ecological characteristics (Ramakrishnan, 1981). The ultimate objective should be to conserve biological diversity at specific and sub-specific levels and to integrate conservation with the existing or altered life styles of the human population.

The relationship between poverty, deforestation and conservation is obvious but this could be reversed through appropriate redevelopment strategies as shown through this interdisciplinary study. Linking conservation with development through adequate understanding of ecological systems (Huntley *et al.*, 1991) would in the ultimate analysis, ensure sustainability of the biosphere (Table 11.1), in a manner that satisfies current needs without compromising the needs or options of future generations.

REFERENCES

Abrahamson, W.G. 1979. Patterns of resource allocation in wildflower populations of fields and woods. *Amer. J. Bot.* **66**: 71–79.

Abrahamson, W.G. 1980. Demography and vegetative reproduction. In : O.T. Solbrig (Ed.) *Demography and Evolution in Plant Populations.* pp. 89–106. Blackwell, Oxford.

Abrahamson, W. G. and Gadgil, M. 1973. Growth form and reproductive effort in goldenrods (*Solidago*, Compositae). *Amer Nat.* **107**: 651–661.

Addicott, F.T. and Lyon, J.L. 1973. Physiological ecology of abscission. In: T.T. Kozlowski (Ed.) *Shedding of Plant Parts.* pp. 85–124. Academic Press, New York.

Ahlgren, I.F. and Ahlgren, C.E. 1960. Ecological effects of of forest fires. *Bot. Rev.* **26**: 483–533.

Ahlgren, I.F. and Ahlgren, C.E. 1965. Effects of prescribed burning on soil microorganisms in a Minnesota Jack pine forest. *Ecology* **46**: 303–310.

Akachukw, A.E. 1985. Cost–benefit analysis of wood and food components of agrisilviculture in Nigerian forest zone. *Agroforestry System* **3**: 307–316.

Akkermans, A.D.L. and Van Dijk, C. 1981. Now-leguminous root-nodule symbioses with actinomycetes and *Rhizobium.* In: W.J. Broughton (Ed.) *Nitrogen Fixation.* Vol.1. Ecology. pp. 57–103. Oxford Univ. Press, Oxford.

Alcorn, J.B. 1984. *Huastec Mayan Ethnobotany.* University of Texas Press, Austin, Texas.

Aldrich, R. J. 1984. *Weed-Crop Ecology : Principles in Weed Management.* Breton Publishers, North Scitnate, Mass, 465 pp.

Allan, W. 1965. *The African Husbandman.* Oliver and Boyd, London. 505 pp.

Allen, S.E. 1964. Chemical aspects of heather burning. J. Appl. Ecol. **1**: 347–367.

Altieri, M.A. 1983. *Agroecology : The Scientific Basis of Alternative Agriculture.* Div. Biol. Control, Univ. of California, Berkeley. 162 pp.

Altieri, M.A. and Letourneau, D.K. 1982. Vegetation management and biological control in agroecosystems. *Crop Protection* **1**: 405–430.

Altieri, M.A. and Liebman, M. (Eds) 1988. *Weed Management in Agroecosystems : Ecological Approaches.* CRC Press, Inc. Boca Raton, Florida, 354 pp.

Alvim, P. de T. 1964. Tree growth periodicity in tropical climate. In: M.H. Zimmermann (Ed.) *The Formation of Wood in Forest Trees.* pp. 479–496. Acad. Press, New York.

Andow, D.A. 1988. Management of weeds for insect manipulation in agroecosystems. In: M.A. Altieri and M. Liebman (Eds) *Weed Management in Agroecosystems: Ecological Approaches.* pp. 265–301. CRC Press, Boca Raton, Florida.

Andreae, B. 1980. *The Economics of Tropical Agriculture.* Commonwealth Agriculture Bureaux, England.

Anonymous, 1971. *Census of India.* Govt. of India Press, New Delhi.

Anonymous, 1977. *National Atlas of India.* National Atlas Organization, Calcutta 234 pp.

Ashton, D.H. 1976. Phosphorus in forest ecosystems in Beenak, Victoria. *J. Ecol.* **64** : 171–186.

Aweto, A.O. 1981. Secondary succession and soil fertility restoration in south-western Nigeria. I. Succession. *J. Ecol.* **69** : 601–607.

Baker, H.G. 1965. Characters and modes of origin of weeds. In : H.G. Baker and G.L Stebbins (Eds) *The Geneties of Colonizing Species*. pp 147–172. Academic Press, New York.

Barley, K.P. and Jennings, A.C. 1959. The influence of earthworms on the availability of nitrogen *Austr. J. Agric. Res.* **19** : 364–370.

Bartholomew, W.V., Meyer, J. and Laudelout, H. 1953. Mineral nutrient immobilization under forest and grass fallow in the Yangambi (Belgian Congo) region. Publications de l' Institut National pour l'Etude Agronomique du Congo Belge, Serie Scientifique, **57**: 1–27.

Baruah, U. and Ramakrishnan, P.S. 1988a. Field germination and establishment patterns of shrubs over a successional gradient. *Proc. Indian natn. Sci. Acad.* B **54**: 343–348.

Baruah, U. and Ramakrishnan, P.S. 1988b. Growth and allocation strategies of *Clerodendron infortunatum* Gaerth. and *Ficus fulva* Reinwdt. in successional environments. *Proc. Indian natn. Sci. Acad.* B **54**: 403–408.

Baruah, U. and Ramakrishnan, P.S. 1989a. Architecture and growth patterns of early versus late successional shrubs of sub-tropical moist forests of north-eastern India. *Proc. Indian Acad. Sci.* (*Plant Sci*). **99** : 353–362.

Baruah, U. and Ramakrishnan, P.S. 1989b. Growth strategies of early and late successional shrubs from a sub-tropical moist forest under two light regimes. *Proc. Indian Acad. Sci.* (*Plant Sci.*) **99**: 343–351.

Baruah, U. and Ramakrishnan, P.S. 1989c. Leaf dynamics of early versus late successional shrubs of sub-tropical moist forests of north-eastern India *Proc. Indian Acad. Sci.* ((*Plant Sci.*) **99**: 431–436.

Baruah, U. and Ramakrishnan, P.S. 1989d. Biomass, production and allocation patterns of early versus late successional shrubs of sub-tropical humid forests in north-eastern India. *Proc. Indian natn. Sci. Acad.* **B55**: 63–68.

Baruah, U. and Ramakrishnan, P.S. 1989e. Phenology of the shrub strata of successional sub-tropical humid forests of north-eastern India. *Vegetatio* **80**: 63–67.

Bazzaz, F.A. 1979. The physiological ecology of plant succession. *Ann. Rev. Ecol. Syst.* **10**: 351–371.

Bazzaz, F.A and Harper, J.L. 1977. Demographic analysis of growth of *Linum usitatissimum. New Phytol.* **78**: 193–208.

Bazzaz, F.A. and Pickett, S.T.A. 1980. Physiological ecology of tropical succession: A comparative review. *Ann. Rev. Ecol. Syst.* **11**: 287–310.

Belser, L.W. and Schmidt, E.L. 1978. Diversity in ammonia oxidizing nitrifying population of a soil. *Appl. Env. Microbiol.* **36**: 548–588.

Berkowitz, A.R. 1988. Competion for resources in weed-crop mixtures. In: Altieri, M.A. and Liebman, M. (Eds) *Weed Management in Agroecosystems : Ecological Approaches* pp. 89–119. CRC Press Inc. Boca Raton, Florida.

Bhadauria, T. and Ramakrishnan, P.S. 1989. Earthworm population dynamics and contribution to nutrient cycling during cropping and fallow phases of shifting agriculture (Jhum) in north-east India. *J. appl. Ecol.* **26**: 505–520.

Bhadauria, T. and Ramakrishnan, P.S. 1990. Population dynamics of earthworms and their activity in forest ecosystems of north-east India. J. Trop. Ecol. **6**: (in press).

Binkley, D. 1981. Nodule biomass and acetylene reduction rates of red alder and Sitka alder on Vancouver Island, B.C. *Can. J. For. Res.* **11**: 281–286.

Birch, H.F. and Friend, M.T. 1956. The organic matter and nitrogen status of east African soils. *J. Soil Sci.* **7**: 156–167.

Black. C.C. 1971. Ecological implications of dividing plants into groups with distinct photosynthetic production capabilities. *Adv. Ecol. Res.* **7**: 87–144.

Blackman, G.E. and Templeton, W.G. 1938. The nature of the competition between cereal crops and annual weeds. *J. Agric. Sci.* **27**: 247–271.

Boojh, R. and Ramakrishnan, P.S. 1981a. Germination of seeds of *Alnus nepalensis* Don. *Nat. Acad. Sci. Letters.* **4**: 53–56.

Boojh, R. and Ramakrishnan, P.S. 1981b. Temperature responses to seed germination in two closely related tree species of *Schima* Reinw. *Curr. Sci.* **50**: 416–418.

Boojh, R. and Ramakrishnan, P.S. 1981c. Phenology of trees in a sub-tropical evergreen montane forest in north-east India. *Geo. Eco. Trop.* **5**: 189–209.

Boojh, R. and Ramakrishnan, P.S. 1982a. Growth strategy of trees related to successional status I. Architecture and extension growth. *Forest Ecol. Manage.* **4**: 359–374.

Boojh, R. and Ramakrishnan, P.S. 1982b. Growth strategy of trees related to successional status. II. Leaf dynamics. *Forest Ecol. Manage.* **4**: 375–386.
Boojh, R. and Ramakrishnan, P.S. 1982c. On bifurcation ratio in *Schima* species. *Curr. Sci.* **51**: 436–437.
Boojh, R. and Ramakrishnan, P.S. 1982d. Growth and architecture of two altitudinal populations of *Schima wallichii. Proc. Indian natn. Sci. Acad.* **B48**: 534–545.
Boojh, R. and Ramakrishnan, P.S. 1982e. Seed germination and seedling establishment of two closely related *Schima* species. *Proc. Indian Acad. Sci. (Plant Sci.)* **91**: 397–407.
Boojh, R. and Ramakrishnan, P.S. 1982f. Litterfall pattern in a sub-tropical evergreen montane forest in north-east India. *Geo. Eco. Trop.* **6**: 33–44.
Boojh, R. and Ramakrishnan, P.S. 1983a. The growth pattern of two species of *Schima. Biotropica.* **15**: 142–147.
Boojh, R. and Ramakrishnan, P.S. 1983b. Sacred groves and their role in environmental conservation. In: *Strategies for Environmental Management – Souvenir Vol.* pp. 6–8. Dept. Science and Environment, Govt. of U.P., Lucknow.
Bor, N.L. 1942. The relict vegetation of the Shillong plateau, Assam. *Indian For. Rec.* N.S. **3**: 152–195.
Borchert, R. 1969. Unusual shoot growth in a tropical tree, *Oreopanax* (Araliaceae) *Amer. J. Bot.* **56**: 1033–1041.
Borchert, R. and Slade, N.A. 1981. Bifurcation ratios and the adaptive geometry of trees. *Bot. Gaz.* **142**: 394–401.
Bormann, F.H. and Likens, G.E. 1979. *Pattern and Processes in a Forest Ecosystem.* Springer-Verlag, New York. 253 pp.
Bormann, F.H., Likens, G.E., Fisher, D.W. and Pierce, R.S. 1968. Nutrient loss accelerated by clear-cutting of a forested ecosystem. *Science* **159**: 882–884.
Borthakur, D.N., Singh, A., Awasthi, R.P. and Rai, R.N. 1978. Shifting cultivation in the north-eastern region. In: Proceedings of the National Seminar – *Resources, Development and Environment in the Himalayan Region.* pp. 330–342. Dept. Sci. Tech., Govt. of India, New Delhi.
Boserup. E. 1965. *The Conditions of Agricultural Growth.* Aldine, Chicago, 123 pp.
Bray, J.R. and Gorham, E. 1964. Litter production in forests of the world. *Adv. Ecol. Res.* **2**: 101–157.
Brown, C.C., McAlpine, R.G. and Kormanick, P.P. 1967. Apical dominance and form in woody plants : a reappraisal. *Amer. J. Bot.* **54**: 153–162.
Brown, R.H. 1978. A difference in N use efficiency in C_3 and C_4 plants and its implication in adaptation and evolution. *Crop Sci.* **18**: 93–98.
Brunig, E.F. 1976. Tree forms in relation to environmental conditions : an ecological view point. In: M.G.R. Cannell and F.T. Last (Eds) *Tree Physiology and Yield Improvement.* pp. 139–156. Academic Press, New York.
Bull, T.A. 1971. The C_4 pathway related to growth rates in sugarcane. In : M.D. Hatch. C.B. Osmond and R.O. Slatyer (Eds) *Phososynthesis and Photorespiration* pp 68–75. Wiley Interscience, New York.
Burley, J. and Stewart, J.L. (Eds) 1985. *Increasing Productivity of Multipurpose Species.* International Union Forestry Research Organizations, Vienna, Austria. 560 pp.
Caldwell, M.M., White, R.S., Moore, R.T. and Camp, L.B. 1977. Carbon balance, productivity and water use of cold winter desert shrub communities dominated by C_3 and C_4 species. *Oecologia* **29**: 275–300.
Carneiro, R.L. 1960. Slash and burn agriculture: A closer look at its implications for settlement patterns. In : A.F.C., Wallace (Ed.) *Men and culture.* pp. 229–234. Philadelphia.
Cavendish, F.H. 1905. A flowering of *Dendrocalamus hamiltonii* in Assam *Indian For.* **31**: 479.
Chacon, J.C. and Gliessman, S.R. 1982. Use of the 'non-weed' concept in traditional tropical agroecosystems of south-eastern Mexico. *Agro-Ecosystems* **8**: 1–10.
Chandra, S., Evenson, J.P. and Deboer, A.J. 1976. Incorporating energetic measures in an analysis of crop production practices in Sigtoka Valley, Fiji. *Agric. Systems* **1**: 301–311.
Chapin, F.S. 1980. The mineral nutrition of plants. *Ann. Rev. Ecol. Syst.* **11**: 233–260.
Chin, S.C. 1982. The significance of rubber as cash crop in Kenyah swidden village in Sarawak. *Federation Museum Journal* **27**: 23–37.

Christensen, N.L. and Kimber, P.C. 1975. Effects of prescribed burning on the flora and fauna of south-west Australian forests. *Proc. Ecol. Soc. Australia* **9**: 85–106.
Clements, F.E. 1916. *Plant succession: Analysis of the Development of Vegetation.* Carnegie Institution of Washington Publication, 242, 542 pp.
Cody, M.L. 1966. A general theory of clutch size. *Evolution* **20**: 174–184.
Conklin, H.C. 1957. *Hanunoo Agriculture.* FAO Forestry Development Paper No. 12, FAO, Rome. 109 pp.
Cook, D.B. 1941. The period of growth in some northeastern trees. *J. Forestry* **39**: 957–959.
Cook, L. 1939. A contribution to our information on grass burning. *South Afr. J. Sci.* **36**: 270–282.
Cornell, J.H. and Slatyer, R.O. 1977. Mechanisms of succession in natural communities and their role in community stability and organization. *American Naturalist* **111**: 1119–1144.
Coulter, J.K. 1950. Organic matter in Malayan soils: A preliminary study of the organic matter content in soil under virgin jungle, forest plantations and abandoned cultivated land. *Malay For.* **13**: 189–202.
Dansereau, P. 1957. *Biogeography: An Ecological Perspective* Ronald Press, New York. 394 pp.
Das, A.K. and Ramakrishnan, P.S. 1985. Litter dynamics in Khasi pine (*Pinus kesiya* Royle ex. Gordon.) of north-eastern India. *Forest Ecol. Manage.* **10**: 135–153.
Das, A.K. and Ramakrishnan, P.S. 1986. Adaptive growth strategy of Khasi pine (*Pinus kesiya* Royle Ex. Gordon). *Proc. Indian Acad. Sci.* (*Plant Sci.*) **96**: 25–36.
Das, A.K. and Ramakrishnan, P.S. 1987. Above-ground biomass and nutrient contents in an age series of Khasi pine (*Pinus kesiya*). *Forest Ecol. Manage.* **18**: 61–72.
Dash, M.C. and Patra, V.C. 1977. Density, biomass and energy budget of a tropical earthworm population from a grassland site in Orissa, India. *Revue D'Ecologie Et De Biologie Du Sol* **14**: 461–471.
Daubenmire, R. 1968. Ecology of fire in grasslands. *Adv. Ecol. Res.* **5**: 209–266.
Dazhong, W. and Pimental, D. 1984. Energy inputs in agricultural systems of China. *Agric. Ecosys. Environ.* **11**: 29–35.
Dean, G.W., Garter, H.O., Wagstaff, H.R., Olayide, S.O., Ronning, M. and Bath, D.L. 1972. *Production Functions and Linear Programming Models for Dairy Cattle Feeding*, Giannini found. Monogr. No. 31, Calif. Exp. Stn., Univ. Calif, Berkeley. 52 pp.
Debell, D.S. and Ralston, C.W. 1970. Release of nitrogen by burning light forest fuels. *Soil Sci. Soc. Am. Proc.* **34**: 936–938.
Deevey, E.S. 1947. Life-tables for natural populations of animals. *Quarterly Rev. Biol.* **22**: 283–314.
Deka, H.K. and Mishra, R.R. 1982. Decomposition of bamboo (*Dendrocalamus hamiltonii* Nees.) leaf litter in relation to age of jhum fallows in north-east India. *Plant and Soil* **68**: 151–159.
Dev, P. and Ramakrishnan, P.S. 1987a. Coexistence of closely related *Eupatorium* species. I. *Eupatorium odoratum* L. versus *Eupatorium adenophorum* Spreng. and *Eupatorium adenophorum* Spreng. versus *Eupatorium riparium* Regel. at different altitudes. *Proc. Indian Acad. Sci.* (*Plant Sci.*) **97**: 165–176.
Dev, P. and Ramakrishnan, P.S. 1987b. Coexistence of closely related *Eupatorium* species. II. *Eupatorium adenophorum* Spreng. and *Eupatorium riparium* Regel. at different altitudes. *Proc. Indian Acad. Sci.* (*Plant Sci.*) **97**: 177–183.
de Wit, C.T. 1960. On Competition. *Versl. laudbonwk. Onderz. Ned.* **66**: 1–82.
Donovan, D.G. 1981. *Fuelwood: How Much Do We Need*? Institute of Current World Affairs, Hanover, N.H. 23 pp.
Doust, L.L. 1981. Population dynamics and local specialization in a clonal perennial (*Ranunculus repens*). I. The dynamics of ramets in contrasting habitats. *J. Ecol.* **69**: 743–755.
Dove, M.R. 1983. Forest preference in swidden agriculture. *Trop. Ecol.* **24**: 122–142.
Drew, W.B. 1974. *The Ecological Role of Bamboos in Relation to the Military Use of Herbicides on Forests in South Vietnam.* National Academy of Science, National Research Council, Washington, DC, 14 pp.
Edwards, C.A. 1983. Earthworm ecology in cultivated soils In : J.E. Satchell (Ed.) *Earthworm Ecology from Darwin to Vermiculture.* pp. 123–136. Chapman and Hall, London.

Edwards, C.A. and Lofty, J.R. 1977. *Biology of Earthworms*. 2nd Edn. Chapman and Hall, London 333 pp.
Edwards, C.A. and Lofty, J.R. 1982. The effect of direct drilling and minimal cultivation on earthworm populations. *J. appl. Ecol.* **19**: 723–724.
Edwards, P.J. 1977. Studies of mineral cycling in a montane rain forest in New Guinea II. Production and disappearance of litter. *J. Ecol.* **65**: 971–992.
Edwards, P.J. and Grubb, P.J. 1982. Studies of mineral cycling in a montane rain forest in New Guinea. IV. Soil characteristics and the division of mineral elements between the vegetation and soil. *J. Ecol.* **70**: 649–666.
Egler, F.E. 1954. Vegetation science concepts. I. Initial floristic composition-a factor in old field vegetation development. *Vegetatio*, **4**: 412–417.
Ehleringer, J.R. and Bjorkman, O. 1977. Quantum yields for CO_2 uptake in C_3 and C_4 plants: dependence on temperature, CO_2 and O_2 concentration. *Plant Physiol.* **59**: 86–90.
Epstein, E. 1972. *Mineral Nutrition in Plants: Principles and Perspectives*. John Wiley, New York. 412 pp.
Eussen, E.H.H. and Wirjahardja, S. 1973. Studies on alang-alang (*Imperata cylindrica* (L.) Beauv. vegetation *Biotrop. Bull.* **6**: 1–24.
Evans, G.C. 1972. *The Quantitative Analysis of Plant Growth*. Blackwell Sci. Pub., Oxford.
Ewel, J. 1971. Biomass changes in early tropical succession. *Turrialba* **21**: 110–112.
Ewel, J.J. 1976. Litter fall and leaf decomposition in a tropical forest succession in eastern Gautemala. *J. Ecol.* **64**: 293–308.
Ewel, J., Berish, C., Brown, B., Price, N. and Raich, J. 1981. Slash and burn impacts on a Costa Rican wet forest site. *Ecology* **62**: 816–829.
Ewel, J., Benedict, F., Berish, C., Brown, B., Gliessman, S., Amador, M., Bermudez, R., Martinez, A., Miranda, R. and Price, N. 1982. Leaf area, light transmission, roots and leaf damage in nine tropical plant communities. *Agro-Ecosystems* **7**: 305–326.
FAO 1978. Shifting cultivation. *Forest News for Asia and the Pacific* **2**: 1–26.
FAO/SIDA, 1974. *Report on Regional Seminar on Shifting Cultivation and Soil Conservation in Africa*. FAO, Rome. 248 pp.
Faucher, D. 1949. *Geographie Agraire*. Genin, Paris.
Finck, A. 1970. Moglichkeiten de Nahrungs produktion in Landon. *Ernahrungs Umschau*, **2**: 47–52.
Fisher, J.B. 1978. A quantitative study of *Terminalia* branching. In: P.B. Tomlinson and M.H. Zimmermann (Eds) *Tropical Trees as Living Systems*. pp. 285–320. Cambridge Univ. Press, New York.
Flinn, M.A. and Wein, R.W. 1977. Depth of underground plant organs and theoretical survival during fire. *Can. J. Bot.* **55**: 2550–2554.
Fox, J. 1984. Firewood consumption in Nepali villages. *Environ. Manage.* **8**: 243–250.
Frankie, W.G., Baker, H.G. and Opler, P.A. 1974. Comparative phenological studies of trees in tropical wet and dry forests in the low-lands of Costa Rica. *J. Ecol.* **62**: 381–913.
Furer-Haimendorf, C. Von. 1962. The Apatanis and their Neighbours. London.
Furer-Haimendorf, C. Von. 1985. *Tribals of India: The Struggle for Survival*. Oxford Univ. Press, Delhi 342 pp.
Gangwar, A.K. 1987. *Cropping and Yield Patterns under Slash and Burn Agriculture (Jhum) in North-East India and Related Ethnobiological Studies*. Ph.D. thesis. North-Eastern Hill Univ., Shillong 248 pp.
Gangwar, A.K. and Ramakrishnan, P.S. 1987a. Cropping and yield patterns under different land use systems of the Khasis at higher elevations of Meghalaya in north-eastern India. *Int. J. Ecol. Environ. Sci.* **13**: 73–86.
Gangwar, A.K. and Ramakrishnan, P.S. 1987b. Agriculture and animal husbandry among the Sulungs and Nishis of Arunachal Pradesh. *Social Action* **37**: 345–372.
Gangwar, A.K. and Ramakrishnan, P.S. 1989a. Ecosystem function in a Khasi village of the desertified Cherrapunji area in north-east India. *Proc. Indian Acad. Sci. (Plant Sci.)* **99**: 199–210.
Gangwar, A.K. and Ramakrishnan, P.S. 1989b. Cultivation and use of lesser-known plants of food value by tribals of north-east India. *Agric. Ecosys. Environ.* **25**: 253–267.
Gangwar, A.K. and Ramakrishnan, P.S. 1990. Ethnobiological notes on some tribes of north-eastern India. *Econ. Bot.* **44**: 94–105.

Gebhardt, H. and Coleman, N.T. 1974. Anion adsorption by allophanic tropical soils. III. Phosphate adsorption. *Soil Sci. Amer. Proc.* **38**: 255–261.
Gill, A.M. 1975. Fire and the Australian flora: a review *Australian Forestry* **38**: 4–25.
Gill, A.M. and Tomlinson, P.B. 1971. Studies on the growth of red mangrove (*Rhizophora mangle* L.). 3. Phenology of the shoot. *Biotropica* **3**: 109–124.
Gleason, H.A. 1926. The individualistic concept of the plant association. *Bull. Torr Bot. Club*, **53**: 7–26.
Gliessman, S.R. 1980. Some ecological aspects of traditional agricultural practices in Tabasco, Mexico: Applications for production. *Biotica* **5**: 93–101.
Gliessman, S.R. 1988. Ecology and management of weeds in traditional agroecosystems. In: M.A. Altieri and M. Liebman (Eds) *Weed Management in Agroecosystems : Ecological Approaches.* pp 237–244. CRC Press Inc. Boca Raton, Florida.
Gliessman, S.R. 1989a. Integrating trees into agriculture : The home garden agroecosystem as an example of agroforestry in the tropics. In: S.R. Gliessman (Ed.) *Agroecology: Researching the Ecological Basis for Sustainable Agriculture.* pp 160–168. Springer-Verlag, New York.
Gliessman, S.R. 1989b. Agroecology: Researching the ecological basis for sustainable development. In: S.R. Gliessman (Ed.) *Agroecology : Researching the Ecological Basis for Sustainable Development.* pp 3–10. Springer-Verlag, New York.
Gliessman, S.R., Garcia, E.R. and Amador, A.M. 1981. The ecological basis for the application of traditional agricultural technology in the management of tropical agroecosystems. *Agro-Ecosystems* **7**: 173–185.
Gokhale, A.M., Zeliang, D.K., Kevichusa, R. and Angami, T. 1985. *Nagaland: The Use of Alder Trees.* Education Dept., Kohima, Nagaland. 26 pp.
Golley, F.B., McGinnis, J.T., Clements, R.G., Child, G.I. and Duever, M.J. 1975. *Mineral Cycling in a Tropical Moist Forest Ecosystem.* Univ. of Georgia Press, Athens, Georgia.
Gomez-Pompa, A., Vasguez-Yanes, C. and Guevara, S. 1972. The tropical rain forest: a non-renewable resource. *Science* **172**: 762–765.
Gonzalez, J.A. 1985. Home gardens in central Mexico. In : I.S. Farrington (Ed.) . *Prehistoric Agriculture in the Tropics.* BAR International Series 232, Oxford, England.
Gopalan, C.B., Ramasastri, V. and Balasubramanian, S.C. 1985. *Nutriative Value of Indian Food.* National Institute of Nutrition, Hyderabad. 204 pp.
Grant, W.C. 1955. Temperature relationships in the megascolecid earthworm, *Pheretima hupeiansis. Ecology* **36**: 412–417.
Greenland, D.J. 1975. Bringing the green revolution to the shifting cultivators. *Science* **190**: 841–844.
Greenland, D.J. 1977. Contribution of microorganisms to the nitrogen status of tropical soils. In : A. Ayanaba and P.J. Dart (Eds) *Biological Nitrogen Fixation in Farming Systems of the Tropics.* pp. 13–25. Wiley, New York.
Greenland, D.J. and Kowal, J.M.L. 1960. Nutrient content of the moist tropical forest of Ghana. *Plant and Soil* **12**: 154–174.
Grigg, D.B. 1974. *The Agriculture Systems of the World : An Evolutionary Approach.* Cambridge Univ. Press, Cambridge. 358 pp.
Grime, J.P. 1974. Vegetation classification by reference to strategies. *Nature, London* **250**: 26–31.
Grime, J.P. and Jeffrey, D.W. 1965. Seedling establishment in vertical gradients of sunlight. *J. Ecol.* **53**: 621–642.
Grubb, P.J. and Edwards, P.J. 1982. Studies of mineral cycling in a montane rain forest in New Guinea. III. The distribution of mineral elements in the above-ground material. *J. Ecol.* **70**: 623–648.
Gupta, K.K. 1972. Flowering in different species of bamboos in Cachar District of Assam in recent times. *Indian For.* **98**: 83–85.
Hadley, E.B. and Kieckhefer, B.J. 1963. Productivity of two prairie grasses in relation to fire frequency. *Ecology* **44**: 389–395.
Haig, I.T., Hubermann, M.A. and Din, U.A. 1958. *FAO Forest and Forest Product Studies*, 13. (*Tropical Sylviculture*) Vol. 1. FAO, Rome, 190 pp.
Hains, B. and Foster, R.B. 1977. Energy flow through litter in a Panamian forest. *J. Ecol.* **65**: 147–155.

Hallé, F. and Oldeman, R.A.A. 1970. *Essai sur l'Architecture et la Dynamique de Croissance de Arbres Tropicaux*. Masson, Paris.
Hallé, F., Oldemann, R.A.A. and Tomlinson, P.B. 1978. *Tropical Trees and Forests: An Architectural Analysis*. Springer-Verlag, New York. 441 pp.
Hare, W.L., Marlowe, J.P., Rae, M.L., Gray, F., Humphries, R. and Ledgar, R. 1990. *Ecologically Sustainable Development*. Australian Conservation Foundation, Fitzroy, Victoria.
Harper, J.L. 1977.*Population Biology of Plants*. Academic Press, London.
Harper, J.L. 1978. The demography of plants with clonal growth. In : A.H.J. Freysen and J.W. Woldendrop (Eds) *Structure and Functioning of Plant Populations*. pp. 27–41. North Holland, Amsterdam.
Harper, J.L. and Ogden, J. 1970. The reproductive strategies of higher plants. I. The concept of strategy with special reference to *Senecio vulgaris* L. *J. Ecol.* **58**: 681–698.
Harper, J.L. and White, J. 1974. The demography of plants. *Ann. Rev. Ecol. Syst.* **5**: 419–463.
Hartnett, D.C. and Bazzaz, F.A. 1985. The integration of neighbourhood effects by clonal genets of *Solidago canadensis*. *J. Ecol.* **73**: 415–422.
Haswell, M.R. 1973. *Tropical Farming Economics*. Longman, London. 173 pp.
Hay, R.K.M., Holmes, J.C. and Hunter, J.A. 1978. The effect of tillage, direct drilling and nitrogen fertilizer on soil temperature under a barley crop. *J. Soil Sci.* **19**: 174–183.
Hayashi, I. and Numata, M. 1968. Ecology of pioneer species of early stages in secondary succession. II. The seed production. *The Botanical Magazine*, Tokyo **81**: 55–66.
Haynes, R.J. and Goh, K.M. 1978. Ammonium and nitrate nutrition of plants. *Biol Rev.* **53**: 465–510.
Hazra, P.K. 1975. *Visitor's Guide to Law-Lyngdoh (Sacred Grove) of Mawphlang*. Forest Dept., Govt. of Meghalaya.
Hickman, J.C. 1975. Environmental unpredictability and plastic energy allocation strategies in *Polygonum cascadense* (Polygonaceae). *J. Ecol.* **63**: 691–698.
Hofstra, J.J. and Steinstra, A.W. 1977. Growth and photosynthesis of closely related C_3 and C_4 grasses as influenced by light intensity and water supply. *Acta Bot. Neerl.* **26**: 63–72.
Holdsworth, N. 1963. Intermittant growth of the mango tree *J. West African Sci. Assoc.* **7**: 163–171.
Holland, P.G. 1969. The maintenance of structure and shape in three Mallee eucalypts. *New Phytol.* **68**: 411–421.
Holttum, R.E. 1940. On periodic leaf-change and flowering of trees of Singapore, *Garden Bulletin*, Singapore **11**: 119–175.
Honda, H. and Fisher, J.B. 1978. Tree branch angle-maximizing effective leaf area. *Science* **199**: 888–890.
Horn, H.S. 1971. *The Adaptive Geometry of Trees*. Princeton Univ. Press, N.J. 144 pp.
Horton, R.E. 1945. Erosional development of streams and their drainage basins : hydrophysical approach to quantitative morphology. *Bull. Geol Soc. Am.* **56**: 275–370.
Hughes, A.P. and Freeman, P.R. 1967. Growth analysis using frequent small harvests. *J. appl. Ecol.* **4**: 553–560.
Hummel, F.E. 1946. The formation of growth rings in *Entandrophragma macrophyllum* A. Chev. and *Khaya grandifolia* D.C. *Empire For. Rev.* **25**: 103–107.
Humphery, R.R. 1958. The desert grassland, a history of vegetational change and an analysis of causes. *Bot Rev.* **24**: 193–252.
Huntley, B.J., Ezcurra, E., Fuentes, E.R., Fujii, K., Grubb, P.J., Haber, W., Harger, J.R.E., Holland, M.M., Levin, S.A., Lubchenco, J., Mooney, H.A., Neronov, V., Noble, I., Pulliam, H.R., Ramakrishnan, P.S., Risser, P.G., Sala, O., Sarukhan, J. and Sombroek, W.G., 1991. A sustainable biosphere: The global imperative. *Ecology International* **20**: 3–14.
Jackson, J.F. 1978. Seasonality of flowering and leaffall in a Brazilian sub-tropical lower montane moist forest. *Biotropica* **10**: 38–42.
Jacobs, P., Gardner, J. and Munro, D.A. 1987 Sustainable and equitable development : An emerging paradigm. In : P. Jacobs and D.A. Munro (Eds) *Conservation with Equity : Strategies for Sustainable Development*. pp. 17–29. IUCN, Gland, Switzerland.
Jakoby, W.D. and Bhat, J.V. 1958. Microbial metabolism of oxalic acid. *Bacteriological Revue* **22**: 75–80.

Janzen, D.H. 1971. Seed predation by animals. *Ann. Rev. Ecol. Syst.* **2**: 465–492.
Janzen, D.H. 1973. Tropical agro-ecosystems. *Science* **182**: 1212–1219.
Janzen, D.H. 1976. Why bamboos wait so long to flower? *Ann. Rev. Ecol. Syst.* **7**: 347–391.
Jenny, H., Gessel, S.P. and Bringham, F.T. 1949. Comparative study of decomposition rates of organic matter in temperate and tropical regions. *Soil Sci.* **68**: 419–432.
Jones, J.M. and Richardson, B.M. 1977. Effect of reforestation on turnover of 15_N-labelled nitrate and ammonium in relation to changes in soil microflora. *Soil Biol. Biochem.* **9**: 383–392.
Jones, M.J. 1976. The significance of crop residues to the maintenance of fertility under continuous cropping at Samaru, Nigeria. *J. Agric. Sci., Camb.* **86**: 117–125.
Jordan, C.F. 1985. *Nutrient Cycling in Tropical Forest Ecosystems.* Wiley, Chichester.
Jordan, C.F. and Kline, J.R. 1972. Mineral cycling : Some basic concepts and their application in a tropical rain forest. *Adv. Ecol. Syst.* **3**: 35–50.
Jordan, C.F., Kline, J.R. and Sasscer, D.S. 1972. Relative stability of mineral cycles in forest ecosystems. *Amer. Naturalist*, **106**: 237–253.
Juo, A.S.R. and Lal, R. 1977. The effect of fallow and continuous cultivation on the chemical and physical properties of an alfisol in western Nigeria. *Plant and soil* **47**: 567–584.
Kapoor, P. and Ramakrishnan, P.S. 1975. Studies on crop-legume behaviour in pure and mixed stands. *Agro- Ecosystems* **2**: 61–74.
Kartawinata, K. and Vayda, A.P. 1984. Forest conversion in East Kalimantan, Indonesia : The activities and impact of timber companies, shifting cultivators, migrant pepper farmers and others. In: F. di Castri, F.W. G. Baker and M. Hadley (Eds) *Ecology in Practice* Vol. 1 pp. 98–126. Tycooly Publ., Dublin.
Keeley, J.E. 1981. Reproductive cycles and fire regimes. In: H.A.Mooney, T.M. Bonnicksen, N.L. Christensen, J.E. Lotan, W.A. Reiners (Eds) *Fire Regimes and Ecosystem Properties* pp. 278–300. Gen. Tech. Rept. WO – 26, Forest Service, USDA, Honolulu, Hawaii.
Keeley, J.E. and Keeley, S.E. 1977. Energy allocation pattern of a sprouting and non-sprouting species of *Arctostaphylos* in the California Chapparal. *Am. Midl. Nat.* **98**: 1–10.
Kempf, J.S. and Pickett, S.T.A. 1981. The role of branch length and angle in branching pattern of forest shrubs along a successional gradient. *New Phytol.* **88**: 111–116.
Khiewtam, R.S. 1986. *Ecosystem Function of Protected Forests of Cherrapunji and Adjoining Areas.* Ph.D. Thesis, North-Eastern Hill Univ., Shillong 202 pp.
Khiewtam, R.S. and Ramakrishnan, P.S. 1989. Socio-cultural studies of the sacred groves at Cherrapunji and adjoining areas in north-eastern India. *Man In India* **69**: 64–71.
Kinerson, R.S., Higginbotham, K.O. and Chapman, R.C. 1974. The dynamics of foliage distribution within a forest canopy. *J. appl. Ecol.* **11**: 347–353.
King, L.L. 1966. *Weeds of the World : Biology and Control.* Leonard Hill Books, London. 526 pp.
Klinge, H. 1973. Root mass estimation in lowland tropical rainforest of central Amazonia, Brazil. I. Fine root masses of a pale yellow latosol and a giant humus podzol. *Trop. Ecol.* **14**: 28–38.
Knight, H. 1965. Loss of nitrogen from the forest floor by burning. *For. Chron.* **42**: 149–152.
Koriba, K. 1958. On the productivity of tree growth in the tropics with reference to the mode of branching, the leaf fall and the formation of the resting bud. *Gardens' Bulletin*, Singapore **17**: 11–81.
Kozlowski, T.T. 1964. Shoot growth in weedy plants *Bot. Rev.* **30**: 335–392.
Kozlowski, T.T. 1971. *Growth and Development of Trees.* Vol. I. Academic Press, New York.
Kozlowski, T.T. 1973. Extent and significance of shedding of plant parts. In: T.T. Kozlowski (Ed.) *Shedding of Plant Parts.* pp. 1–44. Academic Press, New York.
Kozlowski, T.T. and Ahlgren, C.E. (Eds) 1974. *Fire and Ecosystems.* Academic Press, New York. 542 pp.
Kramer, P.J. 1943. Amount and duration of shoot growth in tree seedlings. *Plant Physiol.* **19**: 239–257.
Kramer, P.J. and Kozlowski, T.T. 1979. *Physiology of Woody Plants.* Academic Press, New York.
Krishnamoorthy, R.V. 1985. Nitrogen contribution by earthworm populations from grassland and woodland sites near Bangalore, India. *Revue D'Ecologie Et De Biologie Du Sol* **22**: 463–472.

Kucera, C.L. and Ehrenreich, J.H. 1962. Some effect of annual burning on central Missouri prairie. *Ecology* **43**: 334–336.
Kumar, A. and Ramakrishnan, P.S. 1989a. Village ecosystem function of tribal societies in north-eastern India with special reference to Hill Miris of Arunachal Pradesh. In : J.S. Singh and B. Gopal (Eds) *Perspectives in Ecology*. pp. 353–376. Jagmander Book Agency, New Delhi.
Kumar, A. and Ramakrishnan, P.S. 1989b. Ecological implications of some cash crop ecosystems in north-eastern India. *Proc. Indian Acad. Sci.* (*Plant Sci.*) **99**: 211–221.
Kumar, A. and Ramakrishnan, P.S. 1990. Energy flow through an Apatani village ecosystem of Arunachal Pradesh in north-east India. *Human Ecology*, (in press).
Kumar, Y. and Rao, R.R. 1985. Studies on Balphakram wildlife sanctuary-3 : General account, forest types and fauna. *Indian J. For.* **8**: 300–309.
Kushwaha, S.P.S. and Ramakrishnan, P.S. 1982. Observations on growth of *Eupatorium odoratum* L. and *Imperata cylindrica* (L.) Beauv. var. Major. under different light and moisture regimes. *Proc. Indian natn. Sci. Acad.* **B48**: 689–693.
Kushwaha, S.P.S. and Ramakrishnan, P.S. 1987. An analysis of some agro-ecosystem types of north-eastern India. *Proc. Indian natn. Sci. Acad.* **B53**: 161–168.
Kushwaha, S.P.S., Ramakrishnan, P.S. and Tripathi, R.S. 1981a. Population dynamics of *Eupatorium odoratum* in successional environments following slash and burn agriculture. *J. appl. Ecol.* **18**: 529–535.
Kushwaha, S.P.S., Ramakrishnan, P.S. and Tripathi, R.S. 1981b. Population dynamics of *Scleria tessellata* Willd. – A shade tolerant secondary successional species in slash and burn agriculture (Jhum) fallows. *Proc. Indian natn. Sci. Acad.* **B47**: 768–773.
Kushwaha, S.P.S., Ramakrishnan, P.S. and Tripathi, R.S. 1983a. Competitive relationships of the plants of *Imperata cylindrica* established from rhizomes and from seeds. *Trop. Plant Sci. Res.* **1**: 53–57.
Kushwaha, S.P.S., Ramakrishnan, P.S. and Tripathi, R.S. 1983b. Population dynamics of *Imperata cylindrica* (L.) Beauv. var. Major related to slash and burn agriculture (Jhum) in north-eastern India. *Proc. Indian Acad. Sci.* (*Plant Sci.*) **92**: 313–321.
Kushwaha, S.P.S., Ramakrishnan, P.S. and Tripathi, R.S. 1983c. Competition between *Imperata cylindrica* and *Eupatorium odoratum* at two levels of NPK treatment. *Proc. Indian natn. Sci. Acad.* **B49**: 50–56.
Lal, R., Wilson, G.F. and Okigbo, B.N. 1979. Changes in properties of an Alfisol produced by various cover crops. *Soil Sci.* **127**: 377–382.
Lamb, D. 1980. Soil nitrogen mineralization in a secondary rainforest succession. *Oecologia* **47**: 257–263.
Landelout, H. and Meyer, J. 1954. Mineral element and organic material cycles in the equatorial forest of the Congo. *Oecol. Plant.* **7**: 1–21.
Leach, G. 1976. *Energy and Food Production.* IPC Science and Technology Press, Guildford. 137 pp.
Leburn, J. and Gilbert, G. 1954. *La Classification Ecologique des Forests du Congo.* I.N.E.A.C., Brussels, Publ. No. 63, 89 pp.
Lee, K.E. 1985. *Earthworms : Their Ecology and Relationships with Soils and Land Use.* Academic Press, Sydney.
Lee, R.B. and De Vore, I. (Eds) 1968. *Man the Hunter.* Aldine Publ. Co., Chicago.
Lewis, O. 1951. *Life in a Mexican Village : Tepoztlan Restudied.* University of Illinois Press, Urbana. 512 pp.
Lewis, W.M. Jr. 1974. Effects of fire on nutrient movement in a South Carolina pine forest. *Ecology* **55**: 1120–1127.
Litsinger, J.A. and Moody, K. 1976. Integrated pest management in multiple cropping systems. In : M. Stelly, (Ed.) *Multiple cropping*. pp. 293 316. *Am. Soc. Agron.* Madison, Wisconsin.
Livingston, R.B. and Allessio, M.L. 1968. Buried viable seeds in successional fields and forest stands, Harvard Forests, Massachusetts, *Bull. Torry Bot. Club* **95**: 58–69.
Lloyd, P.S. 1971. Effects of fire on the chemical status of herbaceous communities of the Derbyshire Dales. *J. Ecol.* **59**: 261–273.
Loach, K. 1970. Shade tolerance in tree seedling. II. Growth analysis of plants raised under artifical shade. *New Phytol.* **69**: 273–286.

Longman, K.A. and Jenik, J. 1974. *Tropical Forest and its Environment*. Longman, London.

Loomis, R.S. 1984. Traditional agriculture in America. *Ann. Rev. Ecol. Syst.* **15**: 449–478.

Lovejoy, P.E. 1986. Species leave the Ark one by one. In : B.G. Norton (Ed.) *The Preservation of Species : The value of Biological Diversity*. pp. 16–17. Princeton Univ. Press, Princeton, N.J.

Lovett-Doust, L. 1981. Population dynamics of local specialization in a clonal perennial (*Ranunculus repens*). I. The dynamics of ramets in contrasting habitats. *J. Ecol.* **69**: 743–755.

Lyr, H. and Hoffmann, G. 1967. Growth rates and growth periodicity of tree roots. In: J.A. Romberger and P. Mikola (Eds) International Review of Forestry Research, Vol. 2. pp. 181–236. Academic Press, New York.

MacArthur, R.H. and Wilson, E.O. 1967. *The Theory of Island Biogeography*, Princeton, Univ. Press, Princeton, N.J.

Maikhuri, R.K. and Ramakrishnan, P.S. 1990a. Ecological analysis of a cluster of villages emphasizing land use of different tribes in Meghalaya in north-east India. *Agric. Ecosys. Environ.* **31**: 17–37.

Maikhuri, R.K. and Ramakrishnan, P.S. 1990b. Comparative analysis of the village ecosystem function of different tribes living in the same area in Arunachal Pradesh in north-eastern India. *Agric. Systems* **35**: (in press)

Margalef, R. 1968. *Perspectives in Ecological Theory*. Univ. of Chicago Press, Chicago.

Marks, P.L. 1974. The role of pin cherry (*Prunus pensylvanica* L.) in the maintenance of stability in northern hardwood ecosystems. *Ecol. Monogr.* **44**: 73–88.

Marks, P.L. 1975. On the relation between extension growth and successional status of deciduous trees of the north-eastern United States. *Bull. Torry Bot. Club* **102**: 172–177.

McClure, F.A. 1966. *The Bamboos : A Fresh Perspective*. Harvard Univ. Press, Massachusetts.

Misra, B. 1976. A positive approach to the problem of shifting cultivation in eastern India and a few suggestions to the policy makers. In B. Pakem, J.B. Bhattacharjee, B.B. Dutta, and B. Duttaray. (Eds) *Shifting Cultivation in North-east India*. pp. 80–91. North-East India Council Social Sci. Res., Shillong.

Mishra, B.K. and Ramakrishnan, P.S. 1981. The economic yield and energy efficiency of hill agro-ecosystems at higher elevations of Meghalaya in north-eastern India. *Acta ecologica-Oecol. Applic.* **2**: 369–389.

Mishra, B.K. and Ramakrishnan, P.S. 1982. Energy flow through a village ecosystem with slash and burn agriculture in north-eastern India. *Agric. Systems* **9**: 57–72.

Mishra, B.K. and Ramakrishnan, P.S. 1983a. Secondary succession subsequent to slash and burn agriculture at higher elevations of north-east India. I. Species diversity, biomass and litter production. *Acta Oecologica – Oecol. Applic.* **4**: 95–107.

Mishra, B.K. and Ramakrishnan, P.S. 1983b. Secondary succession subsequent to slash and burn agriculture at higher elevations of north-east India. II. Nutrient cycling. *Acta Oecologica-Oecol. Applic.* **4**: 237–245.

Mishra, B.K. and Ramakrishnan, P.S. 1983c. Slash and burn agriculture at higher elevations in north-eastern India. I. Sediment, water and nutrient losses. *Agric. Ecosys. Environ.* **9**: 69–82.

Mishra, B.K. and Ramakrishnan, P.S. 1983d. Slash and burn agriculture at higher elevations in north-eastern India.II. Soil fertility changes. *Agric. Ecosyst. Environ.* **9**: 83–96.

Mishra, B.K. and Ramakrishnan, P.S. 1984. Nitrogen budget under rotational bush fallow agriculture (Jhum) at higher elevations of Meghalaya in north-eastern India. *Plant and Soil* **81**: 37–46.

Mishra, K.C. and Ramakrishnan, P.S. 1988. Earthworm population dynamics in different jhum fallows developed after slash and burn agriculture in north-eastern India. *Proc. Indian Acad. Sci.* (*Anim.Sci.*) **97**: 309–318.

Mitchell, R. 1979. *An Analysis of Indian Agroecosystem*. Interprint, New Delhi. 180 pp.

Montenegro, G., Aljaro, M.E. and Kummerow, J. 1979. Growth dynamics of Chilean matorral shrubs. *Bot. Gaz.* **140**: 114–119.

Mooney, H.A. and Billings, W.D. 1960. The annual carbohydrate cycle of alpine plants as related to growth. *Amer. J. Bot.* **47**: 594–598.

Mooney, H.A., Bonnicksen, T.M., Christensen, N.L., Lotan, J.E. and Reiners, W.A. (Eds) 1981. *Fire Regimes and Ecosystem Properties.* U.S.D.A. Forest Service. Gen. Tech. Rept. WO-26, Hawaii, Honolulu. 593 pp.

Mott, J.J. 1982. Fire and survival of *Stylosanthes* spp. in dry savanna woodlands of northern territory. *Austr. J. Agric. Res.* **33**: 203–211.

Mutsaers, H.J.W., Mbouemboue, P. and Boyomom, M. 1981. Traditional food crop growing in the Yaounde area (cameron). Part I. Synopsis of the system. *Agro-Ecosystems* **6**: 273–287.

Myers, N. 1980. Conversion of Tropical Moist Forests. National Acad. Sci., Washington, D.C. 205 pp.

Nair, P.K.R. 1984. *Soil Productivity Aspects of Agroforestry.* International Council for Research in Agroforestry (ICRAF), Nairobi, Kenya.

Nakano, K. and Syahbuddin, 1989. Nutrient dynamics in forest fallows in south-east Asia. In: J. Proctor (Ed.) *Mineral Nutrients in Tropical Forest and Savanna Ecosystems* pp. 325–336. Blackwell Scientific Pub., Oxford.

National Remote Sensing Agency, 1983a *Mapping of Forest Cover in India from Satellite Imagery.* 1972–1975 and 1980–1982. Summary Report, NRSA, Hyderabad. 8 pp.

National Remote Sensing Agency, 1983b. *Mapping of Forest Cover in India from Satellite Imagery.* 1972–1975 and 1980–1982. Summary Report, N.E. States and Union Territories, NRSA, Hyderabad.

National Research Council, 1975. *Underexploited Tropical Plants with Promising Economic Value.* National Acad. Sci. Washington, D.C. 189 pp.

National Research Council, 1979. *Tropical Legumes : Resources for the Future.* National Acad. Sci., Washington D.C. 331 pp.

National Research Council, 1982. *Ecological Aspects of Development in the Humid Tropics.* National Academy of Sciences, Washington, D.C. 297 pp.

Nelson, N.D., Burk, J. and Icebrands, J.G. 1981. Crown architecture of short rotation, intensively cultured *Populus.* I. Effects of close spacing on first order branch characteristics. *Can. J. For. Res.* **11**: 73–81.

Newell, S.J. and Tramer, E.J. 1978. Reproductive strategies in herbaceous plant communities during succession. *Ecology* **59**: 228–234.

Newhouse, M.E. and Madgwick, H.A.I. 1968. Comparative seedling growth of four hardwood species. *Forest Science* **14**: 27–30.

Nguu, N.V. and Palis, R.K. 1977. Energy input and output of a modern and traditional cultivation system in lowland rice culture. *Phillippines J. Biol.* **6**: 1–8.

Noble, I.R. and Slatyer, R.O. 1980. The use of vital attributes to predict successional changes in plant communities subject to recurrent disturbances. *Vegetatio* **43**: 5–21.

Norman, M.J.T. 1978. Energy inputs and outputs of subsistence cropping systems in the tropics. *Agro-Ecosystems* **4**: 355–366.

North Eastern Council, 1982. *Shifting Cultivation in North-Eastern Region.* North-Eastern Council Pub. No. 17, Shillong.

North-Eastern Council, 1984. *Basic Statistics of North-Eastern India.* North-Eastern Council Pub., Shillong.

Numata, M., Ikusima, I and Ohga, N. 1974. Ecological aspects of bamboo flowering: Ecological studies of bamboo forests in Japan. XIII. *Bot. Mag.* Tokyo **87**: 271–284.

Nutman, P.S. (Ed.), 1976. *Symbiotic Nitrogen Fixation in Plants.* Cambridge University Press, Cambridge, 584 pp.

Nye, P.H. 1955. Some soil forming processes in the humid tropics. IV. The action of soil fauna. *J. Soil Sci.* **6**: 78–84.

Nye, P.H. 1961. Organic matter and nutrient cycles under moist tropical forest. *Plant and Soil* **13**: 333–346.

Nye, P.H. and Bertheux, M.H. 1957. The distribution of phosphorus in forest and savannah soils of the Gold Coast and its agricultural significance. *J. Agric. Sci.* Camb. **49**: 149–159.

Nye, P.H. and Greenland, D.J. 1960. *The Soil under Shifting Cultivation.* Technical Communication No. 51, Commonwealth Bureau of soils, Harpenden, England. 156 pp.

Nye, P.H. and Greenland, D.T. 1964. Changes in the soil after clearing tropical forest. *Plant and Soil* **21**: 101-112.

Nye, P.H. and Tinker, P.B. 1977. *Solute Movement in the Soil Root System.* California Univ. Press, Berkeley.
Odum, E.P. 1969. The strategy of ecosystem development. *Science*, New York **164**: 262–270.
Ogden, J. 1974. The reproductive strategy of higher plants. II. The reproductive strategy of *Tussilago farfara. J. Ecol.* **62**: 291–324.
Okafor, F.C. 1987. Population pressure and land resource depletion in southeastern Nigeria, *Applied Geography* **7**: 242–256.
Oohata, S. and Shidei, T. 1971. Studies on the branching structure of trees. I. Bifurcation ratio of trees in Horton's law. *Jap. J. Ecol.* **21**: 7–14.
Partap, T. and Kapoor, P. 1985. The Himalayan grain Chenopods. I. Distribution and ethnobiology. *Agric. Ecosyst. Environ.* **14**: 185–199.
Patnaik, S. 1988. *Studies on Land use and Village Ecosystem Function in the Khasi Hills of Meghalaya.* Ph.D. Thesis, North-Eastern Hill Univ., Shillong. 162 pp.
Patnaik, S. and Ramakrishnan, P.S. 1989. Comparative study of energy flow through village ecosystems of two co-existing communities (the Khasis and the Nepalis) of Meghalaya in north-east India. *Agric. Systems* **30**: 245–267.
Patnaik, S. and Ramakrishnan, P.S. 1990. Cation budget under terrace agroecosystem in Meghalaya in north-east India. *Proc. Indian Acad. Sci.* (*Plant Sci.*), **100**: 23–31.
Payne, W.J.A. 1985. A review of the possibilities for integrating cattle and tree crop production systems in the tropics. *For. Ecol. Manage.* **12**: 1–36
Perozzi, R.G. and Bazzaz, F.A. 1978. The response of an early sussessional community to shortened growing season. *Oikos* **31**: 89–93.
Pickett., S.T.A. and Kempf, J.S. 1980. Branching patterns in forest shrubs and understorey trees in relation to habitat. *New Phytol.* **86**: 210–228.
Pimental, D. and Pimental, M. 1979. *Food, Energy and Society.* Edward Arnold, London. 165 pp.
Purdie, R.O. and Slatyer, R.O. 1976. Vegetation succession after fire in woodland sclerophyll communities in south-eastern Australia. *Austr. J. Ecol.* **1**: 223–236.
Puri, G.S. 1960. *Indian Forest Ecology.* Vol. I. Oxford Book and Stationary Co., New Delhi. 318 pp.
Raison, R.J. 1979. Modification of the soil environment by vegetation fires, with particular reference to nitrogen transformations, a review. *Plant and Soil* **51**: 73–108.
Ram, S.C. 1986. *Ecosystem Structure and Function of Seral Communities of Degraded Environment at Cherrapunji and Adjoining Areas.* Ph.D. Thesis. North-Eastern Hill Univ., Shillong. 218 pp.
Ram, S.C. and Ramakrishnan, P.S. 1988a. Growth and allocation strategy of some important species of seral grasslands at Cherrapunji in north-eastern India. *Proc. Indian Acad. Sci.* (*Plant Sci.*) **98**: 13–23.
Ram, S.C. and Ramakrishnan, P.S. 1988b. Hydrology and soil fertility of degraded grasslands at Cherrapunji in north-eastern India. *Environ. Conserv.* **15**: 29–35.
Ram, S.C. and Ramakrishnan, P.S. 1988c. Litter production and decomposition patterns in seral grasslands at Cherrapunji in north-eastern India. *Pedobiologia* **32**: 65–76.
Ram, S.C. and Ramakrishnan, P.S. 1989. Productivity and biomass turnover rates for seral grasslands of Cherrapunji in north-eastern India. *Curr. Sci.* **58**: 523–525.
Ramakrishnan, P.S. 1970. Nutritional requirements of the edaphic ecotypes in *Melilotus alba* Medic. III. Interference between the calcareous and acidic populations in the two soil types. *New Phytol.* **69**: 81–86.
Ramakrishnan, 1972. Individual adaptation and its significance in population dynamics In: V. Puri and Y.S. Murty (Eds) *Biology of Land Plants.* pp 344–355. Sarita Prakashan, Meerut, India.
Ramakrishnan, P.S. 1978. Approaches to studies on forest ecosystem function in the north-eastern hill region. In: D.N. Sen and R.P. Bansal (Eds) *Environmental Physiology and Ecology of Plants.* pp. 97–108. Singh and Singh, Dehra Dun.
Ramakrishnan, P.S. 1980. Ecological impact of jhum (slash and burn agriculture) on forested ecosystem of north-eastern India. *INSA Newsletter* **60**: 3–7.
Ramakrishnan, P.S. 1981. Form and function in ecologic race formation. In : S.P. Sen (Ed). *Recent Advances in Plant Sciences.* pp. 279–294. S.M. Sircar Memomal Volume, Calcutta.

Ramakrishnan, P.S. 1983. Socio-economic and cultural aspects of jhum in the north-east and options for eco-development of tribal areas. In: N.P. Chaubey (Ed.) *Tribal Techniques, Social Organisation and Development: Disruption and Alternates.* pp 12–30. Indian Acad. Social Sci., Allahabad.

Ramakrishnan, P.S. 1984a. The science behind rotational bush fallow agriculture system (Jhum). *Proc. Indian Acad. Sci. (Plant Sci.)* **93**: 397–400.

Ramakrishnan, P.S. 1984b. Let the tribals decide what they want. *Science Age* **2**: 8–11.

Ramakrishnan, P.S. 1984c. Problems and prospects of conservation of plant resources in the north-eastern hill region of India. In: S.K. Jain (Ed.) *Conservation of Tropical Plant Resources.* pp 172–180. Bot. Surv. India., Howrah.

Ramakrishnan, P.S. 1984d. The need to conserve Slient Valley and tropical rain-forest ecosystem in India. *Environ. Conserv.* **11**: 170–171.

Ramakrishnan, P.S. 1985a. Jhum cultivation-prospects for developing countries. *Science Technology and Development,* (ICSU-COSTED) **9**: 1–3.

Ramakrishnan, P.S. 1985b. Humid tropical forests. In: *Research on Humid Tropical Forests.* Regional Meeting National MAB committee of Central and South Asian countries. MAB-India, Ministry Environ. and Forests, New Delhi. pp. 39.

Ramakrishnan, P.S. 1985c. Tribal man in the humid tropics of the north-east. *Man in India* **65**: 1–32.

Ramakrishnan, P.S. 1985d. Conversion. of rain forests in north-eastern India. In : J.S. Singh (Ed.) *Environmental Regeneration in the Himalaya : Concepts and Strategies.* pp. 69–84. Central Environmental Association, Nainital.

Ramakrishnan, P.S. 1985e. Cherrapunji – A wet desert ? *Expanse* **1**: 65–66.

Ramakrishnan, P.S. 1985f. The Eucalyptus controversy: A new angle. *Science Age* **3**: 11–13.

Ramakrishnan, P.S. 1986a. Morphometric analysis of growth and architecture of tropical trees and their ecological significance. pp. 209–222. Naturalia Monspeliensia-Colloque International sur l'Arbore, Montpellier.

Ramakrishnan, P.S. 1986b. Environmental communication. In: V.K. Dubey, K.N. Pandey, and S. Kumar (Eds) *Sampreshan,* National Council of Developmental Communication, Varanasi, India. pp 32–38.

Ramakrishnan, P.S. 1987a. Role of tree architecture in agroforestry. In: P.K. Khosla, and D.K. Khurana (Eds) *Agroforestry for Rural Needs.* pp. 112–130. Indian Soc. Tree Scientists, Solan, India.

Ramakrishnan, P.S. 1987b. Tropical tree architecture and photosynthesis. *Proc. Indian natn. Sci. Acad.* **B53**: 565–571.

Ramakrishnan, P.S. 1987c. Management of biosphere reserves with peoples' participation. In: *Biosphere Reserves.* pp. 220–226. Ministry of Environ. and Forests, Govt. of India, New Delhi.

Ramakrishnan, P.S. 1987d. Shifting agriculture and rain forest ecosystem management. *Biology International* **15**: 17–18.

Ramakrishnan, P.S. 1987e. Energy flow and shifting cultivation. In : T.M. Vinod Kumar and D.K. Ahuja (Eds) *Planning for the Indian Himalaya.* pp. 247–276. Wiley Eastern, New Delhi.

Ramakrishnan, P.S. 1988a. Successional theory: Implications for weed management in shifting agriculture, mixed cropping, and agroforestry systems. In: M.A. Altieri, and M. Liebman (Eds) *Weed Management in Agroecosystems : Ecological Approaches.* pp. 183–196. CRC Press, Inc. Boca Raton, Florida.

Ramakrishnan, P.S. 1988b. Patterns of primary terrestrial productivity and ecosystem function. *Proc. Indian natn. Sci. Acad.* **B54**: 349–360.

Ramakrishnan, P.S. 1988c. Forest farmer and ecosystem redevelopment in the Asian humid tropics. In: S. Adisoemarto (Ed.) *Ecodevelopment Process for Degraded Land Resources in Southeast Asia.* pp. 103–118. UNESCO-MAB (Indonesia), Jakarta, Indonesia.

Ramakrishnan, P.S. 1988d. Ecology of shifting agriculture. In : R.N. Trivedi (Ed.) *Environment Pollution and its Impact on the Organism.* pp. 47–83. Bharati Bhawan, Patna.

Ramakrishnan, P.S. 1989. Nutrient cycling in forest fallows in north-eastern India. In : J. Proctor (Ed.) *Mineral Nutrients in Tropical Forest and Savanna Ecosystems.* pp. 337–352. Blackwell Scientific Publications, Oxford.

Ramakrishnan, P.S. 1990. Agricultural systems of the north- eastern hill region of India. In : S.R. Gliessman (Ed.) *Agroecology- Researching the Ecological Basis for Sustainable Agriculture.* pp. 251–274. Springer-Verlag, New York.

Ramakrishnan, P.S. (ed.) 1991. *Ecology of Biological Invasion in the Tropics.* National Institute of Ecology and International Science Publishers, New Delhi.

Ramakrishnan, P.S. and Das, A.K. 1983. Studies on pine ecosystem function in Meghalaya. *Trop. Plant. Sci. Res.* **1**: 15–24.

Ramakrishnan, P.S. and Gupta, U. 1972a. Nutrient factors influencing the distribution of two closely related species of *Argenone. Weed Res.* **12**: 234–240.

Ramakrishnan, P.S. and Gupta, U. 1972b. Ecotypic differences in *Cynodon dactylon* (L.) Pers. related to weed - crop interference. *J. appl. Ecol.* **9**: 333–339.

Ramakrishnan, P.S. and Jeet, N. 1972. Competitive relationships existing between two closely related species of *Argenone* living in the same area. *Oecologia* **9**: 279–288.

Ramakrishnan, P.S. and Kumar, S. 1971. Productivity and plasticity of wheat and *Cynodon dactylon* (L.) Pers. in pure and mixed stands. *J. appl. Ecol.* **8**: 85–98.

Ramakrishnan, P.S. and Maikhuri, R.K. 1988. Ecology and management of jhum cultivation in the Himalayas. In : *Souvenir Vol. III. International Rangeland Congress.* pp. 26–37. Range Manage, Soc., India, Jhansi, India.

Ramakrishnan, P.S. and Mishra, B.K. 1981. Population dynamics of *Eupatorium adenophorum* Spreng. During secondary succession after slash and burn agriculture. (jhum) in north-eastern India. *Weed Res.* **22**: 77–84.

Ramakrishnan, P.S. and Ram., S.C. 1988. Vegetation, biomass and productivity of seral grasslands of Cherrapunji in north-east India. *Vegetatio* **74**: 47–53.

Ramakrishnan, P.S. and Saxena, K.G. 1984. Nitrification of Cherrapunji. *Curr. Sci.* **53**: 107–109.

Ramakrishnan, P.S. and Shukla., R.P. 1982. On the relation among growth strategies, allocation pattern, productivity and successional status of trees of a sub-tropical forest community. In : P.K. Khosla (Ed.) *Improvement of Forest Biomass.* pp. 403–412. Indian Soc. Tree Scientists, Solan, India.

Ramakrishnan, P.S. and Singh, J. 1983. Distribution of fine roots in a sub-tropical humid forest of north-east India. *Trop. Plant Sci. Res.* **1**: 163–165.

Ramakrishnan, P.S. and Swamy, P.S. 1987. Role of mycorrhizae in the management of tropical rain forest. In: A.K. Varma, A.K. Oka, K.G. Mukherji, K.V.B.R. Tilak and J. Raj (Eds) *Mycorrhiza Round Table : Proceedings of a Workshop.* pp. 471–487. International Develop. Res. Centre and Jawaharlal Nehru Univ., New Delhi.

Ramakrishnan, P.S. and Toky, O.P. 1978. Preliminary observations on the impact of jhum (shifting agriculture) on the forested ecosystem. In : *Seminar on Resources, Development and Environment in the Himalayan Region.* pp. 343–354. Department of Science and Technology, Govt. of India, New Delhi.

Ramakrishnan, P.S. and Toky, O.P. 1981. Soil nutrient status of hill agro-ecosystems and recovery pattern after slash and burn agriculture (jhum) in north-eastern India. *Plant and Soil.* **60**: 41–64.

Ramakrishnan, P.S. and Toky, O.P. 1983. Some aspects of environmental degradation in north-eastern hill areas of India. In: T.V. Singh, and J. Kaur (Eds) *Studies in Eco-development: Himalayas Mountains and Men.* pp. 149–156. Print House (India), Lucknow.

Ramakrishnan, P.S. and Vitousek, P.M. 1989. Ecosystem level processes and the consequences of biological invasions. In : J.A. Drake, H.A. Mooney, F. di Castri, R.H. Groves, F.G. Kruger, M. Rejmanek, and M. Williamson (Eds) *Biological Invasions : A Global Perspective.* pp. 281–300. SCOPE – John Wiley and Sons, Chichester.

Ramakrishnan, P.S., Toky, O.P. and Mishra, B.K. 1981a. Jhum-An ecological assessment. In : A. Singh and P. Wahi (Eds) *Souvenir Volume, Silver Jubilee Symposium.* pp 41–49. *International Society for Tropical Ecology*, Varanasi.

Ramakrishnan, P.S., Toky, O.P., Mishra, B.K. and Saxena, K.G. 1981b. Slash and burn agriculture in north-eastern India. In : H. A. Mooney T.M. Bonnicksen, N.L. Christensen, J.E. Lotan, and W.A. Reiners (Eds) *Fire Regimes and Ecosystem Properties.* pp. 570–586. USDA. Forest Serv. Gen. Tech. Rept., W.O. 26, Washington, D.C.

Ramakrishnan, P.S., Shukla, R.P. and Boojh, R. 1982. Growth strategies of trees and their application to forest management. *Curr. Sci.* **51**: 448–455.

Ranjhan, S.K. 1977. *Animal Nutrition and Feeding Practices in India.* Vikas Publ. House Pvt. Ltd., New Delhi. 339 pp.

Rao, K.S. and Ramkrishnan, P,S. 1987a. Comparative analysis of the population dynamics of two bamboo species, *Dendrocalamus hamiltonii and Neohouzeua dulloa,* in a successional environment. *Forest Ecol. Manage.* **21**: 177–189.

Rao, K.S. and Ramakrishnan, P.S. 1987b. Socio-economic analysis of bamboo resources in East Khasi Hills district of Meghalaya. *Man In India* **67**: 221–231.

Rao, K.S. and Ramakrishnan, P.S. 1988a. Architectural plasticity of two bamboo species (*Neohouzeua dulloa* A Camus. and *Dendrocalamus hamiltonii* Nees. and Arn.) in successional environments in north-east India. *Proc. Indian Acad. Sci.* (*Plant Sci.*) **98**: 121–133.

Rao, K.S. and Ramakrishnan, P.S. 1988b. Leaf dynamics of two bamboo species (*Neohouzeua dulloa* A. Camus and *Dendrocalamus hamiltonii* Nees. and Arn.) in successional environments in north-east India. *Proc. Indian natn. Sci. Acad.* **B54**: 63–69.

Rao, K.S. and Ramakrishnan, P.S. 1989. Role of bamboos in nutrient conservation during secondary succession following slash and burn agriculture (jhum) in north-east India. *J. appl. Ecol.* **26**: 625–633.

Rappaport, R.A. 1971. The flow of energy in an agricultural society. *Sci. Am.* **225**: 117–132.

Raven, P.H. 1981. Tropical rainforests: a global responsibility. *Natural History* **90**: 28–32.

Raynal, D.J. and Bazzaz, F.A. 1973. Establishment of early successional plant populations on forest and prairie soil. *Ecology* **54**: 1335–1343.

Reddy, A.K.N. 1981. An Indian village agricultural ecosystem- Case study of Ungra village. Part II : Discussion. *Biomass* **1**: 77–88.

Reiners, W.A. and Reiners, N.M. 1970. Energy and nutrient dynamics of forest floors in three Minnesota forests. *J. Ecol.* **58**: 497–519.

Revelle, R. 1976. Energy use in rural India. *Science* **192**: 969–975.

Reynold, G.H. and Bohning, J.W. 1956. Effect of burning on a desert grass-shrub range in southern Arizona. *Ecology* **13**: 393–401.

Rice, E.L. and Pancholy, S.K. 1974. Inhibition of nitrification by climax ecosystems. III. Inhibitors other than tannins. *Amer J. Bot.* **61**: 1095–1103.

Richards, P.W. 1952. *Tropical Rain Forest.* Cambridge Univ. Press, Cambridge.

Risi, J.C. and Galwey, N.W. 1984. The *Chenopodium* grains of *Andes. Adv. Appl. Biol.* **10**: 145–195.

Robertson, G.P. and Vitousek, P.M. 1981. Nitrification potentials in primary and secondary succession. *Ecology* **62**: 376–386.

Rodin, L.E. and Bazilevich, N.T. 1967. *Production and Mineral Cycling in Terrestrial Vegetation.* Oliver and Boyd, Edinburgh.

Romberger, J.A. 1963. *Meristems, Growth and Development in Woody Plants.* U.S. Dept. Agric., Forest Service Tech. Bull. 1293.

Roos, F.H. and Quinn, J.S. 1977. Phenology and reproductive allocation in *Andropogon scoparius* (Gramineae) populations in communities of different successional stages. *Amer J. Bot.* **64**: 535–540.

Ross, R. 1954. Ecological studies on the rain forest of southern Nigeria. III. Secondary succession in the Shasha forest reserve. *J. Ecol.* **42**: 259–282.

Rosswall, T. and Panstian, K. 1984. Cycling of nitrogen in modern agricultural systems. *Plant and Soil* **76**: 3–21.

Russell, R.S. 1977. *Plant Root Systems : Their Function and Interaction with the Soil.* McGraw Hill Co. London.

Ruthenberg, H. 1971. *Farming Systems in the Tropics.* Clarendon Press, Oxford. 366 pp.

Sahu, S.D. 1978. *Energy Requirements for Some Selected Cultural Practices in a Hilly Region of Meghalaya.* M.Tech. Thesis. I.I.T., Kharagpur.

Salisbury, E.J. 1942. *The Reproductive Capacity of Plants.* Bell, London 244 pp.

Sanchez, P A 1989. Soils. In: H. Leith and Werger, M.J.A. (Eds) *Tropical Rain Forest Ecosystems.* pp. 73–88. Elsevier, Amsterdam.

Sanchez, P.A. and Cochrane, T.T. 1980. Soil constraints in relation to major farming systems of tropical America. In : *Priorities for Alleviating Soil-Related Constraints to Food Production in the Tropics.* pp. 107–140. IRRI, Los Banos, Phillipines.

Sanchez, P.A. and Salinas, J.G. 1981. Low-input technology for managing oxisols and ultisols in tropical America. *Adv. Agron.* **34**: 279–406.

San Jose, J.J. and Medina, Y.E. 1976. Organic matter production in the *Trachypogon* savanna at Colabozo, Venezuela. *Trop. Ecol.* **17**: 113–124.

Sarukhan, J. and Harper, J.L. 1973. Studies on plant demography : *Ranunculus repens* L., *R. bulbosus* L. and *R. acris* L. I. Population flux and survivorship. *J. Ecol.* **61**: 675–716.

Sarma, S.N. 1985 *Biology of Imperata cylindrica* (L.) Beauv. Ph.D. Thesis, North-Eastern Hill Univ. 178 pp.

Sasson, A. 1987. Conserving plant resources. *World Scientist*, **16**: 50–61.

Sasson, A. 1990. *Feeding Tomorrow's World.* UNESCO, Paris pp. 805.

Satchell, J.E. 1967. Lumbricidae. In : A Burges and F. Raw (Eds) *Soil Biology*. pp. 259–322. Academic Press, London.

Satchell, J.E. 1983. Earthworm ecology in forest soils. In : J.E. Satchell (Ed.) *Earthworm Ecology from Darwin to Vermiculture.* pp. 161–170. Chapman and Hall, London.

Saxena, K.G. and Ramakrishnan, P.S. 1981. Growth strategy and allocation pattern of *Eupatorium odoratum* and *Imperata cylindrica* at different fertility levels of the soil. *Proc. Indian natn. Sci. Acad.* **B47**: 861–866.

Saxena, K.G. and Ramakrishnan, P.S. 1982a. Reproductive efficiency of secondary successional herbaceous populations subsequent to slash and burn of sub-tropical humid forests in north-eastern India. *Proc. Indian Acad. Sci (Plant Sci.).* **91**: 61–68.

Saxena, K.G. and Ramakrishnan, P.S. 1982b. Partitioning of biomass and nutrients in the secondary successional herbaceous populations subsequent to slash and burn. *Proc. Indian natn. Sci. Acad.* **B48**: 807–818.

Saxena, K.G. and Ramakrishnan, P.S. 1983a. Growth and allocation strategies of some perennial weeds of slash and burn agriculture (jhum) in north-eastern India. *Can. J. Bot.* **61**: 1300–1306.

Saxena, K.G. and Ramakrishnan, P.S. 1983b. Growth, resource allocation patterns and nutritional status of some dominant annual weeds of shifting agriculture (Jhum) in north-eastern India. *Acta Oecologica-Oecol. Plant.* **4**: 323–333.

Saxena, K.G. and Ramakrishnan, P.S. 1984a. Growth and patterns of resource allocation in *Eupatorium odoratum* L. in the secondary successional environments following slash and burn agriculture (jhum) *Weed Res.* **24**: 127–134.

Saxena, K.G. and Ramakrishnan, P.S. 1984b. Herbaceous vegetation development and weed potential in slash and burn agriculture (jhum) in N.E. India. *Weed Res.* **24**: 135–142.

Saxena, K.G. and Ramakrishnan, P.S. 1984c. C_3/C_4 species distribution among successional herbs following slash and burn in north-eastern India. *Acta Oecologica-Oecol. Plant.* **5**: 335–346.

Saxena, K.G. and Ramakrishnan, P.S. 1986. Nitrification during slash and burn agriculture (jhum) in north-eastern India. *Acta Oecologica-Oecol. Plant.* **7**: 307–319.

Saxena, K.G. and Ramakrishnan, P.S. 1988. Eco-physiology and organization of the early successional herbaceous communities. *Int. J. Ecol. Environ. Sci.* **14**: 1–19.

Schier, G.A. and Zasda, J.C. 1973. Role of carbohydrate reserves in the development of root suckers in *Populus tremuloides. Can. J. For. Res.* **3**: 243–250.

Schultz, T.W. 1964. *Transforming Traditional Agriculture.* Yale Univ. Press, New Haven, Conn. 212 pp.

Sharma, E. and Ambasht, R.S. 1988. Nitrogen accretion and its energetics in the Himalayan alder. *Functional Ecol.* **2**: 229–235.

Sharpley, A.N. and Syers, J.K. 1976. Potential role of earthworm casts for the phosphorus enrichment of run-off water. *Soil Biol. Biochem.* **8**: 341–346.

Sharrow, S.H. and Wright, H.A. 1977. Effect of fire, ash and litter on soil nitrate, temperature, moisture and Tobosa grass production, turnover and efficiencies of energy capture and water use. *J. Ecol.* **66**: 573–597.

Shinozaki, K., Yoda, K., Hozumi, K. and Kira, T. 1964. A quantitative analysis of plant form the pipe model theory. I. Basic analysis. *Japan. J. Ecol.* **14**: 97–105.

Shukla, R.P. and Ramakrishnan, P.S. 1981a. On photoblastism in seed germination of *Duabanga sonneratioides* Ham. *Proc. Indian Acad. Sci.* (*Plant Sci.*) **90**: 547–553.

Shukla, R.P. and Ramakrishnan, P.S. 1981b. Adaptive significance of seed polymorphism in *Lagerstroemia parviflora* Roxb. *Curr. Sci.* **50**: 685–688.

Shukla, R.P. and Ramakrishnan, P.S. 1981c. The nature and ecological significance of heterophylly in *Artocarpus chaplasha* Roxb. *Proc. Indian natn. Sci. Acad.* **B47** : 551–556.

Shukla, R.P. and Ramakrishnan, P.S. 1982a. Comparative study on field germination and establishment of early versus late successional trees in north-eastern India. *Proc. Indian natn. Sci. Acad.* **B48**: 115–120.

Shukla, R.P. and Ramakrishnan, P.S. 1982b. Phenology of trees in a sub-tropical humid forest in north-eastern India. *Vegetatio* **49**: 103–109.

Shukla, R.P. and Ramakrishnan, P.S. 1984a. Leaf dynamics of tropical trees related to successional status. *New Phytol.* **97**: 697–706.

Shukla, R.P. and Ramakrishnan, P.S. 1984b. Biomass allocation strategies and productivity of tropical trees related to successional status. *Forest Ecol. Manage.* **9**: 315–324.

Shukla, R.P. and Ramakrishnan, P.S. 1986. Architecture and growth strategies of tropical trees in relation to successional status. *J. Ecol.* **74**: 33–46.

Simmon-Sylvestre, G. 1976. Preliminary results on the effects of steam sterilization of soils. *Ann. Agron.* **18**: 243–266.

Sims, P.L. and Singh, J.S. 1978. The structure and function of ten western North American grasslands. III. Net primary production, turnover and efficiencies of energy capture and water use. *J. Ecol.* **66**: 573–597.

Singh, A. 1987. Spectral separability of tropical forest cover classes. *Internat. J. Remote Sensing* **8**: 971–979.

Singh, J. and Ramakrishnan, P.S. 1981. Biomass and nutrient movement through litter in *Shorea robusta* Gaertn. plantations in Meghalaya. *Proc. Indian natn. sci. Acad.* **B47**: 852–860.

Singh, J. and Ramakrishnan, P.S. 1982a. Structure and function of a sub-tropical humid forest of Meghalaya. I. Vegetation, biomass and its nutrients. *Proc. Indian Acad. Sci.* (*Plant Sci.*) **91**: 241–253.

Singh, J. and Ramakrishnan, P.S. 1982b. Structure and function of a sub-tropical humid forest of Meghalaya. II. Litter dynamics and nutrient cycling. *Proc. Indian Acad. Sci.* (*Plant Sci.*) **91**: 255–268.

Singh, J. and Ramakrishnan, P.S. 1982c. Structure and function of a sub-tropical humid forest of Meghalaya III. Nutrient flow through water. *Proc. Indian Acad. Sci.* (*Plant Sci.*) **3**: 269–280.

Singh, J. and Ramakrishnan, P.S. 1983. Growth analysis and productivity of *Shorea robusta* Geartn. plantations in Meghalaya. *Trop. Ecol.* **24**: 260–270.

Singh, J.S. and Yadava, P.S. 1974. Seasonal variation and net primary productivity of a tropical grassland at Kurukshetra, India. *Ecol. Monogr.* **44**: 351–376.

Smith, E.L. 1960. Effects of burning and clipping at various times during the wet season on tropical tall grass range in northern Australia. *J. Range Manage.* **13**: 197–203.

Smith, W., Bormann, F.H. and Likens, G.E. 1968. Response of chemoautotrophic nitrifiers to forest cutting. *Soil Sci.* **106**: 471–473.

Smith, W.K. and Nobel, P.S. 1977. Influence of seasonal changes in leaf morphology on water use efficiency for three desert broad leaf shrubs. *Ecology* **58**: 1033–1043.

Soemarwoto, O. 1975. Rural ecology and development in Java, In : E.H. Van Dobber, and R.H. Lowe McConnel (Eds) *Unifying Concepts of Ecology*. pp. 275–287. Junk Publishers, The Hague,

Spedding, C.R.W. 1975. *The Biology of Agricultural Systems*. Academic Press, London.

Spedding, C.R.W. 1979. *An Introduction to Agricultural Systems*. Applied Science Publishers, London, 169 pp.

Spedding, C.R.W. and Walsingham, J.M. 1976. The production and use of energy in agriculture. *J. Agric. Econ.* **27**: 19–30.

Spencer, J.E. 1966. Shifting cultivation in South-eastern Asia. Publication in Geography, No. 199. University of California, Berkeley. 247 pp.

Stark, N. 1971. Nutrient cycling I. Nutrient distribution in some Amazonian soils. *Trop. Ecol.* **12**: 24–50.

Stark, N. and Jordan, C.F. 1977. Nutrient retention by the root mat of an Amazonian rainforest. *Ecology* **58**: 434–437.

Stebbins, G.L. 1974. *Flowering Plants Evolution Above the Species Level*. Harvard Univ. Press, Cambridge.

Steingraeber, D.A., Kasht, L.J. and Frank, D.H. 1979. Variation of shoot morphology and bifurcation ratio in sugar maple (*Acer saccharum*) seedling. *Amer. J. Bot.* **66**: 441–445.

Steinhart, J.S. and Steinhart, C.E. 1974. Energy use in the U.S. food system. *Science* **184**: 307–316.

Stinner, B.R., Odum, E.P. and Grosseley, D.A. Jr. 1983. Nutrient uptake by vegetation in relation to other ecosystem processes in conventional tillage, no-tillage and old-field systems. *Agric. Ecosyst. Environ.* **10**: 1–13

Stout, B.A., Myers, C.A., Hurand, A. and Faidley, L.W. 1979. *Energy for World Agriculture.* FAO Agriculture Series No. 7. 286 pp.

Strahler, A.M. 1957.Quantitative analysis of watershed geomorphology. *Trans. Am. Geophys. Union* **38**: 913–920.

Stromgaard, C.R.W. 1984. The immediate effect of burning and fertilization. *Plant and Soil* **80**: 307–320.

Struik, G.J. and Bray, J.R. 1970. Root-shoot ratios of native forest herbs and *Zea mays* at different soil moisture levels. *Ecology* **5**: 892–893.

Sundarraj, T.S.P. and Mitchell, R. 1987. Analysis of ecosystem structure and energy flow in a South Indian village. In : F.J. Hitzhusen and R.D. MacGregor (Eds) *A Multi-Disciplinary Approach to Renewable Energy in Developing Countries.* pp. 141–167. Publishing Horizons, Columbus, Ohio.

Swamy, P.S. and Ramakrishnan, P.S. 1987a. Weed potential of *Mikania micrantha* H.B.K., and its control in fallows after shifting agriculture (jhum) in north-east India. *Agric. Ecosyst. Environ.* **18**: 195–204.

Swamy, P.S. and Ramakrishnan, P.S. 1987b. Contribution of *Mikania micrantha* H.B.K. during secondary succession following slash and burn agriculture (jhum) in north-east India. I. Biomass, litterfall and productivity. *Forest Ecol. Manage.* **22**: 229–237.

Swamy, P.S. and Ramakrishnan, P.S. 1987c. Contribution of *Mikania micrantha* H.B.K. during secondary succession following slash and burn agriculture (jhum) in north-east India. II. Nutrient cycling. *Forest Ecol. Manage.* **22**: 239–249.

Swamy, P.S. and Ramakrishnan, P.S. 1987d. Effect of fire on population dynamics of *Mikania micrantha* H.B.K. during early succession after slash and burn agriculture (jhum) in north-eastern India. *Weed Res.* **27**: 397–403.

Swamy, P.S. and Ramakrishnan, P.S. 1988a. Effect of fire on growth and allocation strategies of *Mikania micrantha* under early successional environments. *J. appl. Ecol.* **25**: 653–658.

Swamy, P.S. and Ramakrishnan, P.S. 1988b. Growth and allocation patterns of *Mikania micrantha* in successional environments after slash and burn agriculture. *Can. J. Bot.* **66**: 1465–1469.

Swamy, P.S. and Ramakrishnan, P.S. 1988c. Nutrient budget under slash and burn agriculture (jhum) with different weeding regimes in north-eastern India. *Acta Oecologica-Oecol. Applic.* **9**: 85–102.

Swamy, P.S. and Ramakrishnan, P.S. 1988d. Ecological implications of traditional weeding and other imposed weeding regimes under slash and burn agriculture (jhum) in north-eastern India. *Weed Res.* **28**: 127–136.

Swift, M.J. and Sanchez, P.A. 1984. Biological management of tropical soil fertility for sustained productivity. *Nature and Resources* **20**: 2–10.

Syers, J.K. and Springett, J.A. 1984. Earthworms and soil fertility. *Plant and Soil* **76**: 93–104.

Syers, J.K., Sharpley, A.N. and Keeney, D.R. 1979. Cycling of nitrogen by surface casting earthworms in a pasture ecosystem. *Soil Biol. Biochem.* **8**: 341–346.

Thompson, K. and Stewart, A.J.A. 1981. The measurement and meaning of reproductive efforts in plants. *Amer. Nat.* **117**: 205–211.

Tilman, D. 1985. The resource-ratio hypothesis of plant succession. *Amer. Nat.* **125**: 827–852.

Tinker, P.B. 1977. Economy and chemistry of phosphorus. *Nature* **270**: 103–104.

Tisdell, C.A. 1990. Economics and the debate about preservation of species, crop varieties and genetic diversity. *Ecological Economics* **2**: 77–90.

Todd, R.L., Swank, W.T., Doughlass, J.E., Kerr, P.L., Brockway, D.L. and Monk, C.D. 1975. The relationship between nitrate concentration in the southern Appalachian mountain streams and terrestrial nitrifiers. *Agro-Ecosystems* **2**: 127–132.

Toky, O.P. and Ramakrishnan, P.S. 1981a. Cropping and yields in agricultural systems of the north-eastern hill region of India. *Agro-Ecosystems* **7**: 11–25.

Toky, O.P. and Ramakrishnan, P.S. 1981b. Run-off and infiltration losses related to shifting agriculture (jhum) in north-eastern India. *Environ. Conserv.* **8**: 313–321.

Toky, O.P. and Ramakrishnan, P.S. 1982a. A compartive study of the energy budget of hill agro-ecosystems with emphasis on the slash and burn system (jhum) at lower elevations of north-eastern India. *Agric. Systems* **9**: 143–154.

Toky, O.P. and Ramakrishnan, P.S. 1982b. Role of bamboo (*Dendrocalamus hamiltonii* Nees. and Arn.) in conservation of potassium during slash and burn agriculture (jhum) in north-eastern India. *J. Tree. Sci.* **1**: 17–26.

Toky, O.P. and Ramakrishnan, P.S. 1982c. Forest wealth of the north-eastern India and its conservation. In : G.S. Paliwal, (Ed.), *The Vegetational Wealth of the Himalayas*. pp. 422–432. Puja Publishers, Delhi.

Toky, O.P. and Ramakrishnan, P.S. 1983a. Secondary succession following slash and burn agriculture in north-eastern India. I. Biomass, litterfall and productivity. *J. Ecol.*, **72**: 735–745.

Toky, O.P. and Ramakrishnan, P.S. 1983b. Secondary succession following slash and burn agriculture in north-eastern India. II. Nutrient cycling. *J. Ecol.*, **71**: 747–757.

Toky, O.P. and Ramakrishnan, P.S. 1984. Litter decomposition related to secondary succession and species type under slash and burn agriculure (Jhum) in north-eastern India. *Proc. Indian natn. Sci. Acad.* **B50**: 57–65.

Tomlinson, P.B. and Gill, A.M. 1973. Growth habits of tropical trees : some guiding principles. In : B.J. Meggers, E.S. Ayansu and W.D. Duckworth (Eds). *Tropical Forest Ecosystem in Africa and South America : A Comparative Review*. pp. 129–143. Smithsonian Institution, Washington, D.C.

Trenbath, B.R. 1974. Biomass productivity in mixtures. *Adv. Agron.* **26**: 177–210.

Trenbath, B.R. 1989. The use of mathematical models in the development of shifting cultivation systems. In : J. Proctor (Ed.) *Mineral Nutrients in Tropical Forest and Savanna Ecosystems*. pp. 353–369. Blackwell Sci. Publ., Oxford.

Troup, R.S. 1921. *The Silviculture of Indian Trees*. Clarendon, Oxford. 195 pp.

Udea, K. 1960. *Studies on the Physiology of Bamboo with Reference to Practical Application*. Resource bureau reference data No. 34, Science and Technics Agency, Tokyo. 340 pp.

Uhl, C. and Murphy, P. 1981. A comparison of productivities and energy values between slash and burn agriculture and secondary succession in the upper Rio Negro region of the Amazon basin. *Agro-Ecosystems* **7**: 63–83.

UNESCO, 1983. *Swidden Cultivation in Asia*. Vol. 2. Country Profiles, UNESCO Regional Office, Bangkok. 309 pp.

UNESCO, 1986. *Programmes on Man and the Biosphere (MAB)- General Advisory Panel : Final Report*. UNESCO, Paris. 58 pp.

Van Andel, J. and Vera, F. 1977. Reproductive allocation in *Senecio sylvaticus* and *Chamaenerion angustifolium* in relation to mineral nutrition. *J. Ecol.* **65**: 747–758.

Van den Bergh, J.P. 1968. An analysis of yields of grasses in mixed and pure stands. *Versl. landbouwk Onderz. Ned.* **174**: 1–71.

Van den Driessche, R. 1974. Prediction of mineral nutrient status of trees by foliar analysis. *Bot. Rev.* **40**: 374–394.

Varmah, J.C. and Bahadur, K. 1980. Country report and status of research on bamboos in India. *Indian For Rec.* (Bot.) **6**: 28–88.

Viro, P.J. 1974. Effects of forest fire on soil. In : T.T. Kozlowski and C.E. Ahlgren (Eds) *Fire and Ecosystems* pp. 7–44. Academic Press, New York.

Vitousek, P.M., Gosz, J.R., Grier, C.C., Mellilo, J. M. and Reiners, W.A. 1982. A comparative analysis of potential nitrification and nitrate mobility in forest ecosystems. *Ecol. Monogr.* **52**: 155–177.

Vitousek, P.M. and Stanford, Jr. R.L. 1986. Nutrient cycling in moist tropical forest. *Ann. Rev. Ecol. System.* **17**: 137–169.

Wallwork, J.E. 1983. *Earthworm Biology*. Edward Arnold, London.

Wareing P.F. 1970. Growth and its coordination in trees. In : L.C. Luckwill and C.W. Cutting (Eds) *Physiology of Tree Crops*. pp 1–21. Academic Press, New York.

Watanabe, H. and Ruaysoongnern, S. 1984. Cast production by the megascolecid earthworm *Pheretima* spp. in north-eastern Thailand. *Pedobiologia* **26**: 37–44.

Watson, D.J. 1956. Leaf growth in relation to crop yields. In : F.L. Milthrope (Ed.) *The Growth of Leaves*. pp 178–191. Butterworths, London.

Watson, G.A. 1983. Development of mixed tree and food crop system in the humid tropics : a response to population pressure and deforestation. *Exp. Agric.* **19**: 311–332.

Watters, R.F. 1960. Some forms of shifting cultivation in the south west Pacific. *J. Trop. Geogr.* **14**: 35–50.

Watters, R.F. 1971. *Shifting Cultivation in Latin America.* FAO Forestry Development Paper No. 17. F.A.O., Rome. 305 pp.

Wells, C.G. 1971. Effects of prescribed burning on soil chemical properties and nutrient availability. In: *Prescribed Burning Symposium.* pp 86–97. U.S.D.A. Forest Service Southeastern Forest. Expt. Stn. Asheville, NC.

White, E.M., Thompson, W.W. and Gartner, F.R. 1973. Heat effects on nutrient release from soils under ponderosa pine. *J. Range Manage.* **22**: 22–24.

White, J. 1979. The plant as a metapopulation. *Ann. Rev. Ecol. Syst.* **10**: 109–145.

Whitmore, T.C. 1975. *Tropical Rain Forests of the Far East.* Oxford Univeristy Press, London.

Whitney, C.G. 1976. The bifurcation ratio as an indicator of adaptive strategy in weedy plant species. *Bull. Torrey Bot. Club* **103**: 67–72.

Whittaker, R.H. 1972. Evolution and measurement of species diversity. *Taxon*, **21**: 213–251.

Whittaker, R.H. 1975. *Communities and Ecosystems.* McMillan Publ. Co., Inc., New York. 385 pp.

Wilkes, H.G. 1977. The worlds' crop plant germplasm – an endangered resource. *Bull. Atmos. Sci.* **33**: 8–16.

Williams, R.B. and Bell, K.M. 1981. Nitrogen allocation in Mojave desert annuals. *Oecologia* **48**: 145–150.

Wilson, M.F. 1971. Life history consequences of death rates. *Biologist* (Phil. Sigma Soc.) **53**: 49–56.

Wilson, W.J. 1967. Effects of seasonal variation in radiation and temperature on net assimilation and growth rates in an arid climate. *Ann. Bot.* (London), **31**: 264–267.

Wong, R. 1964. Evidence of the presence of growth inhibitory substances in *Mikania cordata* (Burn. f.) B.L. Robinson. *J. Rubber Res. Inst.*, Malaya **18**: 231–242.

Woodmansee, R.W. and Wallach, L.S. 1981. Effects of fire frequency and intensity on biogeochemical cycling. In : H.A. Mooney, T.M. Bonnicksen, N.L. Christensen, J.E. Lotan and W.A. Reiners. (Eds) *Fire Regime and Ecosystem Properties.* pp 379–400. U.S.D.A., Forest Service, Gen, Tech. Rept. WO-26, Washington, D.C.

Woodwell, G.M. and Whittaker, R.H. 1968. Primary production in terrestrial communities. *American Zoology*, **8**: 19–30.

Woodwell, G.M., Whittaker, R.H. and Houghton, R.A. 1975. Nutrient concentration in plants in the Brookhaven oak-pine forest. *Ecology* **56**: 318–332.

Woolhouse, H.W. 1967. The nature of senescence in plants. *Symp. Soc. Exptl. Biol.* **21**: 179–230.

World Resources Institute, 1988. *World Resources* 1988–89. *An Assessment of the Resources Base that Supports the Global Economy.* Basic Books, Inc. New York. 372 pp.

Yadav, A.W. and Tripathi, R.S. 1981. Population dynamics of the ruderal weed *Eupatorium odoratum* L. and its regulation in nature.*Oikos* **23**: 355–361.

Zimdahl, R.L. 1980. *Weed-Crop Competition – A Review.* Int. Plant Protection Centre, Oregon state Univ., Corvallis, 195 pp.

Zinke, P.J., Sabhasri, S. and Kunstadter, P. 1978. Soil fertility aspects of the 'Lua' forest fallow system of shifting cultivation. In : P. Kunstadter, E.C. Chapman and S. Sabhasri (Eds) *Farmers in the Forest.* pp. 143–159. East-West Centre, Honolulu, Hawaii.

INDEX